Probability and Mathematical Statistics (Continue

LEHMANN • Testing Statistical Hypothese
LEHMANN • Theory of Point Estimation
MATTHES, KERSTAN, and MECKE • Inf
MUIRHEAD • Aspects of Multivariate Stat
PRESS • Bayesian Statistics: Principles, Mo
PURI and SEN • Nonparametric Methods i
PURI and SEN • Nonparametric Methods in Multivariate Analysis
PURI, VILAPLANA, and WERTZ • New Perspectives in Theoretical and Applied Statistics
RANDLES and WOLFE • Introduction to the Theory of Nonparametric Statistics
RAO • Linear Statistical Inference and Its Applications, *Second Edition*
RAO • Real and Stochastic Analysis
RAO and SEDRANSK • W.G. Cochran's Impact on Statistics
RAO • Asymptotic Theory of Statistical Inference
ROBERTSON, WRIGHT and DYKSTRA • Order Restricted Statistical Inference
ROGERS and WILLIAMS • Diffusions, Markov Processes, and Martingales,Volume II: Îto Calculus
ROHATGI • An Introduction to Probability Theory and Mathematical Statistics
ROHATGI • Statistical Inference
ROSS • Stochastic Processes
RUBINSTEIN • Simulation and The Monte Carlo Method
RUZSA and SZEKELY • Algebraic Probability Theory
SCHEFFE • The Analysis of Variance
SEBER • Linear Regression Analysis
SEBER • Multivariate Observations
SEBER and WILD • Nonlinear Regression
SEN • Sequential Nonparametrics: Invariance Principles and Statistical Inference
SERFLING • Approximation Theorems of Mathematical Statistics
SHORACK and WELLNER • Empirical Processes with Applications to Statistics
STOYANOV • Counterexamples in Probability

Applied Probability and Statistics

ABRAHAM and LEDOLTER • Statistical Methods for Forecasting
AGRESTI • Analysis of Ordinal Categorical Data
AICKIN • Linear Statistical Analysis of Discrete Data
ANDERSON and LOYNES • The Teaching of Practical Statistics
ANDERSON, AUQUIER, HAUCK, OAKES, VANDAELE, and WEISBERG • Statistical Methods for Comparative Studies
ARTHANARI and DODGE • Mathematical Programming in Statistics
ASMUSSEN • Applied Probability and Queues
BAILEY • The Elements of Stochastic Processes with Applications to the Natural Sciences
BARNETT • Interpreting Multivariate Data
BARNETT and LEWIS • Outliers in Statistical Data, *Second Edition*
BARTHOLOMEW • Stochastic Models for Social Processes, *Third Edition*
BARTHOLOMEW and FORBES • Statistical Techniques for Manpower Planning
BATES and WATTS • Nonlinear Regression Analysis and Its Applications
BECK and ARNOLD • Parameter Estimation in Engineering and Science
BELSLEY, KUH, and WELSCH • Regression Diagnostics: Identifying Influential Data and Sources of Collinearity
BHAT • Elements of Applied Stochastic Processes, *Second Edition*
BLOOMFIELD • Fourier Analysis of Time Series: An Introduction
BOLLEN • Structural Equations with Latent Variables
BOX • R. A. Fisher, The Life of a Scientist
BOX and DRAPER • Empirical Model-Building and Response Surfaces
BOX and DRAPER • Evolutionary Operation: A Statistical Method for Process Improvement

BOX, HUNTER, and HUNTER • Statistics for Experimenters: An Introduction to Design, Data Analysis, and Model Building
BROWN and HOLLANDER • Statistics: A Biomedical Introduction
BUNKE and BUNKE • Statistical Inference in Linear Models, Volume I
BUNKE and BUNKE • Nonlinear Regression, Functional Relations and Robust Methods: Statistical Methods of Model Building
CHAMBERS • Computational Methods for Data Analysis
CHATTERJEE and HADI • Sensitivity Analysis in Linear Regression
CHATTERJEE and PRICE • Regression Analysis by Example
CHOW • Econometric Analysis by Control Methods
CLARKE and DISNEY • Probability and Random Processes: A First Course with Applications, *Second Edition*
COCHRAN • Sampling Techniques, *Third Edition*
COCHRAN and COX • Experimental Designs, *Second Edition*
CONOVER • Practical Nonparametric Statistics, *Second Edition*
CONOVER and IMAN • Introduction to Modern Business Statistics
CORNELL • Experiments with Mixtures: Designs, Models and The Analysis of Mixture Data
COX • Planning of Experiments
COX • A Handbook of Introductory Statistical Methods
DANIEL • Biostatistics: A Foundation for Analysis in the Health Sciences, *Fourth Edition*
DANIEL • Applications of Statistics to Industrial Experimentation
DANIEL and WOOD • Fitting Equations to Data: Computer Analysis of Multifactor Data, *Second Edition*
DAVID • Order Statistics, *Second Edition*
DAVISON • Multidimensional Scaling
DEGROOT, FIENBERG and KADANE • Statistics and the Law
DEMING • Sample Design in Business Research
DILLON and GOLDSTEIN • Multivariate Analysis: Methods and Applications
DODGE • Analysis of Experiments with Missing Data
DODGE and ROMIG • Sampling Inspection Tables, *Second Edition*
DOWDY and WEARDEN • Statistics for Research
DRAPER and SMITH • Applied Regression Analysis, *Second Edition*
DUNN • Basic Statistics: A Primer for the Biomedical Sciences, *Second Edition*
DUNN and CLARK • Applied Statistics: Analysis of Variance and Regression, *Second Edition*
ELANDT-JOHNSON and JOHNSON • Survival Models and Data Analysis
FLEISS • Statistical Methods for Rates and Proportions, *Second Edition*
FLEISS • The Design and Analysis of Clinical Experiments
FLURY • Common Principal Components and Related Multivariate Models
FOX • Linear Statistical Models and Related Methods
FRANKEN, KÖNIG, ARNDT, and SCHMIDT • Queues and Point Processes
GALLANT • Nonlinear Statistical Models
GIBBONS, OLKIN, and SOBEL • Selecting and Ordering Populations: A New Statistical Methodology
GNANADESIKAN • Methods for Statistical Data Analysis of Multivariate Observations
GREENBERG and WEBSTER • Advanced Econometrics: A Bridge to the Literature
GROSS and HARRIS • Fundamentals of Queueing Theory, *Second Edition*
GROVES, BIEMER, LYBERG, MASSEY, NICHOLLS, and WAKSBERG • Telephone Survey Methodology
GUPTA and PANCHAPAKESAN • Multiple Decision Procedures: Theory and Methodology of Selecting and Ranking Populations
GUTTMAN, WILKS, and HUNTER • Introductory Engineering Statistics, *Third Edition*
HAHN and SHAPIRO • Statistical Models in Engineering

Analysis of Complex Surveys

Analysis of Complex Surveys

Edited by

C. J. SKINNER,
D. HOLT
and
T. M. F. SMITH
University of Southampton

JOHN WILEY & SONS
Chichester · New York · Brisbane · Toronto · Singapore

Wiley Editorial Offices

John Wiley & Sons Ltd, Baffins Lane, Chichester,
West Sussex PO19 1UD, England

John Wiley & Sons, Inc., 605 Third Avenue,
New York, NY 10158-0012, USA

Jacaranda Wiley Ltd, G.P.O. Box 859, Brisbane,
Queensland 4001, Australia

John Wiley & Sons (Canada) Ltd, 22 Worcester Road,
Rexdale, Ontario M9W 1L1, Canada

John Wiley & Sons (SEA) Pte Ltd, 37 Jalan Pemimpin #05-04,
Block B, Union Industrial Building, Singapore 2057

British Library Cataloguing in Publication Data

Analysis of complex surveys.
(Wiley series in probability and
mathematical statistics. Applied section.)
1. Surveys. Methodology
I. Skinner, C. J. II. Holt, D.
III. Smith, T. M. F.
001.4′33

ISBN 0 471 92377 X

Phototypesetting by Thomson Press (India) Limited, New Delhi
Printed and bound in Great Britain by Courier International. Tiptree, Essex

Contents

PART C—DISAGGREGATED ANALYSIS: MODELLING STRUCTURED POPULATIONS

List of Contributors

Dr P. D. Ewings

6 Mountfields Avenue, Taunton, Somerset TA1 3BN, UK.

Professor H. Goldstein

Department of Mathematics, Statistics and Computing, University of London Institute of Education, 20 Bedford Way, London WC1H 0AL, UK.

Dr D. J. Holmes

Department of Social Statistics, University of Southampton, Southampton SO9 5NH, UK.

Professor D. Holt

Department of Social Statistics, University of Southampton, Southampton SO9 5NH, UK.

Dr L. M. LaVange

Research Triangle Institute, Research Triangle Park, NC 27709, USA.

Professor E. A. Molina C.

Departmento de Matematica y Ciencia de la Computacion, Universidad Simon Bolivar, Caracas, Venezuela.

Professor G. Nathan

Department of Statistics, Faculty of Social Sciences, The Hebrew University of Jerusalem, Mount Scopus, Jerusalem 81805 Israel.

Dr D. Pfeffermann

Department of Statistics, Faculty of Social Sciences, The Hebrew University of Jerusalem, Mount Scopus, Jerusalem 81805 Israel.

Professor J. N. K. Rao

Department of Mathematics and Statistics, Carleton University, Ottawa, Ontario K1S 5B6 Canada.

Professor A. J. Scott

Department of Mathematics and Statistics, University of Auckland, Private Bag, Auckland, New Zealand.

Ms R. Silver

Department of Mathematics, Statistics and Computing, University of London Institute of Education, 20 Bedford Way, London WC1H 0AL, UK.

Dr C. J. Skinner

Department of Social Statistics, University of Southampton, Southampton SO9 5NH, UK.

Professor T. M. F. Smith

Department of Mathematics, University of Southampton, Southampton SO9 5NH, UK.

Professor D. R. Thomas

School of Business, Carleton University, Ottawa, Ontario K1S 5B6, Canada.

Dr C. J. Wild

Department of Mathematics and Statistics, University of Auckland, Private Bag, Auckland, New Zealand.

Preface

This book's origins lie in two research programmes on 'The Analysis of Data from Complex Surveys', funded by the UK Economic and Social Research Council between 1977 and 1985. The programmes were based at Southampton University and directed by Tim Holt and Fred Smith. Chris Skinner also became involved with the programmes from 1978. A key factor in the development of the programmes was the support they provided for extended visits to Southampton by a number of international researchers with expertise in complex surveys. These researchers included Gad Nathan, Danny Pfeffermann, Alastair Scott and Jon Rao, each of whom has contributed to this book, as well as others, such as Wayne Fuller, Joe Sedransk and Jeff Wu, who have not contributed directly but whose ideas have, nevertheless, exerted a definite influence. Paul Ewings and Dave Holmes were employed full-time on the programmes and they have been responsible not just for the chapters which bear their names, but also for helping in many ways with the work underlying the other chapters. A number of individuals completed their Ph.D. theses at Southampton on topics connected with the programmes; these include Emiro Molina (1982), Chris Skinner (1982), Paul Ewings (1985), David Steel (1985), Muhammad Pervaiz (1986) and Dave Holmes (1987).

A three-day conference was held in Southampton in March 1986 to disseminate the results of this research to an audience of survey researchers and statisticians, drawn mainly from the UK but also from several European countries. Papers were given at the conference by Tim Holt, Chris Skinner, Fred Smith, Paul Ewings, Dave Holmes, Gad Nathan, Danny Pfeffermann and Jon Rao, as well as by Harvey Goldstein, whose independent work on multilevel modelling had much in common with the Southampton work.

The broad decision to write this book was made at that conference, although ideas about the actual form of the book have evolved since that time. Initially the book was to be divided into two parts, on 'the analysis of continuous data' and 'the analysis of discrete data', as on the conference programme. Later, this division seemed less fundamental than the current division into three parts.

Our intention was never to produce a volume of conference proceedings, but rather to write a text in which the various areas of expertise of the different contributors were welded together into a unified whole. Thus we asked contributors beforehand to be prepared for heavy editing and indeed we are extremely grateful to them for allowing us to propose revisions and for letting us ask them repeatedly to restructure and rewrite their chapters. We have attempted to achieve as much consistency of notation and terminology as possible and also, a more difficult aim, to achieve a consistency of level. The diligent reader may still be able to detect slight differences of emphasis by different authors, say on the role of *p*-weighting, but we do not feel that this conflicts in an adverse way with the general aim of the book, which is to cover a broad range of approaches, discussing the relative merits of each, but without necessarily 'taking sides'.

The 'typical reader' we have in mind for this book is a researcher who wishes to apply statistical methods to survey data and who is familiar with those methods to the level of Bishop *et al.* (1975), say, for contingency table analysis, or Draper and Smith (1981) for regression analysis. Such survey analysts might be statisticians or subject-matter researchers such as econometricians, sociologists or epidemiologists. They may well be secondary analysts of survey data collected by government agencies. A background in survey sampling to the level of a text such as Cochran (1977) is also assumed and most of the authors draw in some way on ideas from the survey sampling tradition. Our basic aim has been to bring together a number of theoretical developments which are currently available only in a dispersed form in recent statistical literature. Not unnaturally these developments are often rather theoretical and this is reflected in the mathematical level of some chapters. We have introduced and discussed ideas, principles and the 'bones' of methods rather than produced a step-by-step manual of how to analyse complex survey data. We debated whether to include discussion of computational considerations but eventually decided against, leaving only occasional references to suitable software. We are aware that computational aspects are vital in practice to survey analysts, but the existence of relevant software is so uneven across the topics discussed in this book and the rate of change of available software so high that we felt such discussion would only hinder the general flow of the book. For a description of a number of software packages for variance estimation see Wolter (1985, Appendix E).

As noted above, the book is primarily aimed at researchers rather than students attending any particular course. The book might, however, usefully provide an addition to a standard text, such as Cochran (1977), on a course on sample surveys for postgraduate statistics students. Sample survey courses for statisticians have often been accused of suffering (i) from isolation from other statistics courses, and (ii) from a lack of interesting statistical issues and principles, because of an emphasis on listing estimator/variance/variance estimator triples for different designs. The discussion of different statistical methods in this book would obviously help to address problem (i) and, with respect to problem (ii), we feel that issues considered here, such as the choice of target parameter, aggregation versus disaggregation, and the role of design in analysis, are interesting topics, worthy of discussion in the training of any statistician.

The preparation of a book of this kind depends upon the active support of many colleagues. In particular we wish to mention Beryl Betts and Margaret Youngs for typing parts of the manuscript. However, we offer special thanks to Anne Owens, not only for typing the majority of the text, but for that sense of calm efficiency and cheerfulness that makes major crises seem manageable.

Southampton
December 1988

CHAPTER 1

General Introduction

1.1 ANALYTIC AND DESCRIPTIVE SURVEYS

This book is concerned with the analytic as opposed to the descriptive uses of sample surveys. Descriptive uses are directed at the estimation of summary measures of the population, such as means and frequencies. In contrast, Deming (1950, p. 249) describes analytic uses as 'directed at the underlying causes that have made the frequencies of various classes of the population what they are, and will govern the frequencies of these classes in time to come'. Thus the analytic use takes us beyond the summary measures embodied in description and into the causal explanation of the processes that underlie the descriptive measures.

Since before the turn of the century, surveys have been an important source of information for both descriptive and analytic purposes. Governments have needed an accurate picture of the population in terms of its location, personal characteristics and associated circumstances, such as the quantity and quality of housing stock. Such requirements are essentially descriptive and provide the framework for the formulation of government policy to meet changing social circumstances. Others have sought to go beyond description and have been more concerned with identifying and understanding the causal mechanisms which underlie the picture which the descriptive statistics portray. This has led to the analytic use of surveys, an early example of which is the work of Snow (1855) in investigating the link between the quality of water-supply and the incidence of cholera in the population of London in the mid-nineteenth century.

For those concerned with official statistics in the first third of this century, a major issue was the idea of the 'representative method': how to select samples that would fairly represent the population and from which valid conclusions could be drawn. In the language of modern sample survey theory the issue concerned survey design, estimation and a framework within which to evaluate the properties of alternative procedures. Neyman (1934) created such a framework: establishing the role of randomized methods in the sampling process, using the resulting randomization distribution as a basis for evaluating the properties of alternative procedures, introducing the ideas of stratification and, via optimal allocation, the use of unequal selection probabilities. This *tour de force* was surely the foundation-stone of modern sample survey theory. From this work flowed the extensions to multi-stage sampling, selection with probability proportional to size (Hansen and Hurwitz, 1943; Midzuno, 1951) culminating with a general theory of unequal probability sampling and probability-weighted estimation of Horvitz and Thompson (1952).

From the earliest days, it was recognized that social survey populations invariably have complex structures with systematic differences between subpopulations which may be based on geographical location, personal characteristics or other factors related to individuals or groups of individuals. Good survey design used this structure and designs were developed to incorporate the complexities of stratification, different levels of clustering and measures of size. The basis of Neyman's theory, through the randomization distribution, is essentially distribution-free. However, knowledge and skilful use of the population structure can yield substantial gains in efficiency in appropriate circumstances. The role of the randomization distribution as a link between design and inference, the use of probability-weighted estimators, and the consequential guarantee of approximate unbiasedness with respect to the sampling distribution, have been vitally important to the survey method and to the creation of a standard of acceptable practice. This point is of particular importance in the context of official statistics. Published results must enjoy public confidence; they must be distribution- and assumption-free in so far as this is possible; they must be based on the highest standards of design and analysis, grounded in a sound methodological base. Indeed almost all the advances in sample survey theory from Neyman onwards have been strongly influenced by the descriptive uses of survey sampling. Given the importance of official statistics this emphasis is understandable and justifiable, but a consequence has been neglect of theoretical developments related to the analytic uses of surveys. Most sample survey texts, such as Cochran (1977), are limited to the estimation of descriptive parameters, like the finite population total and mean, and a small number of additional parameters, such as the ratio of two totals or means and the totals and means of domains of study.

As an illustration, to which we will return throughout this chapter, consider a survey to measure unemployment. The number of unemployed people (which is a finite population total) or the unemployment rate (which is a ratio of two totals) are both extremely important indicators for government policy. However, a change in the unemployment rate is not, of itself, much of a guide for the development of specific policies. At the national level, unemployment must be integrated with other macroeconomic indicators such as inflation, money supply, balance of trade and so on within an explanatory model. More importantly, so far as this book is concerned, microeconomic analysis involves the need to disaggregate the unemployment statistics: to understand the differences in levels of unemployment by region, sections of the community and sectors of the economy. Furthermore, dynamic causal mechanisms, such as the relationship between unemployment and training or education, and the impact of factors such as housing and occupational pensions on job mobility, are analytic objectives which may be explored using appropriate methods applied to individual survey responses.

If Neyman can be viewed as laying the foundations of modern sample survey theory with its emphasis on descriptive surveys then the sociologist Paul Lazarsfeld might be identified as the key figure influencing modern developments in the analysis of surveys. Lazarsfeld viewed survey analysis as being essentially concerned with relationships between variables (e.g. Kendall and Lazarsfeld, 1950). This idea and the implied distinction between descriptive surveys and analytic or explanatory surveys are discussed at length by Hyman (1955), who states that: in descriptive surveys 'the focus of ... analysis is ... precise measurement of one or more dependent variables ...' (p. 68), whereas in explanatory surveys the focus is in 'establishing reliably the nature of the relationship between one or more phenomena, or dependent variables, and one or more causes or independent variables' (p. 66).

Lazarsfeld's influence may be traced through at least two aspects of the development of survey analysis. First, in terms of the practice of survey analysis, he was associated with the Bureau of Applied Social Research at Columbia University, which, along with other organizations such as the Survey Research Center at the University of Michigan and the National Opinion Research Center at the University of Chicago, began to conduct extensive analytic surveys which helped to counterbalance the continuing descriptive surveys carried out by governmental agencies. Some examples of the applications of such analytic surveys to the fields of sociology, political science, psychology, economics, anthropology, education, social work, public health and medicine are described in Glock (1967).

Second, and more importantly for our purposes, Lazarsfeld was a pioneer in applying methods of statistical analysis to survey data. He concentrated, in particular, on methods for the causal analysis of contingency tables. A clear exposition of Lazarsfeld's approach to survey data analysis is given by Rosenberg (1968). Such methodological work was driven by the desire of survey analysts to draw appropriate explanatory inferences from survey data. The scope and potential of such analyses have been expanded enormously over the last 20 years by the increased power of computers and the associated development of generalized software. Hirschi and Selvin (1967, p. 162) predicted that the

> Impact [of the computer] promises to be as revolutionary for research in the behavioural sciences as the microscope in biology.
>
> The major impact of the computer on ... quantitative analysis ... [in survey research] will be the replacement of percentaged contingency tables

> as devices of exploration and causal analysis by one or another member of the family of linear statistical techniques—multiple regression, discriminant analysis, canonical analysis, factor analysis and analysis of variance.

This prediction now seems to understate the impact of the computer. Not only are linear statistical techniques now widely applied to survey data but so too are the non-linear techniques of loglinear and logit analysis of contingency tables (Bishop *et al*, 1975), generalized linear modelling (McCullagh and Nelder, 1983), survival analysis (Cox and Oakes, 1984) as well as many other procedures now implemented in standard computer packages such as SPSS X (1986) and SAS (1985). Such techniques have been drawn from many areas of statistical applications and might broadly be said to share the objective of creating statistical models with structural and stochastic components which mirror the process under investigation. Such an aim is broader than the original concern solely with relationships between variables.

It should be emphasized that the sophistication of the statistical modelling techniques currently employed in survey analysis is quite distinct in character, if not in degree, from that of the statistical procedures employed in descriptive surveys for estimation. As noted earlier, traditional inference procedures for descriptive surveys mirror the complexities of the sampling design and the population structure reflected in that design. Indeed survey statisticians who are responsible for estimation procedures may often also be responsible for designing sampling schemes. Thus the complexity in the sampling design is often intimately connected with the complexity of the estimation procedure. In contrast, statistical techniques employed in survey analysis often take no account of the complexity of the design or the population structure. This is usually a consequence of the fact that the technique has been borrowed from some other field with a different concept of sample and population, the usual assumption being that the sample is independently and identically distributed (IID) from some hypothetical superpopulation. Survey analysts are keen to exploit the analytic potential of new techniques and this can be especially easy when such techniques are implemented in standard computer software. Another reason why survey analysts may tend to ignore complex sampling designs is that they are often somewhat removed from the survey design process. Indeed much analysis of survey data consists of the *secondary analysis* of survey data which has been collected previously for descriptive purposes. In such cases the analyst's knowledge of the survey design may not go beyond brief details contained in published reports. For confidentiality reasons unit level information concerning stratum and cluster identification may not be available on the data tape. Thus the secondary analyst cannot use the population structure within an explanatory model even if he or she wishes to.

We have briefly alluded to one other essential difference between the analytic and descriptive uses of survey data. For descriptive uses the objective is essentially fixed. Target parameters, such as the number of unemployed or the unemployment rate, are objectives determined before the data are collected or analysed.

For analytic uses, just as for statistical modelling in general, the parameters of interest are not fixed in advance but evolve through an adaptive process as the analysis progresses. If, in the analysis of the unemployment rate, systematic differences are exhibited between subgroups then the model is adapted to include this factor. Whether other variables such as age or education should be included as explanatory factors is not predetermined but assessed in the light of the survey data. Thus for analytic purposes the process is an evolutionary one where the final parameters to be estimated and the methods to be employed will be chosen in the light of the population structure exhibited through the survey data.

A final distinction between descriptive and analytic uses of surveys, as emphasized by Deming (1950, Chapter 7), is that the target parameter for descriptive purposes is a characteristic of the actual finite population sampled, whereas for analytic purposes, where the objective is to draw conclusions that have wider generalizability in space and time than for the specific population sampled, the target parameter is a characteristic of the underlying model. For example, for descriptive purposes the total number of unemployed adults in area X at time T might be of interest, whereas for analytic purposes interest might centre on the change in the probability that an unemployed person finds employment, arising from attendence at a training scheme. In the former case no inference is desired beyond area X and time T. In the latter case, it might be hoped that the conclusions drawn would have relevance to the effectiveness of training schemes in areas and times different from the ones from which the sample was drawn.

1.2 PROBLEMS WITH APPLYING STANDARD ANALYTIC PROCEDURES TO SURVEY DATA

The mainstream developments in statistical modelling and the availability of computer software have had a very strong impact on data analysis in general and, in the absence of specific survey-based techniques, have been extensively used by data analysts working with survey data. However, the stochastic assumptions underlying such techniques do not reflect the complexity which is usually exhibited by survey populations. Thus, for example, the distributional assumptions upon which loglinear analyses for contingency tables are based are usually Poisson, multinomial or product multinomial distributions and the observations are assumed to be IID. In contrast, survey practitioners recognize that populations are complex with different cell probabilities in different subgroups of the population (e.g. clusters) and this implies a different vector of cell probabilities conditional on the membership of a specific cluster. Thus, to return to the example of unemployment, a cross-tabulation of employment status by age by level of education would not exhibit the properties of an IID sample but would be much more complex in distributional terms. The same issues apply to other statistical methods and, partly because of the lack of statistical theory and partly because available computer software was based on IID assumptions, survey analysts have tended to treat their data as IID by default.

The inadequacy of this assumption has been well known in the sample survey literature for many years. It was known, for example, that the homogeneity which population clusters exhibited tended to increase the variance of sample estimators and, furthermore, estimates of variance wrongly based on IID assumptions were generally downwards biased. The consequences for descriptive statistics of wrongly using IID assumptions for clustered data were that standard errors would be too small, and confidence intervals too narrow. For analytic purposes test statistics would be based on downwardly biased estimates of variance and the results would appear to be more significant than was really the case. From these issues grew the idea of the design effect (Kish, 1965), and Kish and Frankel (1974) applied the same reasoning to a variety of non-linear estimators such as the correlation coefficient and the regression coefficient. Whilst Kish and Frankel considered complex statistics it would be fair to say that their concern was not really the analytic use of survey data. They emphasized that the same considerations of design effect that had been developed for estimators of the finite population mean and total were relevant for non-linear statistics also. Their main purpose was to show that variance estimation techniques taking into account the clustering and stratification in the survey data were needed. Alternative variance estimation methods such as balanced repeated replication and linearization were compared with IID alternatives. In the tradition of sample survey theory, parameters such as the correlation coefficient or regression coefficient were defined as functions of finite population values. They are more complicated than parameters such as the population total or mean because they are not linear functions of the population values but, this apart, the framework for estimation remains the randomization distribution generated by the survey design. Some other important work within this sample survey tradition is that of Koch and his coworkers, for example Koch *et al.* (1975), in which a comprehensive framework for variance estimation and tests of hypotheses is developed, based on the Wald test and asymptotic theory.

There is much more to the analytic use of survey data than simply defining a set of non-linear statistics based on the finite population. Analytic uses of survey data essentially involve a model-building approach. As displayed in Figure 1.1, adapted from Box and Jenkins (1976, p. 19), the process is an iterative

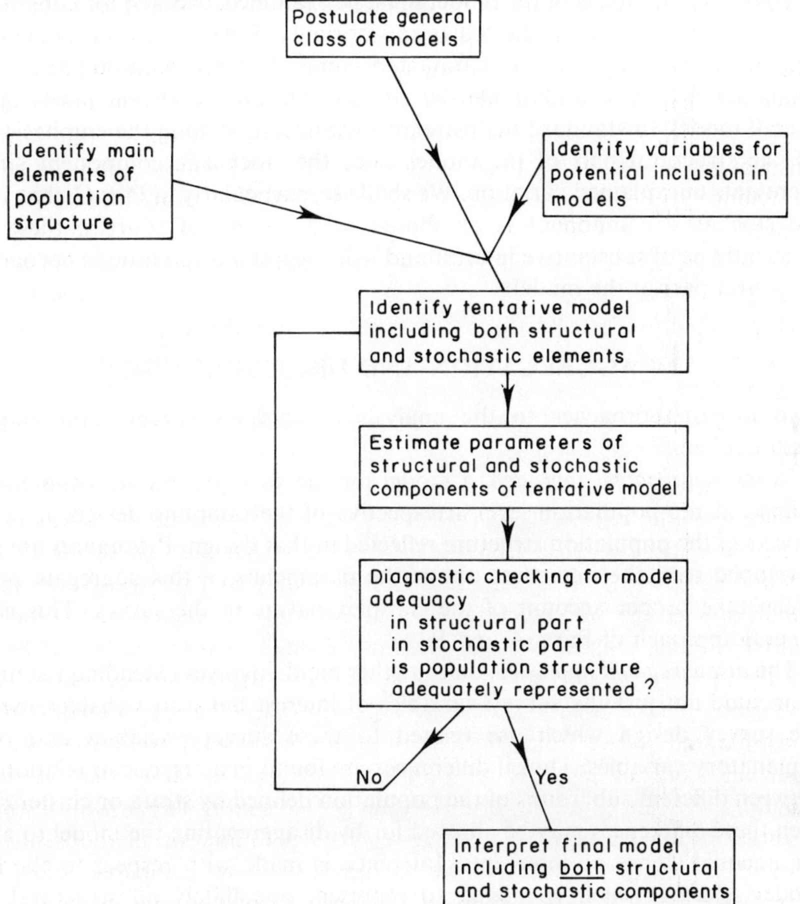

Figure 1.1 Stages in the iterative model-building process.

one in which the objectives are not predetermined but are modified in the light of the structure exhibited by the survey data.

There are several elements to modelling relationships between variables when data are collected from a complex population. The first, as in all statistical modelling procedures, is the choice of variables and the general form of the models to be considered. In addition, essential features of the population structure must be taken into account, including specifying an appropriate variance structure for the observations and considering whether differences between population groupings exist which should be included explicitly in the model. When the model is estimated there is a much larger choice of diagnostic checks and alternative model specifications, since both the structural and stochastic components of the model must be examined, checked for consistency with the survey data and adapted as appropriate. After the final choice of model and the estimation of relevant parameters comes the interpretation phase. Once again the structural and stochastic components are important facets of the overall model. In standard mainstream statistical modelling the emphasis falls on the structural part of the model since the stochastic component simply represents unexplained variation. We shall see, particularly in Part C, that when the stochastic component is attributed to a variety of sources, these will frequently be of substantive interest and will suggest interpretations beyond the structural part of the model.

1.3 AGGREGATION AND DISAGGREGATION

Two broad approaches to the analysis of complex survey data may be distinguished.

In the *aggregated approach*, a model for the survey variables of interest is defined at the population level, irrespective of the sampling design or of any aspects of the population structure reflected in that design. Procedures are then developed to make inferences about the parameters of this aggregate model which take proper account of the complex nature of the survey. This is the general approach of Parts A and B.

The *disaggregated approach*, on the other hand, involves extending the model to include not just the survey variables of interest but also variables used in the survey design which are related to these survey variables and other explanatory variables. Thus if differences are found in a regression relationship between different subgroups of the population defined by strata or clusters, say, then these differences may be allowed for by disaggregating the model to allow for unequal slopes or intercepts. Inference is made with respect to the final model selected which is judged to represent adequately all structural and stochastic variation. This is the general approach of Part C.

For illustration we return to the example of unemployment and suppose that we wish to understand the relationship between the binary dependent variable unemployed/employed and the explanatory variable years of education for individuals in the labour force. In an aggregated approach, one might model the conditional probability of being unemployed given a fixed number of years of education. For example a logistic model might be adopted and the aggregate parameters defined as the coefficients determining this logistic model. The crucial aspect of this approach is that the conditional probabilities are defined across all individuals in the population, and this is done irrespective of whether different relationships might apply in different subgroups of the population. Such population heterogeneity would, however, be allowed for in the inference procedure adopted. In the disaggregated approach, on the other hand, it would be considered important to examine whether the relationship between unemployment and years of education differed between strata, clusters or other population groupings. If, for example, it was found that the relationship differed between regional strata then the logistic model would be modified accordingly to allow for different coefficients in different regions.

Thus the aggregated approach defines the parameters, unconditionally across population subgroups, leading to interpretations of the form 'the average population logistic regression parameters are...', whereas the disaggregated approach defines parameters conditioned on the values of the design variables or other variables which 'explain' the variation in the dependent variable.

The aggregated approach follows the traditional survey approach for descriptive purposes. The parameter of interest is defined as an appropriate population parameter and the survey design is of relevance not for parameter definition, but only in the inference procedure. The disaggregated approach follows the statistical modelling tradition. Differences between subgroups of the population are explored and, when appropriate, extra variables are introduced into the model in order to explain the variation between individual units as far as possible.

There are, however, practical limits to how far this disaggregation process can go. It may be that having allowed for regional effects in both intercept and slope we still recognize that within each region separate clusters show differences in unemployment rate, even after adjusting for education, and furthermore that the logistic regression coefficients vary between clusters. A highly disaggregated approach would be to treat each cluster as a separate group, which would involve calculating separate intercepts and slopes for each cluster and in effect carrying out a very large fixed-effects analysis. A practical objection would be that there would be far too many parameters to make sense of, and a theoretical objection is that inference would be restricted to the sampled clusters. These objections are even more obvious in the extreme case where each individual unit is treated as a separate group, for in this situation no statistical inference from the sample to the population would be possible at all.

There may also be substantive reasons for limiting the extent of disaggregation. For example, the area within which an individual lives is influenced by the individual's economic circumstances, which are themselves influenced by employment status. These effects are represented in the path diagram in Figure 1.2.

For a model relating the dependent variable unemployment to the independent variable years of education, disaggregation to the cluster level involves controlling the relationship for the area of residence, as defined by the clustering. But if individuals' areas of residence are partly influenced by their employment status, so that the unemployed become much more concentrated in some areas than others for reasons unconnected with their educational level, it is likely that the relationship between unemployment and education within clusters will be weaker than in the population as a whole. Thus the within-cluster

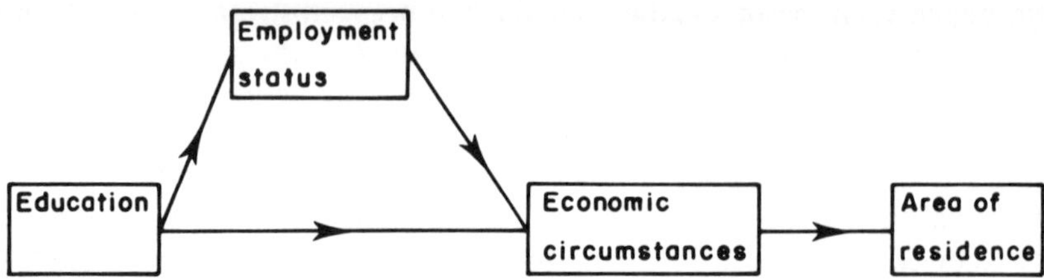

Figure 1.2 Path diagram illustrating how disaggregation by area of residence can lead to 'selection on the dependent variable' in a model relating unemployment to education.

coefficients of the logistic relationship in the disaggregated model are artificially attenuated by 'selection on the dependent variable'. This effect will be discussed at greater length in Part B. In such situations the coefficients of a model aggregated across clusters may be more amenable to causal interpretation than the coefficients of the disaggregated model.

In practice, what is needed is a compromise between disaggregation and aggregation. Some degree of disaggregation will usually lead to better understanding of the underlying phenomena and to parameters which are more likely to possess a 'causal' interpretation. But complete disaggregation may not be desirable for practical reasons or because not all the groupings of the population into clusters or strata are suited to act as 'exogenous' factors in the model for reasons of interpretation. Thus, although for simplicity of exposition Parts A and B adopt a purely aggregated approach and Part C a purely disaggregated approach, it will usually be sensible in practical applications to combine ideas and methods from all three parts of the book.

1.4 SURVEY DESIGN AND POPULATION STRUCTURE

The historical development of sample survey theory and the role of the randomization distribution in creating an essential link between survey design and estimation has had one important side-effect. Since the survey design determines the randomization distribution and this in turn is the basis for evaluating the properties of the estimator, there has been a tendency to ascribe these properties to the design. In particular the variance inflation that is associated with clustered samples is referred to as a consequence of the design, and terms such as 'design effect' and 'design factor' reinforce this interpretation.

For analytic uses of survey data the precise roles of the survey design and the population structure need to be disentangled and clarified. We illustrate the issues by considering the design effect for a two-stage sample with m observations per cluster and an intra-cluster correlation coefficient τ to describe the cluster homogeneity. In standard sample survey theory, the design effect (deff) for the sample mean $\bar{y}$ is defined as the ratio of the variance of $\bar{y}$ under the complex design to the variance of $\bar{y}$ under a simple random sampling (SRS) design of the same size. For the case of two-stage sampling a standard approximate result (see Chapter 2) is that

$$\text{deff}(\bar{y}) = 1 + (m - 1)\tau.$$

The usual interpretation is that the deff measures the extent to which the clustered design has caused the variance of the sample mean to be inflated. The essential comparison which the design effect represents is between two alternative designs: the design used and SRS. For the survey analyst this comparison is not the main point at issue since the survey has already been designed and the data collected and it remains to choose methods of analysis which are appropriate to the given situation. A statistical inference procedure must be proposed which captures the essential features of the sample as it relates to the population structure. It is not a comparison of alternative designs which is of interest but a comparison of alternative inference procedures with their associated assumptions.

Suppose, for example, that the data have been collected by a two-stage design and the clusters exhibit a positive intra-cluster correlation τ. If the survey analyst proposes an inference procedure which makes no provision for this correlation then this may have consequences for both the efficiency of point estimators and the bias of the standard error. If, for example, the analyst carries out a regression analysis under the usual ordinary least squares (OLS) assumptions and these are inappropriate, the estimators of the regression parameters $\boldsymbol{\beta}$ will be unbiased but not fully efficient, whereas the estimator of the covariance matrix of $\hat{\boldsymbol{\beta}}$ will be biased. The latter bias will affect the inference procedure. In a model-based approach the inference procedure will be based upon a specified model. Thus analogous to a design effect is a misspecification effect, which summarizes the impact on an inference procedure of possible misspecification of the model.

This distinction between a design effect and a misspecification effect is discussed in Chapter 2. At this stage we wish simply to emphasize that for the analyst it is not the design so much as the population structure which matters. If a population comprises clusters which exhibit intra-cluster correlation then the analyst must take account of the correlation structure in the statistical model. Even if an SRS design is used, the observations from the same cluster will still be correlated for analysis purposes even though the design effect is one. It is the population structure not the design which generates the correlation structure. Of course the design is important since a clustered sample from the population will guarantee correlated observations, whereas a simple random sample will tend to yield only at most one observation per cluster. Nevertheless, for the analyst it is the outcome of the distribution of sample observations across clusters that matters rather than the intention as expressed by the survey design.

1.5 OBJECTIVES AND STRUCTURE OF THE BOOK

In general terms our aim has been to explore the implications of applying a variety of methods of statistical analysis to complex survey data. Within this, there are three major objectives:

1. To investigate the effects of complex designs and population structure on standard procedures.
2. To explore methods of making simple adjustments to standard procedures to restore their properties.
3. To develop new statistical methods more appropriate to complex survey data.

Objective (1) is addressed by Parts A and B of the book and includes the use of standard procedures such as OLS regression analysis, loglinear models and χ^2 testing procedures for contingency tables. The properties of standard significance tests and standard error estimators, associated with these procedures, are investigated in Part A, whereas the properties of point estimators of parameters, and in particular the extent to which they are biased, are left till Part B. This order does not reflect any assumption about the priority of importance of these two topics. Indeed, point estimation is sometimes considered to be of more fundamental importance than standard error estimation or significance testing. Rather, the order follows that usually taken in the survey sampling tradition, where the sampling bias is generally taken to be of $0(n_0^{-1})$ after suitable p-weighting (where n_0 is the sample size), which is less important in large samples than the standard error of $0(n_0^{-1/2})$ (e.g. Cochran, 1977, p. 160). The extent to which this conventional view holds for both design-based and model-based procedures will be examined in Part B.

Objective (2) is closely linked to objective (1) since results on the impact of complex designs and population structures on standard procedures are often immediately suggestive of modifications to those procedures. Thus objective (2) is addressed side by side with objective (1) through Parts A and B. Likewise no definitive distinction can be drawn between objective (2) and (3) and so some of Parts A and B might also be said to deal with objective (3).

The potential for adjusting standard methods or developing new methods is constrained by the availability of information on the structure of the design and population. In particular, secondary analysts of surveys are often faced with very limited information at the unit level, for reasons, for example, of confidentiality. Thus simple adjustments using minimal accompanying information, are given particular emphasis in Part A.

Part C provides an alternative to standard procedures by modelling explicitly the population structure and hence falls totally under the heading of objective (3). The distinctive aspect of Part C, which has already been discussed in Section 1.3, is the adoption of a disaggregated approach. Again, the position of Part C in the book reflects no assumption about its relative importance. Rather, it appears after Parts A and B since its approach is less familiar in the sample survey tradition. The discussion in Part C makes some references to Parts A and B but is largely self-contained, and hence it should be possible to read chapters in Part C without first having read the whole of Parts A and B.

It will be noted that we have chosen to make the fundamental divisions for the book on the basis of the conceptual issues that underpin the methods. An alternative approach would have been to separate the book into methods suitable for discrete and continuous data. The literature in journals is largely separated on this basis and each has developed a specific form of notation. We have resisted this approach and integrated continuous and discrete variable methods in each part of the book in order to emphasize that the same concepts apply to both cases.

Within the separate parts of the book, chapters on methods for continuous variables, such as linear regression analysis, tend to appear before chapters on methods for discrete variables, such as logistic modelling. One reason for this order is that the criteria of bias and variance, which are used extensively, are most naturally introduced using continuous variables.

1.6 FORMAL FRAMEWORK, TERMINOLOGY AND NOTATION

The aim of this section is to provide a source of reference, for use when reading the remainder of the book, by setting out the basic framework, terminology and notation. The emphasis is on definitions and so discussion is intentionally limited.

1.6.1 Finite Populations, Superpopulation Models and Parameters

A *finite population* is a set of N_o units, where N_o is the *size* of the population. The subscript o emphasizes that N_o is the overall population size. The population is labelled U and the units are labelled from 1 to N_o so that

$$U = \{1, 2, \ldots, N_o\}.$$

The *survey variables*, denoted by the vector $\mathbf{y}$, form the set of variables of interest, the values of which will be recorded for each unit in the sample, but which are not generally known before sampling takes place. The population values of $\mathbf{y}$ are denoted $\mathbf{y}_1, \ldots, \mathbf{y}_{N_o}$.

A *finite population parameter* is some function of the values $\mathbf{y}_1, \ldots, \mathbf{y}_{N_o}$, for example: the *finite population mean*

$$\bar{\mathbf{y}}_U = N_o^{-1} \sum_{t=1}^{N_o} \mathbf{y}_t,$$

the *finite population covariance matrix*

$$\mathbf{V}_U = (N_o - 1)^{-1} \sum_{t=1}^{N_o} (\mathbf{y}_t - \bar{\mathbf{y}}_U)(\mathbf{y}_t - \bar{\mathbf{y}}_U)'.$$

Under a *superpopulation model* the values $\mathbf{y}_1,\ldots,\mathbf{y}_{N_o}$ are assumed to be a joint realized outcome of the random vectors $\mathbf{Y}_1,\ldots,\mathbf{Y}_{N_o}$. The joint distribution of $\mathbf{Y}_1,\ldots,\mathbf{Y}_{N_o}$ is denoted by ξ. Two examples of superpopulation models are:

(a) $\mathbf{Y}_1,\ldots,\mathbf{Y}_{N_o}$ are IID as a multivariate normal random vector with mean vector $\boldsymbol{\mu}$ and covariance matrix $\boldsymbol{\Sigma}$.

(b) Let $\mathbf{y}=(u,\mathbf{v}')'$, where u is a binary dependent variable taking values 0 and 1 and $\mathbf{v}$ is a vector of explanatory variables. Writing $\mathbf{Y}_t=(U_t,\mathbf{V}_t')'$, assume that $\mathbf{Y}_1,\ldots,\mathbf{Y}_{N_o}$ are IID with the logistic conditional distribution:

$$P(U_t=1\,|\,\mathbf{V}_t=\mathbf{v})=\exp(\mathbf{v}'\boldsymbol{\beta})/[1+\exp(\mathbf{v}'\boldsymbol{\beta})].$$

Superpopulation parameters are characteristics of ξ, for example $\boldsymbol{\mu}$ and $\boldsymbol{\Sigma}$ in (a) and $\boldsymbol{\beta}$ in (b).

As mentioned in Section 1.1, superpopulation parameters may often be preferred to finite population parameters as targets for inference in analytic surveys. However, if N_o is large, there will often be little numerical difference between the two since, for example, in (a) above

$$\bar{\mathbf{y}}_U=\boldsymbol{\mu}+O_p(N_o^{-1/2}),\qquad \mathbf{V}_U=\boldsymbol{\Sigma}+O_p(N_o^{-1/2}). \tag{1.1}$$

Hence in much of this book the distinction between finite population and superpopulation parameters will not be emphasized and both will be denoted by Greek letters.

1.6.2 Auxiliary Information and Population Structure

Auxiliary information is information about the finite population known to the survey designers before the sample is selected. This information usually consists of the values $\mathbf{z}_1,\ldots,\mathbf{z}_{N_o}$ of a vector $\mathbf{z}$ of *auxiliary variables*, or else some function of these values such as the vector of population totals $\sum\mathbf{z}_t$. Auxiliary information is important, not only because it usually influences the sampling design, but also because it can be relevant to the choice of target parameter, since it reflects the structure of the population.

Some examples of auxiliary variables are:

(a) *Size measures.* In surveys, where the units are establishments, measures of size z_t are often available. For example, for a survey of schools, the Ministry of Education may provide estimates of the number of children at every school.

(b) *Stratification variables.* Often U is partitioned into H *strata* $h=1,\ldots,H$ with sizes $N_1,\ldots,N_H$, where $\sum N_h=N_o$. Stratum membership could be represented by an $(H-1)\times 1$ vector $\mathbf{z}_t$ of dummy variables. However, we shall more usually use the double suffix ht to represent the tth unit in the hth stratum, $h=1,\ldots,H$; $t=1,\ldots,N_h$ and, for example, label the corresponding $\mathbf{y}$ value by $\mathbf{y}_{ht}$. Note that this represents a slight abuse of notation since the unit label t is used in two ways.

(c) *Multi-stage populations.* Often U has a hierarchical structure, being partitioned into *primary units*, which themselves are partitioned into *secondary units* and so on. Common examples arise from hierarchical geographical classifications. Units in a *two-stage population* containing M primary units of sizes $N_1, \ldots, N_M$ are labelled analogously to stratified populations, with the tth unit in the cth primary unit being denoted ct for $c = 1, \ldots, M$; $t = 1, \ldots, N_c$ and with its associated $\mathbf{y}$ value being denoted $\mathbf{y}_{ct}$. For a *three-stage population* the tth unit in the cth secondary unit in the dth primary unit is labelled dct for $d = 1, \ldots, L$; $c = 1, \ldots, M_d$; $t = 1, \ldots, N_{dc}$ giving $\mathbf{y}$ values $\mathbf{y}_{dct}$. For a *stratified multi-stage population* the label h is added. For example, units in a *stratified three-stage population* are labelled $hdct$ for $h = 1, \ldots, H$; $d = 1, \ldots, L_h$; $c = 1, \ldots, M_{hd}$; $t = 1, \ldots, N_{hdc}$. In this case

$$N_o = \sum_h \sum_d \sum_c N_{hdc}.$$

Given the structure of the population implied by the distribution of such auxiliary variables, it is often natural to define parameters with respect to these variables. For example, for a stratified population, model (a) in Section 1.6.1 might be extended to

$$Y_{ht} \sim N(\mu_h, \boldsymbol{\Sigma}) \qquad h = 1, \ldots, H; \; t = 1, \ldots, N_h.$$

More generally, the values of the auxiliary variables may themselves be viewed as being generated by some superpopulation model. Thus $\mathbf{y}_1, \ldots, \mathbf{y}_{N_o}, \mathbf{z}_1, \ldots, \mathbf{z}_{N_o}$ are assumed to be a joint outcome of the random vectors $\mathbf{Y}_1, \ldots, \mathbf{Y}_{N_o}, \mathbf{Z}_1, \ldots, \mathbf{Z}_{N_o}$ with joint distribution ξ. Parameters of the conditional distribution of $\mathbf{Y}_1, \ldots, \mathbf{Y}_{N_o}$ given $\mathbf{Z}_1 = \mathbf{z}_1, \ldots, \mathbf{Z}_{N_o} = \mathbf{z}_{N_o}$ are relevant to the disaggregated approach discussed in Section 1.3, whereas parameters of the marginal distribution of $\mathbf{Y}_1, \ldots, \mathbf{Y}_{N_o}$ (unconditional on $\mathbf{Z}_1, \ldots, \mathbf{Z}_{N_o}$) are relevant to the aggregated approach (Scott, 1977; Sugden and Smith, 1984; Skinner, 1988).

1.6.3 Sampling Designs

A *sample* is a subset S of U. The *sample size* is the number of distinct units in S. A *sampling design* is a function $p(.)$ defined for every subset S of U, such that

$$p(S) \geqslant 0 \quad \text{for all } S \subset U$$

and

$$\sum_S p(S) = 1.$$

The value $p(S)$ determines the probability that sample S is selected from U. The actual sample selected will be denoted by s and the actual sample size by n_o, where again the suffix o emphasizes that n_o is the overall sample size. A sampling design $p(.)$ has *fixed size* if every sample S such that $p(S) > 0$ has the same size.

The *inclusion probability* π_t of unit t is the probability of selecting that unit, which is

$$\pi_t = \sum_{S \supset t} p(S),$$

where the summation is over all samples S containing t. The *joint inclusion probability* $\pi_{tt'}$ of units t and t' is the probability of selecting both units t and t', which is

$$\pi_{tt'} = \sum_{S \supset t,t'} p(S),$$

where the summation is over all samples S containing both t and t'. An *epsem* (equal probability of selection method) *design* (also called a *self-weighting design*) is one in which $\pi_t = \pi$ is fixed for all $t \in U$.

Some examples of sampling designs are:

(a) *Simple random sampling* (*SRS*).

$$p(S) = \binom{N_o}{n_o}^{-1} \quad \text{if } S \text{ has size } n_o,$$
$$= 0 \quad \text{otherwise.}$$

This is an example of a fixed-size epsem design. The inclusion probabilities are $\pi_t = n_o/N_o$.

(b) *Stratified SRS*. Samples of sizes $n_1, \ldots, n_H$, are selected independently using SRS within each stratum $h = 1, \ldots, H$, giving an overall sample size of $n_o = \sum_1^H n_h$. The inclusion probability is n_h/N_h for units in stratum h and so this design is epsem only under *proportional allocation* when n_h/N_h is fixed for $h = 1, \ldots, H$.

(c) *Multi-stage designs.* A *two-stage design* generates a sample containing m primary units with $n_1, \ldots, n_m$ units sampled within the sampled primary units (labelled without loss of generality $c = 1, \ldots, m$). Again $n_o = \sum n_c$. In general $m < M$ and $n_c < N_c$, $c = 1, \ldots, m$. A *three-stage design* generates a sample containing l primary units with m_d secondary units sampled within the dth sampled primary unit, $d = 1, \ldots, l$, and with n_{dc} units sampled within the dth sampled primary unit. Cochran (1977, Chapters 9–11) gives details of randomization mechanisms used in such designs. One commonly used design involves selecting the primary and higher stage units with *probability proportional to size* (Skinner, 1986c) and selecting a fixed number of units at the final stage by SRS. This provides an epsem design.

1.6.4 Inference

Inference generally follows one of two approaches.

(a) The *design-based approach*: inference proceeds with respect to the sampling distribution of statistics over repeated samples S generated by the sampling design $p(S)$ (with the values $\mathbf{y}_1, \ldots, \mathbf{y}_{N_o}, \mathbf{z}_1, \ldots, \mathbf{z}_{N_o}$ held fixed).
(b) The *model-based approach*: inference proceeds with respect to the sampling distribution of statistics over repeated realizations $\mathbf{y}_1, \ldots, \mathbf{y}_{N_o}$ (and possibly $\mathbf{z}_1, \ldots, \mathbf{z}_{N_o}$) generated by the model ξ (with the selected sample s held fixed).

The *design-bias* or *p-bias* of a point estimator $\hat{\theta}$ of θ is

$$\text{bias}_p(\hat{\theta}) = E_p(\hat{\theta}) - \theta,$$

where $E_p(.)$ is the expectation with respect to the sampling distribution in (a), denoted the *p-distribution*, that is

$$E_p(\hat{\theta}) = \sum_S \hat{\theta} p(S),$$

where $\hat{\theta}$ is treated as a function of S, with $\mathbf{y}_1, \ldots, \mathbf{y}_{N_o}$ (and $\mathbf{z}_1, \ldots, \mathbf{z}_{N_o}$) held fixed. The estimator $\hat{\theta}$ is said to be *design-unbiased* or *p-unbiased* if $\text{bias}_p(\hat{\theta}) = 0$.

The *model-bias* or *ξ-bias* of $\hat{\theta}$ is

$$\text{bias}_\xi(\hat{\theta}) = E_\xi(\hat{\theta} - \theta), \tag{1.2}$$

where, for example, if $\hat{\theta}$ is a function of $\mathbf{y}_1, \ldots, \mathbf{y}_{n_o}$ alone and $\mathbf{Y}_1, \ldots, \mathbf{Y}_{n_o}$ are assumed under ξ to be IID with continuous probability density function f then

$$E_\xi(\hat{\theta} - \theta) = \int \cdots \int (\hat{\theta} - \theta) f(\mathbf{y}_1) \cdots f(\mathbf{y}_{n_o}) \, d\mathbf{y}_1, \ldots, d\mathbf{y}_{n_o}.$$

Note that finite population parameters θ are random variables with respect to ξ, which is why the general expression for $\text{bias}_\xi(\hat{\theta})$ in equation (1.2) contains θ within the parentheses. The estimator θ is *model-unbiased* or *ξ-unbiased* if $\text{bias}_\xi(\hat{\theta}) = 0$.

The *p-variance* and *ξ-variance* of $\hat{\theta}$ are respectively

$$\text{var}_p(\hat{\theta}) = E_p[\hat{\theta} - E_p(\hat{\theta})]^2, \qquad \text{var}_\xi(\hat{\theta}) = E_\xi[\hat{\theta} - E_\xi(\hat{\theta})]^2.$$

The *p-mean square error* (MSE) and *ξ-MSE* are respectively

$$\text{MSE}_p(\hat{\theta}) = E_p(\hat{\theta} - \theta)^2, \qquad \text{MSE}_\xi(\hat{\theta}) = E_\xi(\hat{\theta} - \theta)^2.$$

Similarly, other criteria used in the classical sampling theory approach to statistical inference (Cox and Hinkley, 1974, p. 48) may be defined with respect to either approach. For example, a confidence interval C is said to have *p-confidence level* $1 - \alpha$ if the proportion of the intervals C, obtained from repeated samples S generated by $p(S)$, which contain θ is $1 - \alpha$. Or a procedure for testing a hypothesis H_0 using a test statistic T and a critical region R has a *ξ-significance level* α if the probability that $T \in R$ is α over repeated realizations of $\mathbf{y}_1, \ldots, \mathbf{y}_{N_o}, \mathbf{z}_1, \ldots, \mathbf{z}_{N_o}$ under ξ when H_0 is true.

The extension of classical ideas of asymptotic inference to sample surveys is, however, more complex. Two asymptotic ideas that we shall find useful are those of consistency and of asymptotic distribution.

A point estimator $\hat{\theta}$ is said to be *consistent* for θ if $\hat{\theta} - \theta$ converges to zero as the sample size n_o increases. Various definitions of convergence may be used. We shall require convergence in probability, that is

$$\text{for all } \varepsilon > 0, \quad P(|\hat{\theta} - \theta| < \varepsilon) \to 1 \quad \text{as } n_o \to \infty. \tag{1.3}$$

When the probability measure $P(.)$ in equation (1.3) applies to the p-distribution of $\hat{\theta} - \theta$, $\hat{\theta}$ is said to be *design-consistent for* θ, whereas if equation (1.3) applies to the ξ-distribution, $\hat{\theta}$ is said to be *model-consistent* for θ.

Consistency may be viewed as a necessary condition for an estimator to be 'satisfactory'. It is not a sufficient condition since, amongst consistent estimators, we seek estimators for which $\text{MSE}(\hat{\theta})$ is small. Consistency tends to replace the condition of unbiasedness, familiar in discussions of linear estimators in descriptive surveys (e.g. Cochran, 1977), since in analytic surveys many estimators are non-linear and the property of unbiasedness would be far too stringent to set as a necessary condition.

Results on asymptotic distributions are useful for justifying interval estimators or hypothesis testing procedures. For example, the use of $\hat{\theta} \pm 1.96 v_0^{1/2}$ as an approximate 95% confidence interval for θ, where v_0 is an estimator of $\text{var}(\hat{\theta})$, could be 'justified' by the result that

$$v_0^{-1/2}(\hat{\theta} - \theta) \to N(0, 1) \text{ in law} \quad \text{as } n_o \to \infty. \tag{1.4}$$

Here the asymptotic distribution may be a p-distribution or ξ-distribution according to the approach taken.

A final useful asymptotic concept is the stochastic order of magnitude notation $O_p[a(n_o)]$, where $a(n_o)$ is a given function of n_o such as $a(n_o) = n_o^{-1/2}$ or n_o^{-1}. This notation has indeed already been used in equation (1.1). We write

$$\hat{\theta} - \theta = O_p[a(n_o)]$$

if for all $\varepsilon > 0$ there exists a constant δ and an integer n' such that

$$n_o \geqslant n' \Rightarrow P[|\hat{\theta} - \theta| \leqslant \delta a(n_o)] \geqslant 1 - \varepsilon.$$

The probability measure $P(.)$ may be defined either with respect to the p-distribution or the ξ-distribution.

The problem with expressions such as (1.2)–(1.4) is that their rigorous definition requires the specification not only of a sequence of samples of increasing size, but also of a sequence of finite populations. Furthermore, for both the sequences of samples and populations, it is necessary to specify sequences of sizes of strata, first stage units, etc. This implies that, for any given sample and population, there is no single correct asymptotic framework, the choice is effectively arbitrary. Moreover, different frameworks may imply different results. For example, for a sample of $m = 50$ clusters each containing $n = 50$ units so that $n_o = mn = 2500$, one might consider asymptotic theory as $m \to \infty$ or as $n \to \infty$ or as both m and $n \to \infty$, and $\hat{\theta}$ may be consistent for θ and asymptotically normally distributed as $m \to \infty$ for fixed n but not as $n \to \infty$ for fixed m. There is no way in which we can choose between the 'correctness' of these frameworks. All that we can really ask of asymptotic theory is that it gives us some guidance as to when certain approximations may be expected to work well. As Scott and Wu (1981, p. 99) remark: 'Asymptotic results for finite populations are always rather artificial since there is really only a single population and we could imagine this embedded in a sequence of populations endowed with almost any limiting properties we wanted. The results will only be useful if the conditions point towards those situations in which a normal approximation should work well.' Hence these difficulties should be borne in mind whenever asymptotic results are presented.

1.6.5 Additional Notation

In addition to the notation introduced in Sections 1.6.1–1.6.4, the following conventions are adopted.

(a) *Alphabets*: *Greek lower case*—model parameters; finite population parameters, when they do not need to be distinguished from model parameters (exception: π, inclusion probability).

Greek upper case—parameter matrices; parameter spaces (exceptions: $\sum$, summation, $\mathbf{\Pi}$, matrix of π's).

Roman lower case—values of scalar or vector variables; statistics; indices.

Roman upper case—matrices; random variables; limits to index sets.

Bold face—vectors and matrices.

NB. The convention of using upper-case letters to denote finite population characteristics is avoided.

(b) *Moments*—$\bar{y}$, v and $\mathbf{V}$ denote sample mean, variance and covariance matrix respectively.

$\bar{y}_U$, v_U and $\mathbf{V}_U$ are the corresponding finite population quantities if they need to be distinguished from the (super) population quantities μ, σ^2 and $\mathbf{\Sigma}$. $E(.)$, var(.), cov(.,.) and Var(.) are the expectation, variance, covariance and variance–covariance matrix operators respectively. Cov(.,.) is the covariance operator applied to two random vectors.

(c) *Matrices.* Vectors and matrices are in bold type. All vectors are assumed to be column vectors.

Transpose is denoted $'$.

(d) *Categorical data.* Sample proportions are denoted $\bar{y}$ not p and population proportions μ not π in order to emphasize the link with continuous variables.
(e) *Chi-squared test statistics.* Test statistics referred to the chi-squared distribution are denoted X^2 and suitably subscripted, e.g. X_P^2 (Pearson), X_{LR}^2 (likelihood ratio), X_W^2 (Wald).
(f) *Correlation*—a correlation between two variables is denoted ρ. An intracluster correlation is denoted τ.

PART A

Aggregated Analysis: Standard Errors and Significance Tests

CHAPTER 2

INTRODUCTION TO PART A

C. J. Skinner

2.1 INTRODUCTION

Standard errors and significance tests tend to play a more important role in analytic surveys than in descriptive surveys for at least three reasons.

1. Analytic statistics, such as subclass means, are often based on small subsample sizes and hence have higher standard errors (relative to point estimates) than corresponding statistics, such as overall sample means, in descriptive surveys. Relative standard errors for multivariate techniques such as multiple regression can also be higher due, for example, to multicollinearity.
2. The size of standard errors (reflecting sampling errors and non-sampling variance) relative to bias (such as non-response bias) may be higher for analytic statistics, such as differences in subclass means or regression coefficients, because systematic errors for such statistics may tend to cancel out reducing bias.
3. Significance tests are widely used in analytic surveys to help select models.

The problem addressed by this part of the book is that conventional formulae for standard errors and test procedures, as implemented in standard statistical packages such as SPSS X (1986) or SAS (1985), are based on the assumption of independent and identically distributed (IID) observations, or equivalently that samples are selected by simple random sampling (SRS) with replacement, and this assumption is almost never valid for sample survey data. The aim of Part A is to investigate the consequences for standard errors, interval estimates and significance tests of the kinds of departures from this IID assumption which arise in sample surveys. In this chapter we concentrate on basic ideas and on simple statistics, mainly means. In Chapters 3–5 extensions to more complex statistics are considered.

In Sections 2.2–2.11 we study the impact of complex designs on IID-based procedures. Our general conclusion is that this impact can be very severe, in the sense that the results of IID-based procedures applied to complex survey data can be very misleading. In Sections 2.12–2.14 we consider what alternative procedures may be used which allow for complex designs. In Section 2.15 we introduce Chapters 3–5.

2.2 STANDARD ERRORS, DESIGN EFFECTS AND MISSPECIFICATION EFFECTS

In this section we ask the basic question: what do we mean by the *effect* of a complex design on the standard error of a statistic $\hat{\theta}$?

One way to answer this question is in terms of the difference between the variance of the (randomization) distribution of $\hat{\theta}$ induced by the true complex design, $\text{var}_{\text{true}}(\hat{\theta})$, and the variance of the distribution of $\hat{\theta}$ induced by a hypothetical SRS design of the same sample size n_o, $\text{var}_{\text{SRS}}(\hat{\theta})$, as measured by *Kish's* (1965) *design effect* ($\text{deff}_{\text{Kish}}$) for $\hat{\theta}$:

$$\text{deff}_{\text{Kish}}(\hat{\theta}) = \text{var}_{\text{true}}(\hat{\theta})/\text{var}_{\text{SRS}}(\hat{\theta}). \tag{2.1}$$

This definition is a natural measure of relative efficiency when comparing alternative designs at the design stage of a survey. However, we consider it inappropriate for analysis, since at this stage we have used just one design and comparisons with other designs which might have been used are of little relevance.

Kish's design effect measures the effect of the design on the variance of $\hat{\theta}$. Of more relevance at the analysis stage is the effect of the design on *variance estimators.* Let $v_0 = \widehat{\text{var}}_{\text{IID}}(\hat{\theta})$ be an estimator of the variance of $\hat{\theta}$ derived under the model assumption that observations are IID, or equivalently under the design assumption that the sample is selected by SRS with replacement. For example, v_0 might take the form $\hat{\theta}(1-\hat{\theta})/n_o$ if $\hat{\theta}$ is an estimated proportion. The effect of a complex design on v_0 may be assessed by comparing the distribution of v_0 under the true design (or model) with $\text{var}_{\text{true}}(\hat{\theta})$, the actual variance of $\hat{\theta}$. A conventional summary measure of performance in this setting would be

$$\text{bias}(v_0) = E_{\text{true}}(v_0) - \text{var}_{\text{true}}(\hat{\theta}).$$

However, in order to draw parallels with Kish's design effect, it is convenient to express the bias instead in proportional terms in the following definition of the *misspecification effect* (meff) for the variance estimator v_0 of $\hat{\theta}$:

$$\text{meff}(\hat{\theta}, v_0) = \text{var}_{\text{true}}(\hat{\theta})/E_{\text{true}}(v_0). \tag{2.2}$$

Thus $\text{meff}(\hat{\theta}, v_0)$ measures the extent to which v_0 tends to underestimate (or overestimate) $\text{var}(\hat{\theta})$, so that

$$\begin{aligned} \text{meff}(\hat{\theta}, v_0) &< 1 \quad \text{if bias}(v_0) > 0, \\ &= 1 \qquad\qquad\quad = 0, \\ &> 1 \qquad\qquad\quad < 0, \end{aligned}$$

and the further meff is away from 1, the more the assumptions upon which v_0 is based are judged to be misspecified.

Unlike Kish's design effect, which is a design-based measure, $\text{meff}(\hat{\theta}, v_0)$ may be defined either as a design-based measure or as a model-based measure by taking the moments E_{true} and var_{true} in equation (2.2) to refer to either the randomization distribution induced by the true complex design or to the true model respectively. We now illustrate the distinction between Kish's design effect and our misspecification effect by an example.

Example 1. Cluster Sampling of one cluster of size 2

Consider an infinite population of clusters of size 2 (e.g. husband and wife couples) with mean θ, variance σ^2 and correlation τ between units within clusters (intra-cluster correlation, see Section 2.8) for a variable y. Suppose a single cluster is selected at random and the values y_1 and y_2 are recorded for the two units within the cluster. An unbiased estimator of θ is $\hat{\theta} = (y_1 + y_2)/2$ and, based upon the false assumption that the design is SRS or equivalently that y_1 and y_2 are IID, the variance of $\hat{\theta}$ is $\text{var}_{\text{SRS}}(\hat{\theta}) = \sigma^2/2$ with an unbiased estimator $v_0 = (y_1 - y_2)^2/4$. In fact the actual variance of $\hat{\theta}$ is $\text{var}_{\text{true}}(\hat{\theta}) = \sigma^2(1 + \tau)/2$ and the actual expectation of v_0 is $\sigma^2(1 - \tau)/2$. Hence from equations (2.1) and (2.2):

$$\text{deff}_{\text{Kish}}(\hat{\theta}) = 1 + \tau, \qquad \text{meff}(\hat{\theta}, v_0) = (1 + \tau)/(1 - \tau)$$

and, for example, if $\tau = 0.8$ then $\text{deff}_{\text{Kish}}(\hat{\theta}) = 1.8$ and $\text{meff}(\hat{\theta}, v_0) = 9$, implying that the actual variance of $\hat{\theta}$ is 80% above what it would have been had the hypothetical SRS design of size 2 been used, but is 800% above the average value taken by the IID-based variance estimator v_0.

Thus, in this rather extreme example, the two measures of the effect of cluster sampling take very different values. We shall see, however, that for stratified multi-stage designs with moderate sample sizes $E_{\text{true}}(v_0)$ will often be approximately equal to $\text{var}_{\text{SRS}}(\hat{\theta})$ and hence from equations (2.1) and (2.2) the two measures will often be close. Furthermore, given data from a complex survey it is common to estimate the denominator of $\text{deff}_{\text{Kish}}(\hat{\theta})$ by v_0 (e.g. Kish, 1965, p. 258), which is also the natural estimator of the denominator of $\text{meff}(\hat{\theta}, v_0)$. Thus, even though the population values may be unequal, estimated values of $\text{deff}_{\text{Kish}}$ and meff are often equal:

$$\widehat{\text{deff}}_{\text{Kish}}(\hat{\theta}) = \widehat{\text{meff}}(\hat{\theta}, v_0) = v/v_0, \tag{2.3}$$

where v is a consistent estimator of $\text{var}_{\text{true}}(\hat{\theta})$ (see Section 2.13). Thus the results of empirical studies of Kish's design effects using the estimator in equation (2.3) are relevant to the properties of misspecification effects.

Let us now turn to the use of the term 'misspecification effect'. Within a model-based framework there are at least two reasons why this term seems more natural than 'design effect'.

1. Under the model-based approach $\text{meff}(\hat{\theta}, v_0)$ depends only on the model relationship between the units in the actual sample selected and not on how the sample was selected. Thus in Example 1 the meff of 9 reflects the effect of misspecifying the correlation between the two selected units as zero when in fact it is $\tau = 0.8$. It does not matter how this sample was derived, whether by cluster sampling (the actual design), by SRS in which two units within one cluster were selected by rare chance or by any other hypothetical design. The meff is simply interpreted as the effect of misspecifying the model relationship between the sampled units and so, conditional on the selected sample, the meff is independent of the design.
2. Departures from IID assumptions with consequent bias in v_0 may occur for reasons completely unrelated to the sampling design. For example, in Chapter 3 we shall consider properties of the least-squares regression coefficient $\hat{\beta} = \sum y_t(x_t - \bar{x})/\sum(x_t - \bar{x})^2$ and its conventional variance estimator
$$v_0 = \sum[y_t - \bar{y} - \hat{\beta}(x_t - \bar{x})]^2/[(n_0 - 2)\sum(x_t - \bar{x})^2].$$
We shall see that bias in v_0, as an estimator of the true variance of $\hat{\beta}$, may arise not only from clustering and stratification related to the design but also because of other failures of model assumptions such as homoscedasticity which are unconnected with the design. In such cases any attempt to partition the bias of v_0 into an effect related to the design and an effect due to heteroscedasticity is likely to be arbitrary. We may still, however, interpret $\text{meff}(\hat{\beta}, v_0)$ as an overall measure of the effect of misspecifying the model.

Despite these two points, there are many cases when it does seem to us reasonable to refer to the misspecification effect defined in equation (2.2) as a design effect (understood to be quite distinct from Kish's design effect in equation (2.1)).

With reference to point (1), although a meff is independent of the choice of design conditional on the selected sample, it may be strongly influenced by the design unconditionally. Thus in Example 1, under the actual cluster sampling design the model-based meff equals $(1 + \tau)/(1 - \tau)$ with probability 1 unconditionally across all possible samples generated by the design, whereas under SRS the meff equals 1 with probability 1 (since the probability of selecting two units within the same cluster in an infinite population is zero). Hence, when v_0 is used as an estimator of $\text{var}_{\text{true}}(\bar{y})$, an SRS design tends to make inference very robust to departures from the underlying model assumption that $\tau = 0$, whereas cluster sampling tends to make inference very sensitive to misspecification of this assumption. Because of this strong ultimate influence of the design, it seems reasonable to refer to the meff as a design effect. Indeed the use of the term 'design effect' is now so widespread that it would seem pedantic to avoid it.

Point (2) may be approached in the same way. Any variance estimator v_0 is based upon various assumptions, for example independent observations and homoscedasticity in the regression example above. A complex design may tend to make inference more or less robust to departures from each of these assumptions, for example cluster sampling increases sensitivity to departures from the independence assumption but not necessarily from the homoscedasticity assumption. In cases where bias of v_0 arises from departures from assumptions to which v_0 is made sensitive under the complex design but to which v_0 would have been robust under SRS, it does seem reasonable to refer to the meff as a design effect. In other words, there is a subset of all meffs, which are wholly attributable to the effect of a complex design, which we shall call design effects. Thus in Example 1 the meff of $(1 + \tau)/(1 - \tau)$ may be considered as being wholly attributable to the effect of cluster sampling and hence may be called a design effect.

The above discussion refers to the model-based approach. Under the design-based approach, the design determines the true randomization distributions of $\hat{\theta}$ and v_0 in equation (2.2) and hence determines the meff directly. Hence the term 'design effect' seems generally appropriate for the meff. For many practical stratified multi-stage designs the form of meffs under the design-based and model-based approaches are analogous. Thus in Example 1 the meff has the same form $(1 + \tau)/(1 - \tau)$ under both approaches; the difference lies only in the interpretation of τ. For the model-based approach τ is the model correlation between the pair of values y_{c1} and y_{c2} within any cluster c, whereas for the design-based approach τ is defined (cf. Section 2.8) as the population characteristic

$$\tau = 2\sum_c (y_{c1} - \theta)(y_{c2} - \theta) \Big/ \sum_c \sum_t (y_{ct} - \theta)^2.$$

Such a parallel between the two approaches is shown to hold for two-stage sampling and a broad class of statistics by Skinner (1986b). Because of this general similarity, we shall usually not draw attention to the distinction between the two approaches in Part A of this book.

We have assumed so far that v_0 is based upon IID assumptions. There is, however, no reason why v_0 cannot be based upon any given set of assumptions or model specification. For example, given a stratified clustered design, v_0 might be an estimator of $\text{var}(\hat{\theta})$ which takes account of the clustering but ignores the stratification. In this case $\text{meff}(\hat{\theta}, v_0)$ will measure the effect of stratification after allowing for clustering.

To summarize this section, we shall measure the effect of a complex design on the estimated standard error $v_0^{1/2}$ of a statistic $\hat{\theta}$ by the misspecification effect $\text{meff}(\hat{\theta}, v_0)$. In general, this represents the biasing effect on v_0 of departures from the assumptions upon which v_0 is based. Under the condition that these departures can be attributed to the effect of the sampling design, the misspecification effect may be called a *design effect* and we write

$$\text{deff}(\hat{\theta}, v_0) = \text{meff}(\hat{\theta}, v_0) = \text{var}_{\text{true}}(\hat{\theta})/E_{\text{true}}(v_0). \tag{2.4}$$

Henceforth we shall generally assume this condition holds and use only the term 'design effect' and the notation $\text{deff}(\hat{\theta}, v_0)$. We emphasize that *this design effect is quite distinct from Kish's design effect*, defined in equation (2.1), in particular because it depends upon two arguments $\hat{\theta}$ and v_0 rather than just one. A point estimator of the design effect is

$$\widehat{\text{deff}}(\hat{\theta}, v_0) = v/v_0,$$

where v is a consistent estimator of $\text{var}_{\text{true}}(\hat{\theta})$ as discussed in Section 2.13.

2.3 CONFIDENCE INTERVALS

Suppose that, based upon IID assumptions, $\hat{\theta}$ is a point estimator of θ and, as in Section 2.2, v_0 is an estimator of the variance of $\hat{\theta}$. An IID based confidence interval is conventionally obtained by assuming that

$$t_0 = (\hat{\theta} - \theta)/v_0^{1/2} \tag{2.5}$$

has approximately a standard normal distribution

$$t_0 \underset{\text{IID}}{\sim} N(0, 1) \tag{2.6}$$

so that, for example, a two-sided *IID-based 95% confidence interval* for θ is given by

$$\begin{aligned} C_0 &= \{\theta : |t_0| < 1.96\} \\ &= (\hat{\theta} - 1.96 v_0^{1/2}, \hat{\theta} + 1.96 v_0^{1/2}). \end{aligned} \tag{2.7}$$

What is the effect of a complex design on the properties of this IID-based interval? Under the true complex design (or true model) it may still be reasonable to assume that $\hat{\theta}$ is approximately normal (see Section 2.14) and that $\hat{\theta}$ is approximately unbiased for θ (see Part C for the case where the design induces bias in $\hat{\theta}$) i.e.

$$\hat{\theta} \underset{\text{true}}{\sim} N[\theta, \text{var}_{\text{true}}(\hat{\theta})]. \tag{2.8}$$

In addition, for large samples, v_0 will be close to its expectation $E_{\text{true}}(v_0)$ so that the distribution of t_0 will be close to the distribution of $(\hat{\theta} - \theta)/E_{\text{true}}(v_0)^{1/2}$ and so from equation (2.8) we have approximately

$$\begin{aligned} t_0 &\underset{\text{true}}{\sim} N[0, \text{var}_{\text{true}}(\hat{\theta})/E_{\text{true}}(v_0)] \\ &= N[0, \text{deff}(\hat{\theta}, v_0)], \end{aligned} \tag{2.9}$$

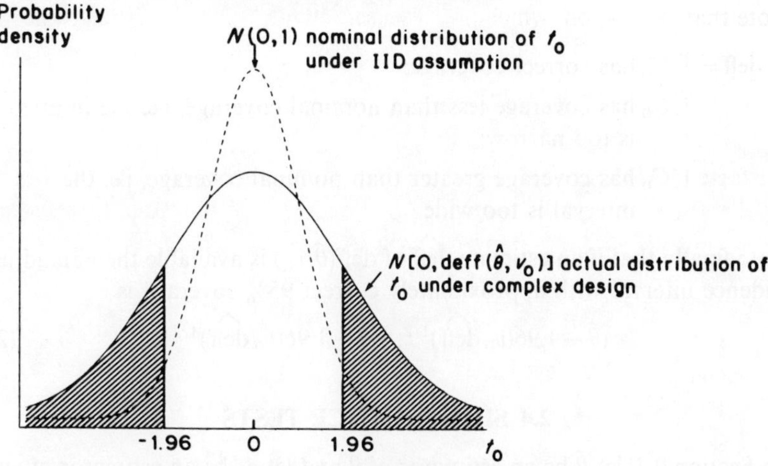

Figure 2.1 Effect of a complex design on the distribution of the pivotal t-statistic $t_0 = (\hat{\theta} - \theta)/v_0^{1/2}$.

from equation (2.4), assuming that it is reasonable to attribute the departures from the IID assumption to the effect of the design. Thus the design effect $\text{deff}(\hat{\theta}, v_0)$ measures the inflation (or deflation) in variance of the IID-based pivotal statistic t_0 due to the design. This is illustrated in Figure 2.1 where the unshaded area under the $N[0, \text{deff}(\hat{\theta}, v_0)]$ curve represents the coverage of the IID-based interval C_0 in equation (2.7). Values of the actual coverage of 95% and 99% IID-based confidence intervals are given in Table 2.1 for various values of deff.

Table 2.1 Coverage of IID-based confidence intervals and significance levels of IID-based significance tests

	Actual coverage		Actual significance level	
Design	Nominal level		Nominal level	
effect	95%	99%	5%	1%
0.9	96	99.3	4	0.7
1.0	95	99	5	1
1.5	89	96	11	4
2.0	83	93	17	7
2.5	78	90	22	10
3.0	74	86	26	14

Note that

if $\text{deff} = 1$, C_0 has correct coverage;

> 1, C_0 has coverage less than nominal coverage, i.e. the interval is too narrow;

< 1, C_0 has coverage greater than nominal coverage, i.e. the interval is too wide.

Note finally that if an estimate $\widehat{\text{deff}}$ of $\text{deff}(\hat{\theta}, v_0)$ is available then an adjusted confidence interval with approximately correct 95% coverage is

$$(\hat{\theta} - 1.96(v_0 \widehat{\text{deff}})^{1/2}, \quad \hat{\theta} + 1.96(v_0 \widehat{\text{deff}})^{1/2}). \tag{2.10}$$

2.4 SIGNIFICANCE TESTS

As in Section 2.3, let $\hat{\theta}$ be an estimator of θ and let v_0 be an estimator of $\text{var}(\hat{\theta})$ based upon IID assumptions. A conventional IID-based normal-theory test of the simple hypothesis $H_0 : \theta = \theta_0$ is obtained by rejecting H_0 if $T_0 = (\hat{\theta} - \theta_0)/v_0^{1/2}$ lies in the critical region of a standard normal distribution. For example, for a two-tailed test with a 95% significance level:

$$H_0 \text{ is rejected if } |T_0| > 1.96. \tag{2.11}$$

What is the effect of a complex design on this test procedure? Under H_0, T_0 is identical to t_0 in equation (2.5) and so from equation (2.9)

$$T_0 \sim N[0, \text{deff}(\hat{\theta}, v_0)] \text{ under } H_0.$$

Hence the actual significance level of the test in equation (2.11) (i.e. the probability that $|T_0| > 1.96$ when H_0 is true) is given by the shaded area in Figure 2.1 and

if $\text{deff} = 1$, actual significance level = nominal significance level

> 1, actual significance level > nominal significance level

< 1, actual significance level < nominal significance level

Values of the actual significance levels for different values of $\text{deff}(\hat{\theta}, v_0)$ and for 5% and 1% nominal significance levels are given in Table 2.1.

In applications it is often considered more desirable, in order to be on the safe side, for a test to be too conservative (actual level < nominal level) than the reverse. Hence when using survey data for significance testing one should be particularly wary of large design effects. A design effect of 'only' 1.5 can more than double a nominal significance level of 5%.

If an estimate $\widehat{\text{deff}}$ of $\text{deff}(\hat{\theta}, v_0)$ is available then the replacement of T_0 by the corrected test statistic

$$T_0' = T_0 / \widehat{\text{deff}}^{1/2} = (\hat{\theta} - \theta_0)/(v_0 \widehat{\text{deff}})^{1/2} \tag{2.12}$$

in the usual IID procedure does lead to a test with approximately the correct significance level.

2.5 DESIGN FACTORS AND EFFECTIVE SAMPLE SIZE

Design factor (deft) and effective sample size (n_e) are two terms derived from the design effect (deff) which each have useful interpretations. The *design factor* is defined as

$$\text{deft} = \text{deff}^{1/2}$$

and is the appropriate inflation factor for standard errors and confidence intervals. This quantity is widely used (see e.g. Verma *et al.*, 1980).

The *effective sample size* (Kish, 1965, p. 259) is defined as

$$n_e = n_o/\text{deff}$$

and has the property that SRS formulae such as (2.7) or (2.11) become 'correct' if n_o is replaced by n_e. This property assumes that the SRS formula for v_0 involves n_o as a denominator, i.e. $v_0 = k/n_0$ so that, when n_o is replaced by n_e, v_0 is replaced by $k/n_e = v_0$ deff and hence, as in equations (2.10) and (2.12), the adjusted confidence interval or test procedure has approximately correct coverage or significance level respectively. The effective sample size may be viewed as measuring the 'amount of information' in a sample. Thus if deff = 1.25 then a given sample of $n_o = 10\,000$ households only contains as much information, for estimating a parameter θ with a given precision, as a SRS of $n_e = 8000$ households. This definition should not be confused with the alternative use of the term 'effective sample size' to denote the number of distinct units in a with-replacement sample (e.g. Cassel *et al.*, 1977, p. 7).

2.6 EXAMPLE 2: CLUSTER SAMPLING OF UNEMPLOYMENT RATES

A numerical example is now presented to illustrate the discussion in Sections 2.2–2.5. Table 2.2 contains unemployment rates in August 1987 for $N_o = 30$ counties (*units*) within Great Britain. These counties are nested within ten regions (*clusters*). To avoid the complexity of unequal cluster sizes, we have taken only the first three counties, by alphabetical order, from within each region.

As an example of a complex design, consider the following *cluster sampling* (CS) design:

Design CS: Select three clusters by SRS without replacement and select all units within each selected cluster, giving a fixed sample size of $n_o = 9$ units.

In order to compute Kish's design effect we also consider an SRS design:

Design SRS: Select $n_o = 9$ units by SRS without replacement.

Table 2.2 Unemployment rates in 30 counties in Great Britain, August 1987

c, Region	t, County	Unemployment rate Y_{ct} (%)
1. South-east	1. Bedfordshire	8.3
	2. Berkshire	5.5
	3. Buckinghamshire	5.9
2. East Anglia	1. Cambridgeshire	7.3
	2. Norfolk	10.2
	3. Suffolk	7.5
3. South-west	1. Avon	9.1
	2. Cornwall	14.3
	3. Devon	11.2
4. West Midlands	1. Hereford and Worcester	10.2
	2. Shropshire	12.1
	3. Staffordshire	10.7
5. East Midlands	1. Derbyshire	11.6
	2. Leicestershire	8.4
	3. Lincoln	11.1
6. Yorkshire and Humberside	1. Humberside	13.8
	2. N. Yorkshire	8.9
	3. S. Yorkshire	16.6
7. North-west	1. Cheshire	11.8
	2. Lancashire	12.1
	3. Greater Manchester	13.7
8. North	1. Cleveland	18.8
	2. Cumbria	9.3
	3. Durham	15.7
9. Wales	1. Clwyd	14.3
	2. Dyfed	15.5
	3. Glamorgan	14.7
10 Scotland	1. Borders	8.8
	2. Central	15.8
	3. Dumfries and Galloway	12.3
Total		$\theta = 11.5$

Source: Department of Employment (1987) *Employment Gazette* **95**.

Let us suppose that we wish to estimate the (unweighted) mean unemployment rate θ across all 30 counties. The true value of θ is 11.5. Consider the sample mean $\bar{y}$ as a point estimator of θ. The sampling distributions of $\bar{y}$ under both designs are illustrated in Figure 2.2, where the histogram for the CS design represents the distribution of $\bar{y}$ over all 120 possible samples of 3 clusters from 10 and the histogram for the SRS design is obtained by generating 1000 independent simple random samples.

Each histogram has approximately a normal shape and is centred at $\theta = 11.5$,

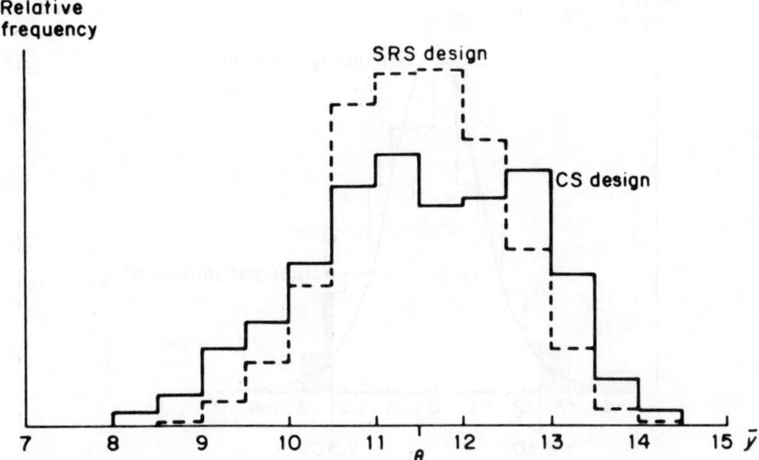

Figure 2.2 Histograms of sampling distributions of $\bar{y}$ for SRS and CS designs.

i.e. $\bar{y}$ is unbiased for θ. The variances of $\bar{y}$ under the two designs are

$$\text{var}_{\text{SRS}}(\bar{y}) = 0.85, \qquad \text{var}_{\text{CS}}(\bar{y}) = 1.59.$$

Hence from equation (2.1) Kish's design effect is

$$\text{deff}_{\text{Kish}}(\bar{y}) = 1.59/0.85 = 1.86,$$

reflecting a substantial inflation of the variance of $\bar{y}$ due to cluster sampling. We consider two estimators of $\text{var}_{\text{CS}}(\bar{y})$ which ignore the cluster sampling: the SRS with replacement estimator

$$v_{\text{WR}} = \sum_s (y_{ct} - \bar{y})^2/[n_o(n_o - 1)]$$

and the SRS without replacement estimator

$$v_{\text{WOR}} = (1 - n_o/N_o)v_{\text{WR}}.$$

Under the CS design we obtain

$$E_{\text{CS}}(v_{\text{WR}}) = 1.13, \qquad E_{\text{CS}}(v_{\text{WOR}}) = 0.79,$$

so using the definition of the design effect in equation (2.4) we have

$$\text{deff}(\bar{y}, v_{\text{WR}}) = 1.59/1.13 = 1.41,$$
$$\text{deff}(\bar{y}, v_{\text{WOR}}) = 1.59/0.79 = 2.01.$$

Thus v_{WOR} tends to underestimate the actual variance of $\bar{y}$ by a factor of 2, whereas v_{WR} tends to underestimate $\text{var}_{\text{CS}}(\bar{y})$ by somewhat less since the

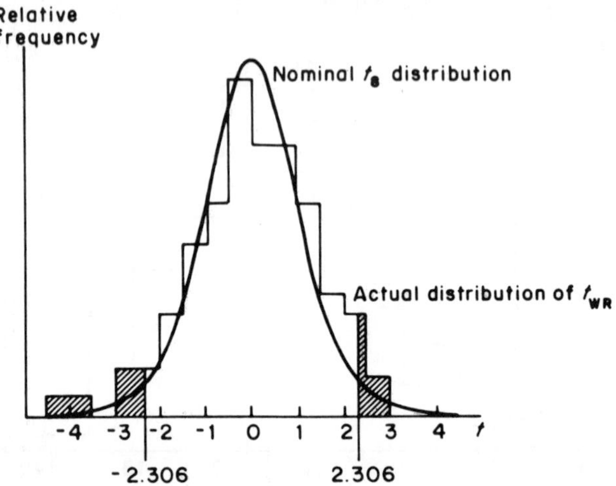

Figure 2.3 Histogram of actual sampling distribution of t_{WR} superimposed on Student's t-distribution with 8 d.f.

variance-inflating effect of cluster sampling is somewhat offset by the variance-reducing effect of without-replacement sampling. This illustrates the point made in Section 2.2 that it is insufficient to refer just to the deff of $\bar{y}$. It is also necessary to specify the choice of variance estimator v_0 since, as we see, $\text{deff}(\bar{y}, v_0)$ can be quite different for different choices of v_0.

In order to assess the effect of the CS design on IID-based confidence intervals and significance tests we evaluate the distribution of the t-statistic (cf. equation (2.5)):

$$t_{WR} = (\bar{y} - \theta)/v_{WR}^{1/2}.$$

A histogram of the sampling distribution of t_{WR} under the CS design is displayed in Figure 2.3 together with the distribution of t_8, Student's t-statistic with 8 d.f. which is the distribution of t_{WR} under the assumption of IID normal observations. Under this assumption a confidence interval for θ with nominal 95% coverage is $\bar{y} \pm 2.306 v_{WR}^{1/2}$, where 2.306 is the 95% point of t_8. This interval contains θ if $|t_{WR}| < 2.306$ which in fact occurs for only 110 or 91.7% of the 120 cluster samples. Thus, as indicated in Section 2.3, a deff of greater than 1 leads to the actual coverage being too low. Theoretically, as in equation (2.9), we might approximate the true distribution of

$$t_{WR}/\text{deff}(\bar{y}, v_{WR})^{1/2} = t_{WR}/1.19$$

by t_8, in which case $|t_{WR}| < 2.306$ if and only if $|t_8| < 2.306/1.19 = 1.94$. From tables of Student's distribution, $\Pr(|t_8| < 1.94) = 0.912$ which is a reasonable approximation to the observed 91.7%.

Similarly an IID-based test with nominal significance level 5% of the true hypothesis $H_0:\theta = 11.5$ is obtained by rejecting H_0 if $|t_{WR}| > 2.306$ and, as indicated above, the actual significance level of this test is 8.3%. Thus, as discussed in Section 2.4, a deff of greater than 1 increases the size of a type I error.

2.7 DESIGN EFFECTS OF WITHOUT-REPLACEMENT SAMPLING AND FINITE POPULATION CORRECTIONS

As a first illustration of the form of a deff for a specific design and a specific parameter, consider the problem of making inference about the mean θ of a finite population U of N_o units using a SRS *without replacement* design of size n_o and the sample mean $\bar{y}$ as the estimator $\hat{\theta}$.

Under a notional SRS *with replacement* design of size n_o we know that

$$E(\bar{y}) = \theta, \qquad \text{var}(\bar{y}) = \sigma^2/n_o,$$

where

$$\sigma^2 = N_o^{-1} \sum_U (y_t - \theta)^2$$

and that an unbiased estimator of σ^2/n_o is

$$v_{WR} = \sum_s (y_t - \bar{y})^2/[(n_o - 1)n_o]. \tag{2.13}$$

What happens to this variance estimator when the true design is without replacement? In this case (see e.g. Cochran, 1977, Chapter 2)

$$E_{\text{true}}(\bar{y}) = \theta, \qquad \text{var}_{\text{true}}(\bar{y}) = \left(1 - \frac{n_o}{N_o}\right)\left(\frac{N_o}{N_o - 1}\right)\frac{\sigma^2}{n_o},$$

$$E_{\text{true}}(v_{WR}) = \left(\frac{N_o}{N_o - 1}\right)\sigma^2/n_o,$$

where $1 - n_o/N_o$ is the *finite population correction* (f.p.c.), and hence from equation (2.4)

$$\text{deff}(\bar{y}, v_{WR}) = 1 - n_o/N_o.$$

Thus the design effect of without-replacement sampling is simply the f.p.c. It should be emphasized that this is only if the parameter of interest is the *finite population* mean. In Chapter 1 it was noted that this is seldom the case in analytic surveys when the parameter of interest is more commonly a model parameter. In this case, if the values $y_1, \ldots, y_n$ are assumed under the model to be IID with mean θ, the target parameter, then the model deff is unity and no f.p.c. is necessary.

2.8 DESIGN EFFECTS OF MULTI-STAGE SAMPLING AND INTRA-CLUSTER CORRELATION

As a second illustration of the form of a deff, consider again the estimation of a finite population mean θ but now using a cluster sampling design. Initially, a design-based approach is adopted. Suppose the population consists of M clusters each of size N so that

$$\theta = \sum_{c=1}^{M} \sum_{t=1}^{N} y_{ct}/(MN),$$

where y_{ct} is the value for the tth unit in the cth cluster. Assume that m clusters are selected by SRS without replacement and y_{ct} is recorded for each $t = 1, \ldots, N$ in each selected cluster labelled, without loss of generality, $c = 1, \ldots, m$. As in Section 2.7, assume for the point estimator of θ:

$$\bar{y} = \sum_{c=1}^{m} \sum_{t=1}^{N} y_{ct}/(mN). \tag{2.14}$$

We consider the deff, $\text{deff}(\bar{y}, v_{\text{WOR}})$, of $\bar{y}$ and the SRS without replacement variance estimator:

$$v_{\text{WOR}} = (1 - m/M) \sum_{c=1}^{m} \sum_{t=1}^{N} (y_{ct} - \bar{y})^2/[(mN - 1)mN], \tag{2.15}$$

which corresponds to equation (2.13) adjusted by the f.p.c. Under the actual design $\bar{y}$ is unbiased for θ and, from Cochran (1977, Theorem 9.2),

$$\text{var}_{\text{true}}(\bar{y}) = \left(\frac{1}{m} - \frac{1}{M}\right) \frac{M\sigma^2(1 + (N-1)\tau)}{N(M-1)}, \tag{2.16}$$

where

$$\sigma^2 = \sum_{c=1}^{M} \sum_{t=1}^{N} (y_{ct} - \theta)^2/MN$$

is the population variance and

$$\tau = \frac{2 \sum_{c} \sum_{t<t'} (y_{ct} - \theta)(y_{ct'} - \theta)}{(N-1)MN\sigma^2} \tag{2.17}$$

is the *intra-cluster correlation*.

It is also straightforward to show that under the given design:

$$E_{\text{true}}(v_{\text{WOR}}) = \left(1 - \frac{m}{M}\right) \frac{\sigma^2}{(mN-1)} \left[1 - \frac{(M-m)(1 + (N-1)\tau)}{mN(M-1)}\right]. \tag{2.18}$$

Substituting equations (2.16) and (2.18) into (2.4) gives

$$\operatorname{deff}(\bar{y}, v_{\mathrm{WOR}}) = \frac{M(mN-1)(1+(N-1)\tau)}{mN(M-1)\{1-[(M-m)/mN(M-1)](1+(N-1)\tau)\}}, \tag{2.19}$$

and if m is large this approximates the well-known formula (Cochran, 1977, p. 242)

$$\operatorname{deff}(\bar{y}, v_{\mathrm{WOR}}) \simeq 1+(N-1)\tau. \tag{2.20}$$

As a numerical illustration, consider the value $\operatorname{deff}(\bar{y}, v_{\mathrm{WOR}}) = 2.01$ in Example 2. The intra-cluster correlation may be computed from Table 2.1 as $\tau = 0.36$, and substituting this value together with $M = 10$, $m = 3$, $N = 3$ into equation (2.19) gives a theoretical deff of 2.00 which differs from 2.01 only because of rounding error in the calculations. Using the approximate formula in equation (2.20) gives a deff of 1.72, but we would not expect the approximation to be very good here since m only equals 3.

From equation (2.20) we see the importance of the intra-cluster correlation τ. This measures the degree to which units within the same cluster are similar with respect to y. The more homogeneous clusters are with respect to y, the greater will be τ. A simple exercise is to show that τ is the Pearson product–moment correlation coefficient for the set of $MN(N-1)$ pairs $(y_{ct}, y_{ct'})$ where $c = 1, \ldots, M$, $t \neq t' = 1, \ldots, N$. In survey applications τ is almost always positive. For example, if y is household income and the clusters are geographical areas then a positive τ would correspond to the fact that households with similar incomes tend to live in similar areas or 'birds of a feather flock together' (Fuller, 1973). Only in very unusual situations is τ negative, for example if y is the 0–1 variable indicating male or female, the clusters are households and the units are adult individuals then τ may be negative because of a tendency for many households to contain couples of one male and one female. In fact it is always true that $\tau \geqslant (N-1)^{-1}$ since we may write

$$\tau = \frac{\sum_c \left[\sum_t (y_{ct} - \theta) \right]^2 - MN\sigma^2}{(N-1)MN\sigma^2} \geqslant \frac{-MN\sigma^2}{(N-1)MN\sigma^2} = \frac{-1}{(N-1)}.$$

The above derivation of the deff of cluster sampling is based on classical randomization theory. We now consider the derivation of corresponding results using a model-based approach. Consider a conventional *one-way random effects model* for y_{ct}:

$$y_{ct} = \theta + \alpha_c + \varepsilon_{ct} \qquad c = 1, \ldots, M, t = 1, \ldots, N, \tag{2.21}$$

where the α_c and ε_{ct} are mutually independent with zero means and

$$\operatorname{var}(\alpha_c) = \tau\sigma_0^2, \qquad \operatorname{var}(\varepsilon_{ct}) = (1-\tau)\sigma_0^2.$$

Note that again τ may be interpreted as an intra-cluster correlation coefficient since under the model:

$$\tau = \frac{\operatorname{cov}(y_{ct}, y_{ct'})}{[\operatorname{var}(y_{ct})\operatorname{var}(y_{ct'})]^{1/2}}, \qquad t \neq t'.$$

Now, of course, τ is bounded below by zero rather than $(N-1)^{-1}$ although the model can be modified to remove this constraint (McHugh and Mielke, 1968; Royall, 1976).

The natural unbiased estimator of θ is again $\bar{y}$, and under IID assumptions, ignoring the clustering and the f.p.c., the natural variance estimator is, as in equation (2.13),

$$v_{\text{IID}} = \sum_{c=1}^{m} \sum_{t=1}^{N} (y_{ct} - \bar{y})^2 / [(mN-1)mN].$$

The actual variance of $\bar{y}$ under the above model is

$$\begin{aligned} \operatorname{var}_{\text{true}}(\bar{y}) &= (mN)^{-2} \operatorname{var}\left(\sum_{c=1}^{m} N\alpha_c + \sum_{c=1}^{m}\sum_{t=1}^{N} \varepsilon_{ct} \right) \\ &= (mN)^{-2}(N^2 m\tau\sigma_0^2 + mN(1-\tau)\sigma_0^2) \\ &= [1 + (N-1)\tau]\sigma_0^2/mN. \end{aligned}$$

Also

$$\begin{aligned} E_{\text{true}}(v_{\text{IID}}) &= \sigma_0^2[mN - (1 + (N-1)\tau)]/[(mN-1)mN] \\ &\simeq \sigma_0^2/mN \text{ if } m \text{ is large,} \end{aligned} \tag{2.22}$$

and hence substituting in equation (2.4) gives

$$\operatorname{deff}(\bar{y}, v_{\text{IID}}) \simeq 1 + (N-1)\tau, \quad \text{as before.}$$

Note that this result is also valid for two-stage sampling where N units are sampled from some larger number of units within each cluster (see Hansen *et al.*, 1953, p. 165 for the corresponding equivalence under randomization theory). The result may also easily be extended to three-stage sampling. Suppose that values y_{dct} are recorded for units $t = 1, \ldots, n$ within each second-stage unit (2SU) $c = 1, \ldots, m$ within each primary sampling unit (PSU) $d = 1, \ldots, l$ and suppose

$$\begin{aligned} \operatorname{cov}(y_{dct}, y_{d'c't'}) &= \sigma_0^2, & d = d', c = c', t = t', \\ &= \tau_2\sigma_0^2, & d = d', c = c', t \neq t', \\ &= \tau_1\sigma_0^2, & d = d', c \neq c', \\ &= 0, & d \neq d', \end{aligned}$$

where τ_2 is the intra-2SU correlation and τ_1 is the intra-PSU correlation. Then

$$\text{var}(\bar{y}) = \text{var}\left[\sum_{d=1}^{l}\sum_{c=1}^{m}\sum_{t=1}^{n} y_{dct}/lmn\right]$$

$$= (1 + (m-1)n\tau_1 + (n-1)\tau_2)\sigma_0^2/lmn$$

and as in equation (2.22)

$$E(v_{\text{IID}}) \simeq \sigma_0^2/lmn \quad \text{if } l \text{ is large}$$

so that

$$\text{deff}(\bar{y}, v_{\text{IID}}) \simeq 1 + (m-1)n\tau_1 + (n-1)\tau_2. \tag{2.23}$$

Thus there are contributions to the deff from both the first and second stages of sampling. Generally, τ_1 is less than τ_2 and this reduces the contribution of the first stage relative to the second stage, but this effect may be more than offset by the factor $(m-1)n$ which is larger than $n-1$ when $m \neq 1$. Empirical examples are given in Section 2.10.

2.9 DESIGN EFFECTS OF STRATIFIED SAMPLING

As a final illustration of the form of a deff, consider the estimation of the finite population mean θ using proportionate stratified sampling. Now

$$\theta = \sum_{h=1}^{H} W_h\theta_h,$$

where W_h is the population proportion and θ_h is the mean in stratum h. If $n_h = W_h n_o$ units are selected by SRS within each stratum then the sample mean $\bar{y}$ remains unbiased for θ and (Cochran, 1977, p. 93)

$$\text{var}(\bar{y}) = (1 - n_o/N_o)\sum_h W_h\sigma_h^2/n_o,$$

where N_o is the population size and σ_h^2 is the variance in stratum h. Using the same variance estimator v_{WOR} as in equation (2.15) and assuming n_o large, it is straightforward to obtain

$$E(v_{\text{WOR}}) \simeq (1 - n_o/N_o)\sum_h W_h[\sigma_h^2 + (\theta_h - \theta)^2]/n_o,$$

so from equation (2.4)

$$\text{deff}(\bar{y}, v_{\text{WOR}}) \simeq \left(\sum_h W_h\sigma_h^2\right) \bigg/ \sum_h W_h[\sigma_h^2 + (\theta_h - \theta)^2],$$

or in terms of the 'analysis of variance':

$$N_o^{-1}\sum_h\sum_t (y_{ht} - \mu)^2 = \sum_h W_h(\theta_h - \theta)^2 + \sum_h W_h\sigma_h^2$$

total variance = between-stratum variance + within-stratum variance we have

$$\text{deff}(\bar{y}, v_{\text{WOR}}) = \frac{\text{within-stratum variance}}{\text{total variance}}.$$

Thus the deff is always less than or equal to 1 and indeed it may be further reduced by the use of optimal disproportionate stratification.

2.10 SOME ACTUAL DESIGN EFFECTS

Some of the preceding theory is now illustrated with estimated deffs from actual surveys. In general a deff will be determined by three factors:

1. The form of statistic and variance estimator;
2. The distribution of the variable(s) in the population;
3. The design.

Note that (1) includes possible variance-reducing effects of post-stratification, ratio-adjustment, etc., and (2) includes possible variance-inflating effects of measurement error such as interviewer variance. In the examples below we shall restrict attention again to means (proportions for 0–1 variables) with the usual variance estimator in equation (2.13) and we explore the effects of (2) and (3). Empirical examples for more complex statistics are given in Chapters 3–5.

Example 1. British Social Attitudes Survey

This survey has a stratified three-stage self-weighting design:

1. One hundred and fourteen parliamentary constituencies (PSUs) are selected from England, Scotland and Wales by stratified probability proportional to size (PPS) sampling.
2. One polling district (2SU) is selected within each selected PSU by PPS.
3. Twenty-two individuals (electors) are selected within each selected 2SU by systematic sampling from the electoral register.

Further details are given by Jowell *et al.* (1986). Of the $114 \times 22 = 2508$ target individuals, interviews were achieved with 1804 (72%). From the preceding theory we expect the deffs to represent the combined effect of a factor less than 1 due to stratification and a factor greater than 1 due to three-stage sampling. Substituting $m = 1$, $n = 1804/114 = 16$ into equation (2.23) suggests that the latter factor will have approximate form

$$1 + 15\tau_2 \tag{2.24}$$

where τ_2 is the intra-polling district correlation. Some estimated deffs for selected variables together with values of τ_2 estimated from equation (2.24) assuming no stratification effects are given in Table 2.3. These deffs were estimated using formula (2.3) and so may be viewed as estimates either of Kish's deff in equation

Table 2.3 Estimated design effects from British Social Attitudes Survey, 1985

Variable	deff	$\hat{\tau}_2$
Completed full-time education by age sixteen	3.06	0.14
Rent house from local authority	5.20	0.28
Identify with Conservative Party	1.99	0.07
Expect unemployment to go up	1.44	0.03

Note: $\hat{\tau}_2 = (\text{deff} - 1)/15$ is estimated from equation (2.24).
Source: Jowell *et al.* (1986, p. 179).

(2.1) or of our own deff in equation (2.4). The largest deff is, not surprisingly, for the housing tenure variable since local authority housing tends to occur in clusters of dwellings and this induces a high positive intra-polling district correlation. Conversely the lowest deff is for unemployment expectations indicating that the heterogeneity of these expectations within polling districts is close to the heterogeneity in the general population. In summary, Jowell *et al.* (1986) report that deffs for classification variables are in the range 1–2.25 with the exception of housing tenure, whilst deffs for behavioural and attitudinal variables lie in the range 1.7–3.2 with deffs for attitudinal variables tending to lie at the lower end of this range.

Example 2. World Fertility Survey

In this example two similar surveys of fertility in Nepal and Colombia are compared. Both surveys involve a stratified three-stage sample of women (Verma *et al.*, 1980) although the design characteristics are somewhat different. In the notation of Section 2.8 the average values of l, m and n are approximately:

	Nepal	*Colombia*
l = number of sampled PSUs	40	35
m = number of sampled 2SUs per PSU	2.5	18
n = number of sampled women per 2SU	60	3.2

Design effects averaged across several variables within each of five classes are given in Table 2.4. Estimates of τ_1 and τ_2 in equation (2.23) are also given. Design effects are generally higher for the Nepal survey. This is mainly because of the larger number of sampled women per 2SU and per PSU than in the Colombian survey. Design effects vary between classes of variables but are

Table 2.4 Estimated design effects from World Fertility Survey

Variable class	Nepal					Colombia				
	deff	% red. from strata[a]	% from PSUs[b]	$\hat{\tau}_1$	$\hat{\tau}_2$	deff	% red. from strata[a]	% from PSUs[b]	$\hat{\tau}_1$	$\hat{\tau}_2$
Nuptuality behaviour	3.61	10	52	0.02	0.02	1.96	41	87	0.02	0.06
Fertility behaviour	3.69	0	68	0.02	0.01	1.77	27	95	0.01	0.02
Fertility preferences	7.62	24	56	0.04	0.05	1.59	19	86	0.01	0.04
Contraceptive knowledge	10.30	19	47	0.05	0.08	9.30	14	91	0.14	0.33
Contraceptive use	4.97	−18	51	0.02	0.03	5.02	12	94	0.07	0.12
All	5.38	12	57	0.03	0.03	3.06	23	91	0.03	0.09

[a] These columns give the percentage reduction of the deff due to stratification e.g. the deff of 3.61 would have been $4.00 = 3.61/(1-0.10)$ without stratification.
[b] These columns give the percentage of deff -1 attributable to PSU differences rather than to 2SU differences, i.e. $[1-(\text{deff}_2-1)/(\text{deff}-1)] \times 100$, where deff_2 is the deff computed ignoring PSU differences.
Source: Verma *et al.* (1980).

noticeably high for contraceptive knowledge which may be expected to be clustered geographically. This clustering seems to be especially high within 2SUs in Colombia ($\hat{\tau}_2 = 0.33$).

The percentage reduction in variance due to stratification is relatively small (12 and 23%) compared with the percentage increase due to multistage sampling (438 and 206% even after allowing for stratification). This relative importance of multi-stage sampling compared to stratification in determining deffs is characteristic of many surveys.

The relative contribution of the PSUs to the deff compared with the contribution of the 2SUs is much greater for the Colombian survey. According to formula (2.23) the approximate contribution of the PSUs is

$$\frac{(m-1)n\tau_1}{(m-1)n\tau_1 + (n-1)\tau_2}.$$

If $\tau_1 = \tau_2$, this is $(m-1)n/(mn-1)$ which is 60% for Nepal and 96% for Colombia. The actual average contributions are 57% for Nepal and 91% for Colombia which is because τ_1 tends to be slightly less than τ_2 for Nepal and quite a bit less for Colombia. Hence we see the important influence on the deffs of the relative values of m and n combined with the relative values of τ_1 and τ_2.

Further discussion of deffs in the World Fertility Survey is given by Kish *et al.* (1976), Verma *et al.* (1980) and Little (1982b).

2.11 MULTIVARIATE DESIGN EFFECTS

So far we have only considered inference about a scalar parameter θ. Suppose now that $\hat{\boldsymbol{\theta}}$ is a $p \times 1$ vector and that $\mathbf{V}_0$ is an estimator of the $p \times p$ covariance matrix of $\hat{\boldsymbol{\theta}}$ based upon IID assumptions (or some other set of null assumptions). Then we may generalize equation (2.4) to define the *design effects matrix* of $\hat{\boldsymbol{\theta}}$ and $\mathbf{V}_0$ as

$$\text{deff}(\hat{\boldsymbol{\theta}}, \mathbf{V}_0) = \boldsymbol{\Delta} = E_{\text{true}}(\mathbf{V}_0)^{-1}\,\text{Var}_{\text{true}}(\hat{\boldsymbol{\theta}}).$$

The eigenvalues of this matrix $\delta_1 \geqslant \cdots \geqslant \delta_p$ are called *generalized design effects* and have the property that δ_1 and δ_p define bounds for the univariate deffs for any linear combination $\mathbf{c}'\hat{\boldsymbol{\theta}}$ of $\hat{\boldsymbol{\theta}}$:

$$\delta_1 = \max_{\mathbf{c}} \text{deff}(\mathbf{c}'\hat{\boldsymbol{\theta}}, \mathbf{c}'\mathbf{V}_0\mathbf{c}),$$

$$\delta_p = \min_{\mathbf{c}} \text{deff}(\mathbf{c}'\hat{\boldsymbol{\theta}}, \mathbf{c}'\mathbf{V}_0\mathbf{c}).$$

This follows from a standard matrix theory result (e.g. Mardia *et al.*, 1979, Corollary A.9.2.1). In the special case where $\text{deff}(\hat{\boldsymbol{\theta}}, \mathbf{V}_0)$ is the $p \times p$ identity matrix, $\delta_1, \ldots, \delta_p$ are all equal to unity so that the univariate deffs of all linear combinations of $\hat{\boldsymbol{\theta}}$ are also unity. A numerical illustration is given in Section 3.3.6.

The generalized design effects are also relevant to the properties of test procedures based upon the IID-based Wald statistic

$$X_0^2 = (\hat{\boldsymbol{\theta}} - \boldsymbol{\theta})'\mathbf{V}_0^{-1}(\hat{\boldsymbol{\theta}} - \boldsymbol{\theta}), \tag{2.25}$$

which arise in several applications (see Chapters 3–5). Under IID assumptions X_0^2 has asymptotically a chi-squared distribution with p degrees of freedom when $\hat{\boldsymbol{\theta}}$ is asymptotically normal (Mardia *et al.*, 1979, Corollary 3.2.1.1), but under the actual design the asymptotic distribution of X_0^2 is that of the linear combination of χ_1^2 variables

$$\sum_{i=1}^{p} \delta_i Z_i^2, \tag{2.26}$$

where $Z_1, \ldots, Z_p$ are independent standard normal random variables (e.g. Mardia *et al.*, 1979, Theorem 3.4.4). Note that, in the special case $p = 1$, X_0^2 is equal to t_0^2, defined in equation (2.5), and result (2.26) follows from equation (2.9).

In applications it is often desirable to obtain critical values K_α such that $\Pr(X_0^2 > K_\alpha) = \alpha$. Computer algorithms for deriving K_α from $\boldsymbol{\Delta}$ and α are available (e.g. Sheil and O'Muircheartaigh, 1977), but a simple *first-order correction* is obtained by approximating the distribution of $X_0^2/\bar{\delta}$ by χ_p^2 where

$\bar{\delta} = p^{-1}\sum\delta_i$ is the mean generalized deff. This approximation is exact when the δ_i are equal. For unequal δ_i a more accurate *second-order correction*, due to Satterthwaite (1946), is obtained by approximating the distribution of $X_0^2/\bar{\delta}(1+a^2)$ by χ^2 with $p/(1+a^2)$ d.f. where a is the coefficient of variation of the δ_i. Both corrections may be obtained directly from $\mathbf{\Delta}$ without computation of the eigenvalues since

$$p\bar{\delta} = \mathrm{tr}(\mathbf{\Delta}), \qquad p\bar{\delta}^2(1+a^2) = \mathrm{tr}(\mathbf{\Delta}^2),$$

where 'tr' denotes the trace operator.

2.12 ALLOWING FOR COMPLEX DESIGNS

To summarize our main conclusion so far: confidence intervals and significance tests based on IID assumptions can be highly misleading in the presence of complex designs. The reader should, by now, be becoming impatient to move on to the obvious next question: if procedures based on IID assumptions are so bad under complex designs, what alternative procedures should we use?

The choice of an appropriate procedure involves three main stages (assuming large sample normal-theory inference), at each of which considerations of sampling design may be relevant:

1. The definition of the target parameter θ;
2. The choice of point estimator $\hat{\theta}$;
3. The choice of standard error estimator $\widehat{\text{s.e.}}(\hat{\theta})$.

Definition of Target Parameter θ

We have already discussed the definition of θ in Chapter 1. We distinguished between the traditional survey sampling approach, in which θ is assumed to be a given population parameter defined prior to data analysis, and the modelling approach, in which θ is only determined following an iterative model-building process. For now, we just assume that θ is well defined.

Choice of Point Estimator $\hat{\theta}$

Part B of this book deals with the choice of point estimator when the complex design introduces bias or inconsistency into the estimator based on IID assumptions. We assume here that such inconsistency has already been corrected for in our estimator $\hat{\theta}$.

There still remains, however, the issue of efficiency, that is amongst consistent estimators of θ we should like to choose a point estimator with relatively small (if not minimum) variance. We have already shown that the variance of an estimator under a complex design is generally different from its variance under

IID assumptions and so it follows that a point estimator which is efficient under IID assumptions need no longer be efficient under a complex design. We shall, however, again leave this issue of choice of optimal point estimator until Part B, although we do note that, unless auxiliary population information is available, the gains in efficiency that may be made are often minor (see, for example, Scott and Holt's (1982) discussion of generalized least squares (GLS) versus OLS estimation in regression).

Choice of Standard Error Estimator

Two broad approaches may be distinguished.

A. Use a non-parametric 'survey-sampling-type' variance estimation procedure, such as will be described in Section 2.13, which allows for the complex design.
B. Extend the parametric statistical model to accommodate the structure of the population used in the design (e.g. strata, clusters) and make inference with respect to this model.

Approach (A) is likely to seem the one most natural to survey samplers. A non-parametric estimator v of the true variance $\text{var}_{\text{true}}(\hat{\theta})$ is chosen, which is consistent under the true design or model. A new pivotal statistic:

$$t = v^{-1/2}(\hat{\theta} - \theta) \tag{2.27}$$

is then defined using the estimated standard error $v^{1/2}$ and inference about θ is carried out by approximating the distribution of t by the standard normal distribution $N(0, 1)$. For example, for confidence intervals, t replaces t_0 in equation (2.7).

Approach (B) is likely to seem most natural when θ has been defined itself as the product of a parametric model-building process. Model building involves checking the adequacy of alternative model specifications, using not only informal procedures such as graphical displays but also formal goodness-of-fit tests. Such tests are generally based on the specified models and it seems natural to assess the sampling variability in the final parameter estimates $\hat{\theta}$ in the same way.

In practice, it is often sensible to view approaches (A) and (B) as complementary rather than as competing alternatives. Some form of modelling approach is often almost essential in order to define relevant parameters θ. But given θ and a consistent estimator $\hat{\theta}$, the non-parametric approach (A) can often provide a model-robust estimator of the standard error of $\hat{\theta}$. For example, if $\boldsymbol{\theta}$ is defined by a fitted regression model $E(Y|\mathbf{X} = \mathbf{x}) = \mathbf{x}'\boldsymbol{\theta}$ and if it is assumed that the OLS estimator $\hat{\boldsymbol{\theta}}$ is consistent for $\boldsymbol{\theta}$ then, as is shown in Chapter 3, the non-parametric linearization estimator of $\text{Var}(\hat{\boldsymbol{\theta}})$ tends to be more robust to departures from the homoscedasticity assumption $\text{var}(Y|\mathbf{X} = \mathbf{x}) = \text{constant}$, than the conventional OLS procedure.

On the other hand, non-parametric variance estimators can break down in some circumstances. For example, the estimators can become very unstable if they are based on few degrees of freedom, such as occurs when there are few PSUs or replicates. This effect can be exacerbated for estimates for subclasses, especially those which tend to be highly clustered, for example, ethnic groups with area-based PSUs. For example, Kovar and Johnson (1986) computed variance estimates for the Hispanic Health and Nutrition Survey using eight pairs of PSUs and found, for example, estimated deffs for one parameter of 3.0, 0.3, 0.7 and 7.0 for adjacent age-groups. There was no substantive explanation for this variability; the only plausible explanation was the instability of the non-parametric variance estimator used to estimate the numerator of the deff. Similar instability was found for alternative non-parametric variance estimators. Rust (1987) reports similar results from a second survey involving 32 pairs of PSUs.

In analytic surveys a further problem concerns the estimation of the inverses of high-dimensional covariance matrices, such as are required for Wald tests in contingency table analysis (see Chapter 4) or Hotellings T^2-tests (see Chapter 3). If the degrees of freedom are less than the dimension of the matrix, as may often happen in practice, then the conventional non-parametric survey sampling estimator of the covariance matrix will be singular and the test statistic will not be defined (e.g. Koch and Lemeshow, 1972).

In such circumstances, some smoothing by modelling is desirable if not essential (cf. Kalton, 1977; Rust, 1987). At its simplest this may mean averaging deffs across similar subclasses or variables. A slightly more sophisticated approach (Wolter, 1985, Chapter 5) is to fit an empirical relationship between parameter estimates $\hat{\theta}$ and their variance estimates $v(\hat{\theta})$, using for example a model of form $v(\hat{\theta}) = \alpha\hat{\theta} + \beta\hat{\theta}^2$, and then to estimate $\text{var}(\hat{\theta})$ by the fitted value $\hat{\alpha}\hat{\theta} + \hat{\beta}\hat{\theta}^2$. More generally, assumptions may be made and indeed tested within a model framework to achieve more stable variance estimators. Examples of such assumptions and estimators will appear in Chapters 3–5. One way of viewing such use of model assumptions is in terms of the trade-off between bias and variance. Traditional non-parametric survey sampling variance estimators are generally unbiased, or at least consistent, but may have very high variances. Estimators smoothed by model assumptions, which are approximately valid, may have a slight bias but a greatly reduced variance.

2.13 NON-PARAMETRIC VARIANCE ESTIMATION

In this section we describe some of the simplest and most widely used non-parametric methods of variance estimation in survey sampling. A more comprehensive account is given by Wolter (1985). Methods will be presented for stratified three-stage designs although generalization to further stages is straightforward.

Linear Statistics

Consider first *linear statistics* of the form

$$\hat{\theta} = \sum_{h=1}^{H} \sum_{d=1}^{l_h} \sum_{c=1}^{m_{hd}} \sum_{t=1}^{n_{hdc}} u_{hdct}, \tag{2.28}$$

where u_{hdct} is a value associated with the tth unit in the cth second stage unit in the dth PSU in stratum h. For example, $\hat{\theta}$ may be the estimator of the mean of a variable y_{hdct} where

$$u_{hdct} = y_{hcdt}/(N_o \pi_{hdct}),$$

N_o is the finite population size and π_{hdct} is the inclusion probability of unit $hdct$.

A simple unbiased estimator of the variance of $\hat{\theta}$ with respect to the randomization distribution may be obtained under the following assumptions.

A1. Samples selected within different strata are independent.
A2. The l_h sample PSUs within each stratum h are selected with replacement. (At each of the l_h draws there is a fixed probability p_{hd} that PSU d is selected, $d = 1, \ldots, L_h$, $\sum_d p_{hd} = 1$.)
A3. $l_h \geqslant 2$.

We may rewrite equation (2.28) as

$$\hat{\theta} = \sum_{h=1}^{H} \sum_{d=1}^{l_h} u_{hd}, \qquad \text{where } u_{hd} = \sum_c \sum_t u_{hdct},$$

and, from A2, $u_{h1}, \ldots, u_{hl_h}$ are IID within strata and so, using also A1,

$$\text{var}(\hat{\theta}) = \sum_{h=1}^{H} l_h \, \text{var}(u_{hd}).$$

Hence from A1 to A3 an unbiased estimator of $\text{var}(\hat{\theta})$ is

$$v(\hat{\theta}) = \sum_{h=1}^{H} \frac{l_h}{(l_h - 1)} \sum_{d=1}^{l_h} (u_{hd} - \bar{u}_h)^2, \tag{2.29}$$

where $\bar{u}_h = \sum_d u_{hd}/l_h$.

The great partical advantage of this estimator is that, provided A1–A3 hold, it does not matter how subsampling occurs within PSUs. Whether there are two, three or more stages of sampling or there is systematic sampling of units within PSUs say, $v(\hat{\theta})$ remains unbiased for $\text{var}(\hat{\theta})$. The estimator is simply computed from the aggregate quantities u_{hd} formed from the *ultimate clusters*, consisting of all the sampled units within each sampled PSU.

The estimator $v(\hat{\theta})$ becomes even simpler in the common case of two PSUs per stratum, i.e. $l_h = 2$, $h = 1, \ldots, H$, when

$$v(\hat{\theta}) = \sum_{h=1}^{H} (u_{h1} - u_{h2})^2. \tag{2.30}$$

Even in surveys where $l_h > 2$ but H is large, ultimate clusters are sometimes combined randomly into two groups within each stratum and the above formula applied for reasons of simplicity. This method is used, for example, in the Current Population Survey (Bureau of the Census, 1978), where the groups are called *Keyfitz clusters* after Keyfitz (1957) who first proposed equation (2.30). The grouping of clusters does, however, lead to some loss of efficiency (Choudhry and Lee, 1987).

Whilst $v(\hat{\theta})$ is unbiased for the variance of $\hat{\theta}$ with respect to its randomization distribution, it is also unbiased for the variance of $\hat{\theta}$ with respect to a model in which, corresponding to A1 and A2, the u_{hd} in different strata are independent and the u_{hd} within strata are IID. Such a model is often plausible (Skinner, 1986b). Hence $v(\hat{\theta})$ may be viewed as either a design-based or a model-based estimator.

Of the three assumptions, A1 is usually valid. Assumption A3 is quite often violated when, in order to maximize efficiency, samplers select only $l_h = 1$ PSU per stratum. In this case, unbiased estimation of $\text{var}(\hat{\theta})$ (in the randomization sense) is impossible and similar strata need to be collapsed (Cochran, 1977, pp. 138–40; Wolter, 1985, pp. 47–54) to form new strata for which $l_h \geqslant 2$. Defining $v(\hat{\theta})$ with respect to the new strata then gives a conservative variance estimator.

Assumption A2 is almost always violated since PSUs are usually selected within strata by some form of without-replacement sampling. In this case unbiased estimation of $\text{var}(\hat{\theta})$ remains possible, but generally involves complex formulae with components for each stage of sampling (e.g. Raj, 1966; Rao, 1975a). Some simplified procedures for the case $l_h = 2$, $h = 1, \ldots, H$ have been proposed by Durbin (1967) and Rao and Lanke (1984). One widely used approximatòn (e.g. Kalton, 1977), which effectively assumes that the l_h ultimate clusters within stratum h form a simple random sample without replacement from the stratum, $h = 1, \ldots, H$, is obtained by inserting a finite population correction (see Section 2.7) into equation (2.29) giving

$$v_{\text{WOR}}(\hat{\theta}) = \sum_{h=1}^{H} \left(1 - \frac{l_h}{L_h}\right) \frac{l_h}{(l_h - 1)} \sum_{d=1}^{l_h} (u_{hd} - \bar{u}_h)^2. \tag{2.31}$$

For example, this is the estimator used in the computer package SUPERCARP (Hidiroglou *et al.*, 1980).

Often the sampling fractions l_h/L_h are small and the difference between $v(\hat{\theta})$ and $v_{\text{WOR}}(\hat{\theta})$ or alternative without replacement formulae is negligible (cf. Durbin, 1953). In any case $v(\hat{\theta})$ is generally a conservative variance estimator. In analytic surveys there are in addition two reasons why $v(\hat{\theta})$ may actually be preferable to $v_{\mathrm{WOR}}(\hat{\theta})$:

1. The parameter of interest is often a superpopulation parameter, in which case the finite population correction is inappropriate (see Section 2.7).
2. The estimator $v(\hat{\theta})$ may be preferable as an estimator of total variance, comprising both sampling variance and response variance. To illustrate this point, suppose

$$u_{hd} = U_{hd} + \varepsilon_{hd},$$

where the U_{hd} are the 'true' values and the ε_{hd} are independent response (measurement) errors arising, for example, from random effects of interviewers nested within PSUs. Then

$$\mathrm{var}(\hat{\theta}) = \mathrm{var}(\sum\sum U_{hd}) + \mathrm{var}(\sum\sum \varepsilon_{hd})$$

and if PSUs are selected by SRS without replacement and the ε_{hd} have fixed variances $\sigma_{\varepsilon h}^2$ then

$$\mathrm{var}(\hat{\theta}) = \sum_h [(1 - l_h/L_h)l_h S_h^2 + l_h \sigma_{\varepsilon h}^2],$$

where

$$S_h^2 = \sum_d (U_{hd} - \bar{U}_h)^2/(L_h - 1), \qquad \bar{U}_h = \sum_d U_{hd}/L_h$$

$$E[v(\hat{\theta})] = \sum_h l_h(S_h^2 + \sigma_{\varepsilon h}^2),$$

$$E[v_{\mathrm{WOR}}(\hat{\theta})] = \sum_h (1 - l_h/L_h)l_h(S_h^2 + \sigma_{\varepsilon h}^2).$$

Hence

$$E[v_{\mathrm{WOR}}(\hat{\theta})] \leqslant \mathrm{var}(\hat{\theta}) \leqslant E[v(\hat{\theta})]$$

and $v(\hat{\theta})$ is preferred since it is a conservative estimator.

Both $v(\hat{\theta})$ and $v_{\mathrm{WOR}}(\hat{\theta})$ may be extended to multivariate linear estimators

$$\hat{\boldsymbol{\theta}} = \sum_h \sum_d \sum_c \sum_t \mathbf{u}_{hdct}. \tag{2.32}$$

Corresponding to $v(\hat{\theta})$, the covariance matrix of $\hat{\boldsymbol{\theta}}$ may be estimated under assumptions A1–A3, by

$$\mathbf{V}(\hat{\boldsymbol{\theta}}) = \sum_{h=1}^{H} \frac{l_h}{(l_h - 1)} \sum_{d=1}^{l_h} (\mathbf{u}_{hd} - \bar{\mathbf{u}}_h)(\mathbf{u}_{hd} - \bar{\mathbf{u}}_h)', \tag{2.33}$$

where

$$\mathbf{u}_{hd} = \sum_c \sum_t \mathbf{u}_{hdct}, \qquad \bar{\mathbf{u}}_h = \sum_d \mathbf{u}_{hd}/l_h.$$

Non-linear Statistics

Various methods are now described for statistics not of form (2.28).

(*a*) *Linearization Method*

Many non-linear statistics $\hat{\phi}$ may still be expressed as non-linear functions of linear statistics, i.e.

$$\hat{\phi} = g(\hat{\boldsymbol{\theta}}), \tag{2.34}$$

where

$$\hat{\boldsymbol{\theta}} = (\hat{\theta}_1, \ldots, \hat{\theta}_p)' = \sum_h \sum_d \mathbf{u}_{hd}, \qquad \mathbf{u}_{hd} = \sum_c \sum_t \mathbf{u}_{hdct} \tag{2.35}$$

and $g(.)$ is a given non-linear function.

For example, consider the ratio $\hat{\phi} = \bar{y}/\bar{x}$ where

$$\bar{y} = \sum y_{hdct}/n_o, \bar{x} = \sum x_{hdct}/n_o.$$

Then $\hat{\phi}$ may be expressed as $g(\hat{\boldsymbol{\theta}})$ where

$$\hat{\boldsymbol{\theta}} = \sum_h \sum_d (u_{1hd}, u_{2hd})' = (\bar{y}, \bar{x})'$$

is a 2×1 vector,

$$u_{1hd} = \sum_c \sum_t y_{hdct}/n_o, \qquad u_{2hd} = \sum_c \sum_t x_{hdct}/n_o \quad \text{and} \quad g[(\theta_1, \theta_2)'] = \theta_1/\theta_2.$$

The idea behind the *linearization method* (also called *Taylor series method* or *δ method*) is to approximate $\hat{\phi}$ by a linear statistic. Assuming $\hat{\boldsymbol{\theta}}$ is consistent for $\boldsymbol{\theta}$, say, as $n_o \to \infty$ and $g(.)$ is suitably differentiable, this is achieved using the first-order Taylor series approximation:

$$\hat{\phi} = g(\hat{\boldsymbol{\theta}}) \simeq g(\boldsymbol{\theta}) + \sum_{j=1}^{p} g_j(\boldsymbol{\theta})(\hat{\theta}_j - \theta_j), \tag{2.36}$$

where $g_j(\boldsymbol{\theta}) = \partial g(\boldsymbol{\theta})/\partial \theta_j$ is the jth partial derivative of $g(\boldsymbol{\theta})$ and $\boldsymbol{\theta} = (\theta_1, \ldots, \theta_p)'$.

Using equation (2.35) and ignoring the constant terms in equation (2.36), the variance of $\hat{\phi}$ may then be approximated by the variance of

$$\sum_{j=1}^{p} g_j(\boldsymbol{\theta}) \sum_h \sum_d u_{jhd} = \sum_h \sum_d \tilde{w}_{hd},$$

where

$$\mathbf{u}_{hd} = (u_{1hd}, \ldots, u_{phd})', \qquad \tilde{w}_{hd} = \sum_{j=1}^{p} g_j(\boldsymbol{\theta}) u_{jhd}, \tag{2.37}$$

which is now a linear statistic of form (2.28) (cf. Woodruff, 1971). The linearization variance estimator is then obtained by substituting $\tilde{w}_{hd}$ for u_{hd} in equation (2.29) and replacing $\boldsymbol{\theta}$ by $\hat{\boldsymbol{\theta}}$, i.e.

$$v_L(\hat{\phi}) = \sum_{h=1}^{H} \frac{l_h}{(l_h - 1)} \sum_{d=1}^{l_h} (w_{hd} - \bar{w}_h)^2, \tag{2.38}$$

where

$$w_{hd} = \sum_{j=1}^{p} g_j(\hat{\boldsymbol{\theta}}) u_{jhd}. \tag{2.39}$$

Thus in the ratio example where $g(\boldsymbol{\theta}) = \theta_1/\theta_2$, the partial derivatives are $g_1(\boldsymbol{\theta}) = 1/\theta_2$, $g_2(\boldsymbol{\theta}) = -\theta_1/\theta_2^2$ and substituting into equation (2.39) gives

$$w_{hd} = (u_{1hd} - \hat{\phi} u_{2hd})/\bar{x}. \tag{2.40}$$

Finally substituting w_{hd} into equation (2.38) gives the usual formula for the variance estimator of a ratio (e.g. Cochran, 1977, pp. 169–71; Kish and Hess, 1959).

One limitation of the linearization method, not shared by the methods that follow, is the requirement for analytic expressions for the partial derivatives. Some examples of these derivatives for various analytic statistics will be given in Chapters 3 and 5.

(b) Simple Replication

Mahalanobis (1946) and Deming (1956) recommend that, for the purpose of variance estimation, the design of a survey should involve r independent replications of the same basic design. This gives a final sample consisting of r *replicates* or *interpenetrating samples.* Any statistic $\hat{\phi}$ for the whole sample can be recomputed for each of the replicates giving $\hat{\phi}_1, \ldots, \hat{\phi}_r$ and $\text{var}(\hat{\phi})$ may be estimated by

$$v_R(\hat{\phi}) = \sum(\hat{\phi}_i - \hat{\phi})^2/r(r-1).$$

In practice, pure replicated designs are seldom used since greater efficiency (of point estimation) is usually achievable using non-replicated designs with a finer degree of stratification. Approximate replicates can usually be constructed, however, for most designs (Wolter, 1985, p. 31). In a stratified multi-stage design, r replicates might be formed by partitioning the l_h PSUs in each stratum h at random into r groups of approximately equal size l_h/r. Combining these groups randomly across strata will give the required r replicates from which the variance estimator v_R may again be calculated. This procedure is sometimes called the *random groups method* to distinguish it from pure replication at the design stage.

The simple replication method has a number of important advantages for analytical surveys.

1. No specialist computer software is necessary. Even for complex estimators,

involving iteration say, the $\hat{\phi}_i$ may simply be computed by applying the same computer algorithm to the r subsets of the data.

2. In analytic surveys there are many multiparameter or even non-parametric estimation problems. For example, Gilks (1986) fits spline curves to birth history data from the World Fertility Survey where $\boldsymbol{\phi}$ determines the shape of the curve. In such cases studying the different $\hat{\boldsymbol{\phi}}_i$, and, the corresponding spline curves for different replicates, may give the analyst a better understanding of sampling variation than by inspecting a high-dimensional covariance matrix as generated by other methods of variance estimation. This is especially true for the statistically unsophisticated analyst.
3. The cross-validatory splitting of a data set into two parts, one for exploration and selection of possible models and the other for confirmation or validation of the selected models is often recommended (e.g. Mosteller and Tukey, 1977). For complex survey data a natural choice of split is to divide the data into two (or more) replicates. Pfeffermann and Nathan (1985) discuss the choice of split in the context of regression and loglinear models.

The main disadvantage of simple replication is that r is often small and this leads to imprecise estimates of variances and wider confidence intervals than could be obtained by other methods such as linearization. The length of a $100(1-\alpha)\%$ confidence interval is on average multiplied by a factor $t_{r-1}(\alpha)/Z(\alpha)$, where $t_v(\alpha)$ is the αth point of Student's t-distribution with v degrees of freedom and $Z(\alpha) = t_\infty(\alpha)$. Some numerical values of this factor for two-sided confidence intervals are given in Table 2.5.

We see that at least ten replicates are need to obtain variance estimators with confidence intervals no more than 20 or 30% longer than those obtainable by other methods. The number of replicates r can always be increased, for example

Table 2.5 Inflation factor for length of confidence intervals for simple replication compared to linearization

Number of replicates, r	Confidence level $(1-\alpha)$	
	0.95	0.99
2	6.5	24.7
3	2.2	3.9
4	1.6	2.3
5	1.4	1.8
10	1.2	1.3
20	1.1	1.1
∞	1.0	1.0

by collapsing strata, but this generally leads to upward bias in the variance estimator which needs to be balanced against its improved precision.

(c) *Balanced Repeated Replication*

The method of *balanced repeated replication* (BRR) or *balanced half-samples* overcomes the problem of low precision of simple replication and has been developed especially for designs with a large number of strata H and with few PSUs, ideally two, per stratum.

The method proceeds, as for simple replication with $r = 2$, by dividing the sample within each stratum h into two replicates; if there are two PSUs per stratum these form the replicates. There are 2^H possible half-samples which may be formed consisting of the union of one replicate from each stratum. Estimates $\hat{\phi}_i$ can be constructed for each half-sample and used to estimate $\text{var}(\hat{\phi})$. It would be computationally expensive to evaluate all 2^H possible $\hat{\phi}_i$, but it is possible to select a 'balanced' set of only k half-samples, where k is the minimum multiple of 4 greater than H, so that the estimator

$$v_{\mathbf{BRR}}(\hat{\phi}) = \sum_{i=1}^{k} (\hat{\phi}_i - \hat{\phi})^2/k$$

has equal asymptotic precision to the same estimator evaluated over all 2^H half-samples (McCarthy, 1969). The estimator $v_{\mathbf{BRR}}$ may also be shown to have equivalent precision to v_{L} to first order (Rao and Wu, 1985). The gain in precision of $v_{\mathbf{BRR}}$ compared to simple replication needs, however, to be balanced against the increased computation required and the loss of the other advantages mentioned earlier. McCarthy (1976) argues that BRR can be extended to cross-validation studies and can actually be preferable for this purpose to simple replication.

(d) *Jackknife Methods*

The jackknife approach is similar to BRR but does not require the definition of just two replicates per stratum. The basic estimator is

$$v_{\mathrm{J}}(\hat{\phi}) = \sum_{h=1}^{H} \frac{(l_h - 1)}{l_h} \sum_{d=1}^{l_h} (\hat{\phi}^{hd} - \bar{\phi}^h)^2,$$

where $\hat{\phi}^{hd}$ is the estimator of ϕ based on the sample with the dth PSU from stratum h omitted and $\bar{\phi}^h = \sum_d \hat{\phi}^{hd}/l_h$. Various alternative forms for v_{J} have also been suggested although Rao and Wu (1985) show these are asymptotically equivalent to second order and that v_{J} is asymptotically equivalent to $v_{\mathbf{BRR}}$ and v_{L} to first order. The number of computations of $\hat{\phi}$ for the jackknife estimator, $\sum l_h$, is generally rather greater than for $v_{\mathbf{BRR}}$, which is at most $H + 4$.

(*e*) *Bootstrap Methods*

This method involves more computation than other approaches and preliminary evidence (Rao and Wu, 1988; Kovar *et al.*, 1988) suggests that the bootstrap variance estimator is less stable than the linearization and jackknife estimators. The direct estimation of confidence intervals via the bootstrap histogram method does, however, avoid the assumption of normality and may be especially useful for constructing one-sided intervals with asymmetric sampling distributions (see Section 2.14 and Rao and Wu, 1988).

Covariance Matrix Estimation

All the methods above extend naturally to the estimation of the covariance matrix of a vector statistic $\hat{\boldsymbol{\phi}}$. For the linearization estimator v_{L}, the partial derivative g_j becomes a vector and hence so does w_{hd} in equation (2.39). The estimator v_{L} in equation (2.38) then becomes

$$\mathbf{V}_{\mathrm{L}}(\hat{\boldsymbol{\phi}}) = \sum_{h=1}^{H} \frac{l_h}{(l_h - 1)} \sum_{d=1}^{l_h} (\mathbf{w}_{hd} - \bar{\mathbf{w}}_h)(\mathbf{w}_{hd} - \bar{\mathbf{w}}_h)'. \tag{2.41}$$

Similarly, for the other methods, $(\hat{\phi}_i - \hat{\phi})^2$ is replaced by $(\boldsymbol{\phi}_i - \hat{\boldsymbol{\phi}})(\boldsymbol{\phi}_i - \hat{\boldsymbol{\phi}})'$.

Discussion

Amongst methods for non-linear statistics, simple replication has several advantages for analytic surveys but tends to be imprecise when there are few replicates. The remaining methods have equivalent precision asymptotically to first order for statistics of form (2.34) (but not necessarily for other statistics such as quantiles, Rao and Wu, 1987). The results of many empirical and simulation studies confirm that none of these methods is uniformly best across all statistics, designs, populations and optimality criteria. Readers who are particularly concerned about precision of variance estimation or confidence interval coverage should consult reviews of these results, such as Rust (1985) or Wolter (1985, Chapter 8), in order to assess the relative performance of the different methods in the context of their own individual type of research. Otherwise the main considerations in choosing between v_{L}, v_{BRR} and v_{J} are likely to be non-statistical, such as the availability of computer software.

One important consideration facing secondary data analysts, as already mentioned in Chapter 1, is that complete design information on the stratum and cluster membership of each unit may often be deliberately concealed by primary data collectors, in order to prevent possible breaches of confidentiality. This may make it impossible to carry out some of the methods of variance estimation described. For stratified multistage designs it would be sufficient for stratum and PSU identifiers to be available, and indeed this information would very rarely permit identification of individuals. As a minimum, the identification of replicates would permit the use of simple replication. There are indeed advantages to the secondary analyst for the primary data collector to reduce the amount of design information to just replicate identifiers, say. Many large-scale government surveys involve such complexity of design, for example both temporal and spatial stratification, rotation of units, etc., that it may be impossible, without 'inside' knowledge of the survey, for the secondary analyst to allow adequately for this complexity in variance estimation. In such cases it may be preferable for the primary data collectors to use their inside knowledge beforehand to 'clean' the design information, for example by collapsing strata, to make the data ready for use with the kinds of variance estimation techniques described in this section. Dippo *et al.* (1984) and Fay (1984) describe such an approach where r 'replicate weights' are added to each record.

2.14 NORMAL APPROXIMATIONS AND CONFIDENCE INTERVALS

The variance estimators of Section 2.13 may be used to construct confidence intervals for θ by approximating the distribution of the pivotal statistic t in equation (2.27) by the standard normal distribution. The approximation $t \sim N(0, 1)$ is usually justified asymptotically by assuming the following two conditions:

C1. $\hat{\theta}$ obeys a central limit theorem, i.e.

$$n_o^{1/2}(\hat{\theta} - \theta) \to N(0, \sigma^2) \text{ in law.} \tag{2.42}$$

C2. v is consistent for the (asymptotic) variance of $\hat{\theta}$, i.e.

$$n_o v \to \sigma^2 \text{ in probability.} \tag{2.43}$$

For, given C1 and C2, it follows that

$$t = v^{-1/2}(\hat{\theta} - \theta) \to N(0, 1) \text{ in law.} \tag{2.44}$$

It is assumed here that the limit is taken as the sample size n_o increases, although, as noted in Section 1.6.4, asymptotic arguments for complex surveys are severely complicated by the requirement also to specify sequences of sizes of finite populations, strata, clusters and so on.

Various results are available in the literature providing conditions under which C1 and C2 hold for alternative statistics $\hat{\theta}$, variance estimators v and sampling designs. Most results apply only to C1 and almost all results concern the randomization distributions induced by a probability sampling design rather than model distributions.

Central limit theorems of form C1 for the sample mean are given by Madow (1948) and Hajek (1960) for SRS without replacement, by Hajek (1964), Rosen (1972, 1974), Holst (1973) and Ohlsson (1986) for various unequal probability designs and by Fuller (1975, Appendix A) for two-stage sampling. Central limit theorems for non-linear statistics of form (2.34) are given by Krewski and Rao (1981), Fuller (1984) and Skinner (1986b) for stratified and two-stage sampling. Binder (1983), Fuller (1984) and Chambless and Boyle (1985) give results for further classes of statistics.

Consistency results of form C2 for linearization, BRR and jackknife variance estimators are given by Krewski (1978), Krewski and Rao (1981), and Wolter (1985, Appendix B).

When judging the adequacy of the approximation $t \sim N(0, 1)$, it is useful to consider not only the results of asymptotic theory but also finite sample empirical studies. Studies concerning the estimation of a mean are reported by Stenlund and Westlund (1975), Barrett and Goldsmith (1976), Sukhatme *et al.* (1984, pp. 32–8) and Dalen (1986) for SRS without replacement and by Bayless (1968) (see Rao, 1978) and Stenlund and Westlund (1976) for unequal probability sampling and stratified sampling. Frankel (1971), Kish and Frankel (1974), Mellor (1973), Bean (1975) and Campbell and Meyer (1978) report studies concerning various complex statistics and complex designs.

These theoretical and empirical results show how the adequacy of the approximation depends on the form of $\hat{\theta}$, the choice of v, the population structure and the design. Although this dependence is complex, some broad conclusions may be drawn. First, the approximations may break down for statistics, such as medians, which are not expressible as smooth functions of linear statistics (e.g. Woodruff, 1952; Rao and Wu, 1987). For linear statistics $\hat{\theta} = \sum u_t$, the normality of the sampling distribution of $\hat{\theta}$ depends mainly on the skewness of the population distribution of u_t. Cochran (1977, p. 42) suggests that the normal approximation will be good, in the sense that a 95% confidence interval should have at least 94% coverage, under SRS with sample size n_o if $n_o > 25G_1^2$, where G_1 is the skewness coefficient of u_t, for example for the exponential distribution $G_1 = 2$ and a sample of size $n_o = 100$ would be required to meet this condition (see also Dalen, 1986). A broad empirical finding (e.g. Frankel, 1971; Rao and Wu, 1987) is that the coverage of two-sided confidence intervals tends to be far more robust to skewness than coverage of one-sided intervals.

For non-linear statistics which may be expressed as smooth functions of linear statistics $\hat{\phi} = g(\hat{\boldsymbol{\theta}}), \hat{\boldsymbol{\theta}} = \sum \mathbf{u}_t$, the normality of $\hat{\phi}$ depends mainly on the skewness of the linearized variables $\sum_j g_j(\boldsymbol{\theta}) u_{jt}$ (cf. equation (2.36)) rather than the original $\mathbf{u}_t$ variables. For example, even if u_{1t} and u_{2t} are very skew, the distribution of the ratio $\sum u_{1t} / \sum u_{2t}$ may be close to normal (Mellor, 1973; Campbell and Meyer, 1978). In addition to the linearized variable being non-skew, it is necessary for the function $g(\boldsymbol{\theta})$ to be approximately linear within the region of sampling error of $\hat{\boldsymbol{\theta}}$ about $\boldsymbol{\theta}$.

For complex designs it is natural to replace conditions on n_o by conditions on the total number of clusters, $\sum l_h$, and conditions on the u_t by conditions on the corresponding cluster quantities, say the u_{hd} in equation (2.29). For the purpose of normal approximation, it is desirable that no clusters have very isolated and hence influential values of u_{hd}, either because they differ greatly from other clusters with respect to the variables under study or because they carry a large weight as a consequence of their size or their small selection probability.

As a crude generalization, the normal approximation for $\hat{\theta}$ tends to work least well in surveys of establishments or organizations where severe skewness and influential observations may occur. In moderate to large-scale household surveys the normal approximation for $\hat{\theta}$ usually works quite well (except perhaps for estimators for small subclasses) and the main problems with the normal approximation of $t = (\hat{\theta} - \theta)/v^{1/2}$ are likely to concern the denominator rather than the numerator. Problems with the denominator may occur because extensive stratification may leave too few degrees of freedom to obtain a sufficiently precise estimator of $\text{var}(\hat{\theta})$. One way of allowing for the imprecision of v as an estimator of $\text{var}(\hat{\theta})$ in obtaining a confidence interval for $\hat{\theta}$ is to approximate the distribution of t by a Student's t-distribution with ν degrees of freedom. For the simple replication variance estimator with r replications it is natural to take $\nu = r - 1$. For the other variance estimators it is common to take $\nu = \sum(l_h - 1)$ (e.g. Frankel, 1971) although the justification for this choice depends upon the assumption of equal cluster variances, an assumption which is rarely tenable in surveys (Fuller, 1984). The problem of few degrees of freedom for v and its consequent imprecision is accentuated in multivariate methods for which the inverse of v is required (see Chapters 3–5).

One method of adjusting for an asymmetric sampling distribution of $\hat{\theta}$ is to use a transformation $g(\hat{\theta})$ of $\hat{\theta}$. Letting v_g be a variance estimator of $g(\hat{\theta})$, a confidence interval for θ may be obtained by approximating the distribution of $[g(\hat{\theta}) - g(\theta)]/v_g^{1/2}$ by the standard normal distribution. Some common choices of transformation for various statistics are given by Wolter (1985, Appendix C). If g is a non-linear function, the implied confidence interval for θ no longer need be symmetric about $\hat{\theta}$. Another way of obtaining an asymmetric confidence interval is to use the bootstrap histogram (Rao and Wu, 1987). A similar but simpler idea is to use the interval between the smallest and largest of the estimates $\hat{\theta}_1, \ldots, \hat{\theta}_r$ obtained by simple replication to provide an approximate $1-(0.5)^{r-1}$ confidence interval (Murthy and Roy, 1975).

2.15 INTRODUCTION TO CHAPTERS 3–5

In this chapter we have attempted to explain how complex designs can affect standard errors, confidence intervals and significance tests and to describe some broad approaches to adjustment for the effects of complex designs. The examples have been kept deliberately simple for the purpose of illustrating ideas rather than for their value in applications. In the following chapters we consider a number of statistical methods which are more widely applicable to survey data.

In Chapter 3 we begin by looking at inference for domain (or subclass) means, perhaps the simplest type of analytic statistic. Comparisons of domain means arise in investigations of the effect of a categorical variable X on a dependent variable Y and form the simplest case of regression analysis, which is covered next. The unifying theme of Chapter 3 is a focus on methods for *continuous* variables and there is additional discussion of multivariate methods for covariance matrices and the general parametric modelling techniques of pseudo maximum likelihood and generalized least-squares inference for moment structure models.

Chapters 4 and 5 are concerned with methods for *discrete* variables, the most common kind of data in social surveys. Chapter 4 focuses on testing procedures, from the simple and widely used chi-square tests of goodness of fit and independence to the use of tests for fitting loglinear models. In addition to testing there is discussion of smoothing contingency tables, residual analysis and some detailed examples from the Canada Health Survey.

The large sample sizes of many sample surveys lead very often to tests rejecting the hypothesis of independence in two-way contingency tables. Chapter 5 focuses on how association in such tables can be measured and on how to compare association in contingency tables arising from separate populations.

Each of Chapters 3–5 follows the common structure:

1. IID-based procedures;
2. The effect of complex designs on IID-based procedures;
3. Alternative procedures which allow for complex designs.

Some might argue that, since it has already been established that IID-based procedures are misleading in the presence of complex designs, discussion should concentrate solely on (3). We would suggest, however, that (1) and (2) remain of interest. First, there are practical advantages to developing methods which involve only simple adjustments to standard procedures rather than completely new procedures, and the investigation of (2) assists the construction of the former type of method. Secondly, as long as some researchers continue to conduct IID-based analyses of survey data, for reasons of software availability or whatever, it helps to be aware of which aspects of these analyses are most sensitive to complex designs, for example to know that one-sample t-tests tend to be more sensitive than two-sample t-tests (Chapter 3) or that chi-squared goodness-of-fit tests may be expected to be more sensitive than F-tests of nested hypotheses in loglinear models (Chapter 4).

CHAPTER 3

Domain Means, Regression and Multivariate Analysis

C. J. Skinner

3.1 INTRODUCTION

In Chapter 2 we investigated the effect of complex designs on inference for means. In this chapter we extend this investigation to other statistical procedures for continuous variables, namely domain mean estimation, regression analysis and multivariate analysis. Regression methods are widely applied to survey data, in particular to control for the effects of confounding variables when investigating 'causal' relationships (Anderson *et al.*, 1980). The comparison of domain means is the simplest instance of such regression analysis. Multivariate methods are also frequently applied to survey data for a wide range of purposes (e.g. Ferber, 1980).

The effect of complex designs on procedures based on IID assumptions will be studied in Section 3.3. Effects will be illustrated numerically using data from the Family Expenditure Survey described in Section 3.2. Methods of allowing for complex designs will be considered in Section 3.4.

3.2 FAMILY EXPENDITURE SURVEY DATA

The 1975 United Kingdom Family Expenditure Survey (FES) is a household survey in which many of the key variables, concerning expenditure on various commodities and income, are continuous. For our purposes we assume the FES involves a standard stratified multistage design with seven strata (each of which consists of a rotation group, defined according to which of the four quarters of the year a PSU appears in the sample) each containing 42 PSUs making 294 PSUs in total. After non-response, the total number of households in the sample is $n_0 = 6923$. In fact the design employs additional geographical stratification to the level of one PSU per stratum but we have ignored this for simplicity. Further details of the design are given in Kemsley *et al.* (1980). Variances and design effects were estimated using SUPERCARP (Hidiroglou *et al.*, 1980).

3.3 IMPACT OF DESIGN ON IID PROCEDURES

3.3.1 Introduction

In Chapter 2 the *design effect* (deff) was defined in equation (2.4) as a measure of the impact of a complex design on an IID-based procedure and the form of deffs for means was investigated. In particular the deff of cluster sampling with clusters of equal size N for a sample mean $\bar{y}$ and an SRS variance estimator v_0 was given in equation (2.20) as

$$\text{deff}(\bar{y}, v_0) \simeq 1 + (N-1)\tau. \tag{3.1}$$

This investigation of the form of deffs is now extended to other statistics.

3.3.2 Domain Means

A *domain* is a subset of a population. A *domain mean* is the mean of a variable within a domain, for example the mean income amongst individuals aged 25–29. Comparisons of domain means arise in analytic surveys when assessing the effect of a categorical variable, X, which defines the domains, on a continuous variable Y, for example the effect of age on income. The means $\mu_1, \ldots, \mu_k$ within the domains $a = 1, \ldots, k$ of X may be represented as linear regression coefficients by writing

$$E(Y \mid X = a) = \mathbf{x}(a)'\boldsymbol{\mu}, \tag{3.2}$$

where $\mathbf{x}(a)$ is the $k \times 1$ vector with ath element one and other elements zero and $\boldsymbol{\mu} = (\mu_1, \ldots, \mu_k)'$ so that $\mathbf{x}(a)'\boldsymbol{\mu} = \mu_a$. Alternatively, equation (3.2) may be expressed with an intercept and differences of domain means as coefficients, for example

$$E(Y \mid X = a) = \alpha + \mathbf{x}^*(a)'\boldsymbol{\beta}, \tag{3.3}$$

where $\alpha = \mu_k$, $\mathbf{x}^*(a)$ is the vector containing the first $k-1$ elements of $\mathbf{x}(a)$ and $\boldsymbol{\beta} = (\mu_1 - \mu_k, \ldots, \mu_{k-1} - \mu_k)'$.

For a self-weighting design, μ_a is usually estimated by the simple unweighted mean $\bar{y}_a$ of Y across all sample units falling into the domain. Kish (1987) proposes extending equation (3.1) for $\bar{y}_a$ under a multi-stage design to

$$\text{deff}(\bar{y}_a, v_{0a}) = 1 + (\bar{n}_a - 1)\tau_a, \tag{3.4}$$

where $\bar{n}_a$ is the mean number of sample units per PSU falling into domain a, i.e.

$$\bar{n}_a = \pi_a \bar{n}, \tag{3.5}$$

where π_a is the overall proportion of sample units falling into domain a, $\bar{n}$ is the mean overall sample size per PSU, τ_a is a measure of intra-PSU correlation in domain a, v_{0a} is the usual IID-based estimator of $\text{var}(\bar{y}_a)$.

In particular, when $\pi_a = 1$ and $\bar{y}$ is the overall sample mean we have

$$\text{deff}(\bar{y}, v_0) = 1 + (\bar{n} - 1)\tau. \tag{3.6}$$

Formulae (3.4) and (3.6) are not derived as analytic results but are used rather to define τ_a and τ implicitly from the deffs. The practical implications of these formulae arise only when the further assumption is made that $\tau_a = \tau$, an approximation which Kish (1987, p. 43) suggests holds 'fairly well in thousands of empirical computations across many kinds of surveys'. Given $\tau_a = \tau$ we can predict deff$(\bar{y}_a)$ (the second argument v_{0a} is now dropped to simplify notation) from deff$(\bar{y})$, using equation (3.4)–(3.6), by

$$\widehat{\text{deff}}(\bar{y}_a) = 1 + \frac{(\pi_a \bar{n} - 1)}{(\bar{n} - 1)} [\text{deff}(\bar{y}) - 1]. \tag{3.7}$$

Some estimated deffs for various domain means for the FES data together with the line predicted by equation (3.7) are plotted against π_a in Figure 3.1. The y variable is the logarithm of expenditure on housing (the log function is

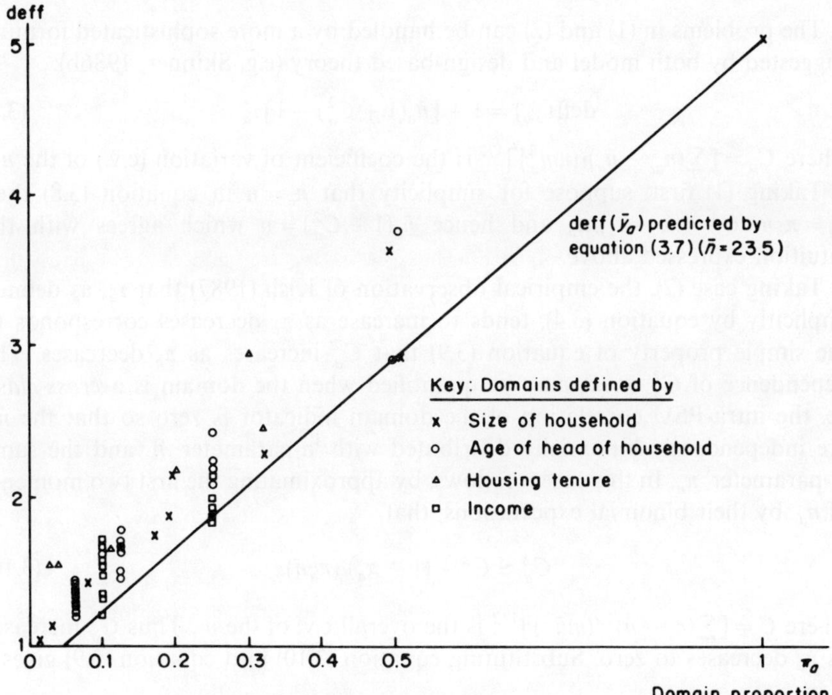

Figure 3.1 Scatterplot of estimated deffs for mean log (housing expenditure) amongst various domains plotted against domain proportion.

used to achieve approximate normality and to reduce the influence of some large values). A strong linear dependence of $\text{deff}(\bar{y}_a)$ on π_a is evident, although the line predicted by equation (3.7) tends to underestimate the deffs. Kish (1987) points out two cases where underestimation from equation (3.7) may be expected.

1. The more unequal the domain sample sizes $n_{a1},\ldots,n_{am}$ in different PSUs, the more equation (3.7) underestimates $\text{deff}(\bar{y}_a)$. The most extreme case occurs when the domain consists of a union A of PSUs, as for example in a regional domain with area PSUs

$$n_{ac} = n_c \text{ for } c \in A, \qquad n_{ac} = 0 \text{ otherwise,} \tag{3.8}$$

   where n_c is the sample size in PSU c, $c = 1,\ldots,m$.

   In this case one would intuitively expect that $\bar{n}_a$ in equation (3.4) should refer to the mean value of n_{ac} across only PSUs in A, rather than the mean across all PSUs, as implied by equation (3.5), which may be much smaller.
2. The smaller is π_a, the more equation (3.7) tends to underestimate $\text{deff}(\bar{y}_a)$. In the extreme case when $\pi_a < 1/\bar{n}$, equation (3.5) implies that $\bar{n}_a < 1$ and hence $\text{deff}(\bar{y}_a) < 1$, although intuitively one would expect $\text{deff}(\bar{y}_a)$ to be bounded below by 1.

The problems in (1) and (2) can be handled by a more sophisticated formula suggested by both model and design-based theory (e.g. Skinner, 1986b):

$$\text{deff}(\bar{y}_a) = 1 + [\bar{n}_a(1 + C_a^2) - 1]\tau_a, \tag{3.9}$$

where $C_a = [\sum(n_{ac} - \bar{n}_a)/(m\bar{n}_a^2)]^{1/2}$ is the coefficient of variation (c.v.) of the n_{ac}.

Taking (1) first, suppose for simplicity that $n_c = n$ in equation (3.8) then $\bar{n}_a = \pi_a n$, $C_a^2 = (1 - \pi_a)/\pi_a$ and hence $\bar{n}_a(1 + C_a^2) = n$ which agrees with the intuition expressed above.

Taking case (2), the empirical observation of Kish (1987) that τ_a, as defined implicitly by equation (3.4), tends to increase as π_a decreases corresponds to the simple property of equation (3.9) that C_a^2 increases as π_a decreases. The dependence of C_a^2 on π_a may be quantified when the domain is a *cross-class*, i.e. the intra-PSU correlation of the domain indicator is zero so that the n_{ac} are independently binomially distributed with 'n-parameter' n_c and the same 'p-parameter' π_a. In this case it follows, by approximating the first two moments of n_{ac} by their binomial expectations, that

$$C_a^2 \simeq C^2 + (1 - \pi_a)/(\pi_a \bar{n}), \tag{3.10}$$

where $C = [\sum(n_c - \bar{n})^2/(m\bar{n}^2)]^{1/2}$ is the overall c.v. of the n_c. Thus C_a^2 increases as π_a decreases to zero. Substituting equation (3.10) into equation (3.9) gives

$$\text{deff}(\bar{y}_a) = 1 + [\bar{n}(1 + C^2) - 1]\pi_a\tau_a, \tag{3.11}$$

which is bounded below by 1 if $\tau_a \geqslant 0$, unlike equation (3.4). Assuming $\tau_a = \tau$,

(3.11) implies the imputation formula:

$$\widehat{\text{deff}}(\bar{y}_a) = 1 + \pi_a[\text{deff}(\bar{y}) - 1]. \tag{3.12}$$

This formula is very simple to use, requiring only knowledge of the domain proportion π_a, which is almost always available, in contrast to formula (3.7) which also requires $\bar{n}$. Note, however, that the difference between these two formulae tends to zero as $\bar{n}$ increases.

None of the above formulae make any allowance for stratification. Under stratified sampling $\text{var}(\bar{y}_a)$ includes a between-stratum component if $\pi_a < 1$ unlike $\text{var}(\bar{y})$ (Cochran, 1977, pp. 142–4). As a consequence the deff due to stratification (see Section 2.9) tends to increase to 1 as π_a decreases to 0. Thus when imputing $\text{deff}(\bar{y}_a)$ from $\text{deff}(\bar{y})$ using either equation (3.7) or (3.12) the presence of stratified sampling may lead to slight underestimation, although this effect is likely to be negligible in typical household surveys compared to the effect of multi-stage sampling.

In Table 3.1 we give predicted deffs from equations (3.7) and (3.12) and also

Table 3.1 Comparison of three formulae for predicting deffs of domain means from deffs of overall means

Variable defining domains	Domain	Domain proportion	$\text{deff}(\bar{y}_a)$ predicted from equation			'Actual'
	a	π_a	(3.7)	(3.13)	(3.12)	$\text{deff}(\bar{y}_a)$
Number of persons in household	1	0.19	1.63	1.82	1.77	1.87
	2	0.32	2.18	2.30	2.30	2.29
	3	0.17	1.54	1.69	1.69	1.75
	4	0.19	1.64	1.83	1.79	2.15
	5	0.08	1.16	1.34	1.33	1.42
	6	0.03	0.95	1.13	1.12	1.15
	7+	0.02	0.89	1.07	1.07	1.04
Housing tenure	Council rented	0.32	2.17	2.70	2.29	2.46
	Owner-occupied (with mortgage)	0.30	2.09	2.41	2.22	2.95
	Owner-occupied (no mortgage)	0.20	1.69	1.95	1.83	2.16
	Rented unfurnished	0.11	1.28	1.59	1.44	1.65
	Rented furnished	0.04	1.01	1.37	1.18	1.55
	Rent free	0.03	0.94	1.20	1.11	1.55

Note: $\text{deff}(\bar{y}) = 5.06$, $\bar{n} = 23.5$, $y = \log$ (housing expenditure).

from the following formula implied by equation (3.9) with $\tau_a = \tau$:

$$\widehat{\text{deff}}(\bar{y}_a) = 1 + \frac{[\pi_a \bar{n}(1 + C_a^2) - 1]}{[\bar{n}(1 + C^2) - 1]}[\text{deff}(\bar{y}) - 1]. \tag{3.13}$$

This formula is less useful in practice for imputing deffs since it depends on C_a^2 and C^2 which are not readily available. However, it provides a benchmark for studying the effect of the binomial approximation. 'Actual' deffs estimated using SUPERCARP are also included. Several comments may be made about Table 3.1:

1. Predictions (3.12) and (3.13) are much closer for household size domains than tenure domains. This is because household size is much more of a cross-class than housing tenure and hence approximation (3.10) works better.
2. Although formula (3.13) tends to under-predict the actual deff in Table 3.1, we have found that across other y variables it displays no systematic under- or over-prediction. What systematic differences exist are often explicable in terms of differences between τ_a and τ. For example, amongst domains of household tenure one would expect council rented accommodation to have the lowest τ_a on housing expenditure, since it is least affected by local 'market forces', and hence for formula (3.13) to over-predict as it does.
3. Our experience is that formula (3.7) is often fairly close to formula (3.12), unless $\bar{n}$ is very low, but on average it does tend to under-predict the actual deff.

In conclusion, we recommend the use of equation (3.12) as a simple method of imputing deffs for domain means from overall deffs, although caution should be exercised under major departures from the cross-class assumption, when underestimation may tend to occur. Further discussion of deffs for domain means and proportions is given by Lepkowski and Landis (1980) and Rust (1987).

3.3.3 Differences of Domain Means

If $\bar{y}_a$ and $\bar{y}_b$ are means of y in mutually exclusive domains a and b then under SRS assumptions, $\bar{y}_a$ and $\bar{y}_b$ are uncorrelated and the usual estimate of $\text{var}(\bar{y}_a - \bar{y}_b)$ is denoted v_{0ab}, where

$$v_{0ab} = v_{0a} + v_{0b}$$

in the notation of equation (3.4).

The deff of $\bar{y}_a - \bar{y}_b$ and v_{0ab} is thus

$$\text{deff}[\bar{y}_a - \bar{y}_b, v_{0ab}] = \frac{\text{var}(\bar{y}_a) + \text{var}(\bar{y}_b) - 2\,\text{cov}(\bar{y}_a, \bar{y}_b)}{E(v_{0a}) + E(v_{0b})}.$$

If $\bar{y}_a$ and $\bar{y}_b$ are uncorrelated under the actual design (or model) then this is simply a weighted mean of

$$\text{deff}(\bar{y}_a, v_{0a}) = \text{var}(\bar{y}_a)/E(v_{0a}),$$

and

$$\text{deff}(\bar{y}_b, v_{0b}) = \text{var}(\bar{y}_b)/E(v_{0b}).$$

In practice, however, just as intra-cluster correlations are almost always positive, it is usually found that $\bar{y}_a$ and $\bar{y}_b$ are positively correlated, in which case the deff of the difference $\bar{y}_a - \bar{y}_b$ tends to be less than the deffs of the individual means.

For example, the correlation matrix of mean housing expenditure within the six domains of household tenure defined in Table 3.1 is

$$\begin{bmatrix} 1 & & & & & \\ 0.32 & 1 & & & & \\ 0.11 & 0.24 & 1 & & & \\ 0.17 & 0.27 & 0.11 & 1 & & \\ 0.19 & 0.36 & 0.16 & 0.31 & 1 & \\ 0.12 & 0.24 & 0.15 & 0.10 & 0.16 & 1 \end{bmatrix}$$

The correlations are positive because local economic factors tend to influence housing costs in the same direction for different tenure types.

3.3.4 Linear Regression

Conventional linear regression analysis assumes a model relating a variable Y to a vector of variables $\mathbf{X} = (X_1, \ldots, X_k)'$ in which

$$E(Y \mid \mathbf{X} = \mathbf{x}) = \alpha + \mathbf{x}'\boldsymbol{\beta}. \tag{3.14}$$

The *ordinary least squares* (*OLS*) *estimator* of $\boldsymbol{\beta}$ is

$$\hat{\boldsymbol{\beta}} = \mathbf{V}_{xx}^{-1}\mathbf{V}_{xy}, \tag{3.15}$$

where

$$\mathbf{V}_{xx} = n_o^{-1}\sum_s (\mathbf{x}_t - \bar{\mathbf{x}})(\mathbf{x}_t - \bar{\mathbf{x}})', \qquad \mathbf{V}_{xy} = n_o^{-1}\sum_s (\mathbf{x}_t - \bar{\mathbf{x}})y_t, \qquad \bar{\mathbf{x}} = n_o^{-1}\sum_s \mathbf{x}_t.$$

The deff of $\hat{\boldsymbol{\beta}}$ depends fundamentally on the choice of estimator of $\text{Var}(\hat{\boldsymbol{\beta}})$. The standard *OLS variance estimator* is

$$\mathbf{V}_{\text{OLS}}(\hat{\boldsymbol{\beta}}) = [n_o(n_o - k)]^{-1}\sum_s e_t^2 \mathbf{V}_{xx}^{-1}, \tag{3.16}$$

where

$$e_t = y_t - \bar{y} - (\mathbf{x}_t - \bar{\mathbf{x}})'\hat{\boldsymbol{\beta}}. \tag{3.17}$$

It is well known, however, that this estimator may be inconsistent for $\text{Var}(\hat{\boldsymbol{\beta}})$ even under SRS if heteroscedasticity is present. For if $\text{var}(Y \mid \mathbf{X} = \mathbf{x}) = \sigma^2(\mathbf{x})$

depends on $\mathbf{x}$, assuming equation (3.14) and independent y_t observations, then

$$E[\mathbf{V}_{\text{OLS}}(\hat{\boldsymbol{\beta}})|\mathbf{x}] \simeq [n_o(n_o - k)]^{-1} \sum_s \sigma^2(\mathbf{x}_t)\mathbf{V}_{xx}^{-1}, \tag{3.18}$$

$$\text{Var}(\hat{\boldsymbol{\beta}}|\mathbf{x}) = n_o^{-2}\mathbf{V}_{xx}^{-1} \sum_s (\mathbf{x}_t - \bar{\mathbf{x}})\sigma^2(\mathbf{x}_t)(\mathbf{x}_t - \bar{\mathbf{x}})'\mathbf{V}_{xx}^{-1}, \tag{3.19}$$

and in general these are unequal. For example, if $k = 1$ then the (conditional) misspecification effect of V_{OLS}, defined in equation (2.2), is

$$\text{meff}(\hat{\beta}, V_{\text{OLS}}) = \text{var}(\hat{\beta}|x)/E[V_{\text{OLS}}|x] \simeq 1 + \rho C_\sigma C_x, \tag{3.20}$$

where C_σ and C_x are the coefficients of variation of $\sigma^2(x_t)$ and $(x_t - \bar{x})^2$ respectively and ρ is their mutual correlation. This misspecification effect measures the inconsistency in V_{OLS} arising from departures from the homoscedasticity assumptions. Inconsistency occurs even under SRS and hence it would be inappropriate to call equation (3.20) a design effect.

An estimator of $\text{Var}(\hat{\boldsymbol{\beta}})$ which is consistent under SRS (assuming no f.p.c.) in the presence of heteroscedasticity is the *SRS linearization estimator*

$$\mathbf{V}_{\text{L,SRS}}(\hat{\boldsymbol{\beta}}) = n_o^{-2}\mathbf{V}_{xx}^{-1} \sum_s (\mathbf{x}_t - \bar{\mathbf{x}})e_t^2(\mathbf{x}_t - \bar{\mathbf{x}})'\mathbf{V}_{xx}^{-1}, \tag{3.21}$$

which is unbiased for equation (3.19) in large samples. The large-sample misspecification effect of this estimator is thus $\text{meff}(\hat{\boldsymbol{\beta}}, \mathbf{V}_{\text{L,SRS}}) = \mathbf{I}_k$, the identity matrix, reflecting no inconsistency under SRS. The estimator $\mathbf{V}_{\text{L,SRS}}$ is implemented in the computer package SAS (1985, p. 660) and is discussed further by White (1980).

In order to estimate deffs for $\hat{\boldsymbol{\beta}}$ and $\mathbf{V}_{\text{OLS}}$ or $\mathbf{V}_{\text{L,SRS}}$ it is necessary to consider estimators of $\text{Var}(\hat{\boldsymbol{\beta}})$ which allow for complex designs.

A modification of $\mathbf{V}_{\text{OLS}}$, to take account of multi-stage sampling, may be based on the following homoscedastic model with *nested error* structure:

$$y_{ct} = \alpha + \mathbf{x}'_{ct}\boldsymbol{\beta} + \varepsilon_{ct}, \tag{3.22}$$

where

$$\varepsilon_{ct} = u_c + v_{ct}, \tag{3.23}$$

$$\begin{aligned} E(\varepsilon_{ct}) = 0, \qquad E(\varepsilon_{ct}\varepsilon_{c't'}) &= \sigma_u^2 + \sigma_v^2, && \text{if } c = c', t = t', \\ &= \sigma_u^2, && \text{if } c = c', t \neq t', \\ &= 0, && \text{if } c \neq c'. \end{aligned}$$

Here y_{ct} and $\mathbf{x}_{ct}$ are the values of Y and $\mathbf{X}$ associated with unit t within PSU c. Under the model the OLS estimator remains unbiased for $\boldsymbol{\beta}$ and an estimator $\mathbf{V}_{\text{NE}}$ which is consistent for $\text{Var}(\hat{\boldsymbol{\beta}})$ under this model may be derived as in Fuller and Battese (1973).

A numerical illustration of the above three variance estimators together with the linearization estimator $\mathbf{V}_{\text{L}}$ which allows for the complex design (to be discussed in Section 3.4.2) is given in Table 3.2. Results are given for various

Table 3.2 Comparison of four variance estimators for an OLS regression coefficient

Variables		OLS coefficient	Variance estimates ($\times 10^6$)			
			Assuming SRS		Assuming complex design	
y	x	$\hat{\beta}$	V_{OLS}	$V_{L,SRS}$	V_{NE}	V_L
Housing	Income	0.09	4.79	117.9	4.80	120.4
Fuel	Income	0.02	0.77	3.1	0.78	3.1
Food	Income	0.13	3.20	34.1	3.24	34.8
Tobacco	Income	0.02	0.86	3.2	0.86	3.3
Clothing	Income	0.13	7.40	257.7	7.51	259.1
Log (housing)	Log (income)	0.55	111	162	102	254
Log (fuel)	Log (income)	0.40	118	139	119	155
Log (food)	Log (income)	0.71	48	70	49	73
Log (tobacco)	Log (income)	0.44	553	537	555	603
Log (clothing)	Log (income)	1.31	751	826	759	831

Note: The y variables measure total expenditure on various commodity types or the logarithm thereof.

simple linear regressions with a single x variable using the FES data. In the first five rows y measures expenditure on various commodities and x is income; in the last five rows log transformations are taken of both y and x.

Scatterplots of y against x suggest that linearity is a reasonable approximation in both cases, although strong heteroscedasticity is present in the unlogged relationship with high positive values of ρ and C_σ in equation (3.20). This implies a strong downward bias of V_{OLS} as reflected in the dramatically different values of V_{OLS} and $V_{L,SRS}$ in the first five rows of Table 3.2. This biasing effect of heteroscedasticity dwarfs the effect of the complex design represented by comparisons between V_{OLS} and V_{NE} or between $V_{L,SRS}$ and V_L. For the logged data, where no obvious heteroscedasticity was present in the scatterplots, the relative differences between V_{OLS} and $V_{L,SRS}$ and between V_{NE} and V_L are not as great, although we would still recommend the use of the linearization methods or other robust alternatives such as jackknife procedures (see Section 2.13). Note, however, that under homoscedasticity V_L may be biased downwards in finite samples because e_t^2 is biased downwards for $\sigma^2(x_t)$ for observations t with large $(x_t - \bar{x})^2$ (Fuller, 1984) and so the linearization methods may themselves be sensitive to influential observations.

Estimates of deffs are given in Table 3.3. The usual estimator of deff$(\hat{\beta}, V_{L,SRS})$ is $V_L/V_{L,SRS}$, values of which are seen to be appreciably lower than the corresponding estimated deffs of the mean of the y variable given in the final column. This is a common empirical finding (e.g. Kish and Frankel, 1974; Hill, 1981) although in our example the degree of understatement is even stronger than

Table 3.3 Comparison of estimated deffs for OLS regression coefficient and deff for mean

Variables		$\widehat{\text{deff}}(\hat{\beta}, V_{\text{OLS}})$	$\widehat{\text{deff}}(\hat{\beta}, V_{\text{L,SRS}})$	$\widehat{\text{deff}}(\bar{y})$
y	x	$V_{\text{NE}}/V_{\text{OLS}}$	$V_{\text{L}}/V_{\text{L,SRS}}$	
Housing	Income	1.00	1.02	2.59
Fuel	Income	1.01	1.00	1.41
Food	Income	1.01	1.02	1.71
Tobacco	Income	1.00	1.04	1.45
Clothing	Income	1.02	1.01	1.12
Log(housing)	Log(income)	1.00	1.57	5.07
Log(fuel)	Log(income)	1.01	1.11	1.86
Log(food)	Log(income)	1.01	1.03	1.63
Log(tobacco)	Log(income)	1.00	1.12	1.39
Log(clothing)	Log(income)	1.01	1.01	1.25

usual. The ratios $V_{\text{NE}}/V_{\text{OLS}}$ are also included and are seen to be smaller still.

A common intuitive explanation for reduced deffs in regression is that regressing on appropriate x variables can control for cluster effects. For example, the deff of 1.63 for the mean log expenditure on food indicates that area of residence 'influences' food spending. But this may only arise because income influences both area of residence and expenditure on food and, once we control for income by regression, the deff falls to 1.03.

A formalization of this argument is suggested by Campbell (1977) and Scott and Holt (1982). They assume model (3.22), and for PSUs of equal sample size N and scalar x, they obtain the result

$$V_{\text{NE}}/V_{\text{OLS}} \simeq 1 + (N-1)\tau_\varepsilon \tau_x \tag{3.24}$$

where τ_ε and τ_x are the intra-PSU correlations of ε_{ct} and x_{ct} respectively. Since both τ's will usually be small, their product will be very small. This is illustrated empirically in Table 3.3 where the ratios $V_{\text{NE}}/V_{\text{OLS}}$ are all very close to 1. Skinner (1986b) argues, however, that $V_{\text{NE}}/V_{\text{OLS}}$ may grossly underestimate $\text{deff}(\hat{\beta}, V_{\text{L.SRS}})$ because this will be dominated by the effect of different regression slopes in different PSUs and this is not allowed for in model (3.22). As an illustration of this argument consider the case where $\hat{\beta}$ is a difference between domain means

$$\begin{aligned}\hat{\beta} = \bar{y}_a - \bar{y}_b &= \sum w_{ac}\bar{y}_{ac} - \sum w_{bc}\bar{y}_{bc} \\ &= \sum w_{ac}(\bar{y}_{ac} - \bar{y}_{bc}) + \sum (w_{ac} - w_{bc})\bar{y}_{bc}\end{aligned} \tag{3.25}$$

where w_{ac} and $\bar{y}_{ac}$ are the proportion and mean value of y respectively in domain a in PSU c. The key assumption of model (3.22) that the PSU effect u_c is constant across X (i.e. domains) implies that

$$\bar{y}_{ac} = \bar{Y}_a + u_c + \bar{v}_{ac}, \qquad \bar{y}_{bc} = \bar{Y}_b + u_c + \bar{v}_{bc}$$

and hence u_c cancels out in the leading term of equation (3.25). This leads to underestimation of deff($\hat{\beta}$, $V_{\mathrm{L.SRS}}$) as reflected in a comparison of columns one and two of Table 3.3 for the logged variables (the only case where model (3.22) is at least plausible).

Multicollinearity is a well-known problem concerning standard errors in conventional regression. If two x variables are collinear then the standard errors of their coefficients will be inflated. In addition, their deffs will become similar and it often aids interpretation if the x variables are redefined to be approximately orthogonal.

Consider, for example, the simple model $y = \alpha + \beta x + \varepsilon$ where y is log (housing expenditure) and x is log(income). The fitted model for the FES data is

$$\begin{array}{lll} & y = 2.75 & + \ 0.55x + \varepsilon. \\ \text{(s.e.)} & (0.17) & (0.016) \\ \text{(deff)} & (1.58) & (1.57) \end{array} \tag{3.26}$$

where standard errors ($V_{\mathrm{L}}^{1/2}$) and estimated deffs ($V_{\mathrm{L}}/V_{\mathrm{L.SRS}}$) are given below the estimated OLS coefficients. In this parametrization the vector of x values is highly collinear with the vector of 1's (the intercept term) and hence the standard error of $\hat{\alpha}$ is dominated by the effect of the sampling error of $\hat{\beta}$ and the deffs for $\hat{\alpha}$ and $\hat{\beta}$ are approximately equal. If instead the model is redefined as $y = \alpha + \beta(x - \bar{x}) + \varepsilon$ so that the column of $x - \bar{x}$ values is orthogonal to the vector of 1's the fit is

$$\begin{array}{ll} y = 8.66 & + \ 0.55(x - 10.71) + \varepsilon. \\ (0.015) & (0.016) \\ (5.07) & (1.57) \end{array} \tag{3.27}$$

Now the standard error of $\hat{\alpha}$ has been greatly reduced and its deff equals the deff for $\bar{y}$ (since $\hat{\alpha} = \bar{y}$).

Similarly, if $z_1, \ldots, z_6$ are 0–1 indicator variables representing six categories of housing tenure the deffs for $\hat{\boldsymbol{\beta}}$ in the model $y = (z_1, \ldots, z_6)\boldsymbol{\beta} + \varepsilon$ will be deffs for domain means as in Table 3.1.

$$\begin{array}{llllll} y = 8.52z_1 + & 8.98z_2 + & 8.70z_3 + & 8.21z_4 + & 8.84z_5 + & 8.19z_6 + \varepsilon \\ (0.02) & (0.02) & (0.03) & (0.04) & (0.05) & (0.06) \\ (2.46) & (2.95) & (2.16) & (1.65) & (1.55) & (1.55) \end{array} \tag{3.28}$$

whereas in a model with an intercept $y = \alpha + \beta_2 z_2 + \cdots + \beta_6 z_6 + \varepsilon$ the deffs for $\hat{\beta}_2, \ldots, \hat{\beta}_6$ will be deffs for differences of domain means and hence tend to be smaller as demonstrated in equation (3.29):

$$\begin{array}{llllll} y = 8.52 & + \ 0.46z_2 + & 0.18z_3 - & 0.31z_4 + & 0.32z_5 - & 0.33z_6 + \varepsilon. \\ (0.02) & (0.03) & (0.03) & (0.04) & (0.05) & (0.06) \\ (2.46) & (2.13) & (1.86) & (1.35) & (1.61) & (1.55) \end{array} \tag{3.29}$$

If the term $x\beta$ ($x = \log(\text{income})$) is added into equation (3.28) or (3.29) the effect is to inflate the standard errors of the coefficients of the z_i as in equation (3.26) and to shrink all deffs towards the deff of the coefficient of x. On the other hand, if the term $(x - \bar{x})\beta$ is added then the magnitudes of the deffs will be only slightly affected, e.g. for equation (3.29):

$$\begin{array}{llllllll} y = 8.60 + & 0.23z_2 + & 0.18z_3 - & 0.24z_4 + & 0.29z_5 - & 0.31z_6 + & 0.49(x - 10.71) + \varepsilon. \\ & (0.02) & (0.03) & (0.03) & (0.04) & (0.06) & (0.016) \\ & (2.02) & (1.82) & (1.45) & (1.45) & (1.73) & (1.48) \end{array}$$

3.3.5 Other Statistics

Before commenting on patterns of deffs for other statistics, we emphasize again that the effects of complex designs cannot always be viewed in isolation from the effects of other kinds of misspecification of assumptions. For example, if we wish to assess the effect of a complex design on the inverse information as an estimator of the variance of a maximum likelihood estimator then we need also to take account of the sensitivity of this estimator to departures from distributional assumptions even under SRS.

Skinner (1986b) considers a general theory of deffs of statistics which may be expressed as smooth functions $\hat{\theta} = g(\bar{\mathbf{y}})$ of a mean vector $\bar{\mathbf{y}}$ and which estimate a parameter $\theta = g(\boldsymbol{\mu})$, where $\boldsymbol{\mu} = E(\bar{\mathbf{y}})$. He suggests that the deff for $\hat{\theta}$ with the linearization variance estimator will be mainly determined by the heterogeneity of the corresponding within PSU parameters $g(\boldsymbol{\mu}_c)$, $\boldsymbol{\mu}_c = E(\mathbf{y}_t | t \in \text{PSU } c)$. If θ reflects some underlying causal mechanism constant across the population, what econometricians call a structural parameter, then the $g(\boldsymbol{\mu}_c)$ may be expected to be similar and the deff may tend to be relatively low, as for the regression slopes in Section 3.3.4. In general, however, there is no reason to suppose that the deff for one statistic of this type should be higher or lower than any other. For example, the deff for a correlation coefficient is mainly determined by the heterogeneity of correlation coefficients in different PSUs and this need not be related at all to the deffs for the means of the individual variables which will be mainly determined by the heterogeneity of PSU means.

As an empirical illustration of the fact that deffs for general complex statistics may be higher or lower than deffs for means, consider the following deffs averaged across statistics of a given type for a 30-strata design, reported by Frankel (1971, pp. 172–3).

Statistic	Mean deff
Mean	2.08
Regression coefficient	1.23
Correlation coefficient	2.00
Partial correlation coefficient	2.10
Multiple R^2	4.83

Only the mean deff for regression coefficients is appreciably lower than the deff for means. This study is discussed further in Kish and Frankel (1974). Bebbington and Smith (1977) present an empirical study of the effect of complex designs on correlation analysis and principal component analysis.

3.3.6 Hypothesis Testing

The effect of a complex design on a *one-sample t-test* has already been described in Section 2.4 with a summary of the effect presented in Table 2.1. The effect for a *two-sample paired t-test*, which is simply a one-sample t-test of differences, follows analogously.

For two-sample t-tests with 'independent' samples, two methods are usually distinguished: pooled variances or separate variances. For survey data with reasonably large but usually unequal sample sizes, the separate variance method is usually to be recommended since it is more robust to unequal variances. In this case the effect of a complex design may again be calibrated using Table 2.1 with the deff of $\bar{y}_a - \bar{y}_b$ discussed in Section 3.3.3.

The hypothesis of equality of more than two domain means $H_0{:}\mu_1 = \cdots = \mu_k$ is conventionally tested using the *F-test* of *one-way ANOVA*. Like the pooled variance two-sample t-test however, this test can be very sensitive to unequal variances in different domains (Box, 1954). This sensitivity is least (in the sense that $E(F) \simeq 1$ under H_0) when the sample sizes in different domains are equal, but this is rarely the case in sample surveys unlike controlled experiments. An alternative test which is robust to unequal variances may be obtained by reformulating the hypothesis as $H_0{:}\boldsymbol{\beta} = \mathbf{0}$ for the linear regression model (3.3) and using the procedure described below.

The linear hypothesis $H_0{:}\mathbf{A}\boldsymbol{\beta} = \mathbf{c}$ for the $k \times 1$ regression coefficient vector $\boldsymbol{\beta}$ in equation (3.14) is conventionally tested by referring the F-test statistic

$$(\mathbf{A}\hat{\boldsymbol{\beta}} - \mathbf{c})'(\mathbf{A}\mathbf{V}_{\mathrm{OLS}}\mathbf{A}')^{-1}(\mathbf{A}\hat{\boldsymbol{\beta}} - \mathbf{c})/q \tag{3.30}$$

to the F-distribution with $(q, n_o - k)$ degrees of freedom (d.f.), where $\mathbf{A}$ is a $q \times k$ matrix of rank $q \leqslant k$, $\hat{\boldsymbol{\beta}}$ is the OLS estimator in equation (3.15) and $\mathbf{V}_{\mathrm{OLS}}$ is given in equation (3.16). The poor performance of this test for survey data is illustrated by Shah *et al.* (1977). As shown in Section 3.3.4, this test will be very sensitive to heteroscedasticity. A more robust SRS test is obtained by substituting $\mathbf{V}_{\mathrm{L,SRS}}$ for $\mathbf{V}_{\mathrm{OLS}}$ in equation (3.30). For large n_o, this statistic will be distributed as $q^{-1}\chi^2_q$ under H_0 and SRS. This is an example of a Wald test. Under a complex design the large-sample distribution will become (see equation (2.26)) the weighted sum of q χ^2_1 random variables with weights given by the eigenvalues of

$$\boldsymbol{\Delta} = q^{-1}[\mathbf{A}E(\mathbf{V}_{\mathrm{L,SRS}})\mathbf{A}']^{-1}(\mathbf{A}\,\mathrm{Var}(\hat{\boldsymbol{\beta}})\mathbf{A}'). \tag{3.31}$$

For those cases, such as in Table 3.3, where the deff matrix for $\hat{\boldsymbol{\beta}}$ is close to $\mathbf{I}_k$ the eigenvalues will be close to q^{-1} and the test should perform reasonably well.

In multivariate analysis, hypotheses about the mean $\boldsymbol{\mu}$ of a $q \times 1$ vector of variables $\mathbf{y}$, are conventionally tested using *Hotelling's T^2 statistic*

$$T^2 = n_o(\bar{\mathbf{y}} - \boldsymbol{\mu})'\mathbf{V}_{yy}^{-1}(\bar{\mathbf{y}} - \boldsymbol{\mu}),$$

where

$$\bar{\mathbf{y}} = n_o^{-1}\sum_s \mathbf{y}_t, \qquad \mathbf{V}_{yy} = (n_o - 1)^{-1}\sum_s (\mathbf{y}_t - \bar{\mathbf{y}})(\mathbf{y}_t - \bar{\mathbf{y}})',$$

which is distributed as $(n_o - 1)q/(n_o - q)$ times an F random variable with $(q, n_o - q)$ d.f. if the $\mathbf{y}_t$ are IID normal with mean $\boldsymbol{\mu}$. Even if the $\mathbf{y}_t$ are not normal, T^2 is asymptotically distributed as χ_q^2 as $n_o \to \infty$ (Anderson, 1984, p.163). Under a complex design, however, it follows from Section 2.11 that T^2 is approximately distributed as a weighted sum of q χ_1^2 random variables with weights $\delta_1, \ldots, \delta_q$, the eigenvalues of

$$\boldsymbol{\Delta} = \boldsymbol{\Sigma}_{\mathrm{SRS}}^{-1}\boldsymbol{\Sigma},$$

where

$$\boldsymbol{\Sigma}_{\mathrm{SRS}} = E(\mathbf{V}_{yy}/n_o), \qquad \boldsymbol{\Sigma} = \mathrm{Var}(\bar{\mathbf{y}}).$$

If the eigenvaues are equal, $\delta_1 = \cdots = \delta_q = \delta$, then the actual significance level of the IID-based test is approximately

$$P(\chi_q^2 > \chi_q^2(\alpha)/\delta), \tag{3.32}$$

where $\chi_q^2(\alpha)$ is the critical point of the χ_q^2 distribution with level α. Values of this probability for $\alpha = 0.05$ and various values of δ and q are tabulated in Table 3.4. Note that the first column of this table is identical to the third column of Table 2.1 since for $q = 1$ the T^2-test is identical to a univariate t-test. For fixed δ the actual level increases markedly as q increases so that, in this situation, multivariate tests can be more misleading than univariate tests. This point is also made by Holt *et al.* (1980) and Fellegi (1980) in the context of categorical data.

If the eigenvalues are unequal then the values in Table 3.4 still retain an interpretation. Tests are often carried out by calculating a p-value for the given data and then rejecting $H_0{:}\boldsymbol{\mu} = \boldsymbol{\mu}_0$ if the p-value is less than α. The *nominal p-value* for a given (fixed) value of $\bar{\mathbf{y}}$ is approximately

$$pval_{\mathrm{SRS}}(\bar{\mathbf{y}}) = P[\chi_q^2 > (\bar{\mathbf{y}} - \boldsymbol{\mu}_0)'\boldsymbol{\Sigma}_{\mathrm{SRS}}^{-1}(\bar{\mathbf{y}} - \boldsymbol{\mu}_0)]$$

and H_0 is rejected using the IID-based test if $pval_{\mathrm{SRS}}(\bar{\mathbf{y}}) < \alpha$. The *actual p-value* can be defined analogously as

$$pval_{\mathrm{true}}(\bar{\mathbf{y}}) = P[\chi_q^2 > (\bar{\mathbf{y}} - \boldsymbol{\mu}_0)'\boldsymbol{\Sigma}^{-1}(\bar{\mathbf{y}} - \boldsymbol{\mu}_0)]. \tag{3.33}$$

Then, amongst values of $\bar{\mathbf{y}}$ in the critical region of the IID-based test, i.e. for which $pval_{\mathrm{SRS}}(\bar{\mathbf{y}}) < \alpha$, the maximum value of $pval_{\mathrm{true}}(\bar{\mathbf{y}})$ can be shown to be

Table 3.4 Actual significance levels (%) of T^2-test for nominal level of 5%

Eigenvalue (δ)	Degrees of freedom (q) 1	2	3	4
0.9	4	4	3	3
1.0	5	5	5	5
1.5	11	14	16	19
2.0	17	22	27	32
2.5	22	30	37	44
3.0	26	37	46	53

Note: The entries in the table are actual significance levels for the T^2-test when all the eigenvalues of $\mathbf{\Delta}$ equal δ. When the eigenvalues are unequal the entry gives the maximum possible actual p-value (%) for values of T^2 nominally significant at 5% where δ is the maximum eigenvalue of $\mathbf{\Delta}$.

$P(\chi_q^2 > \chi_q^2(\alpha)/\delta_1)$, where δ_1 is the maximum eigenvalue of $\mathbf{\Delta}$. But this is just as in equation (3.32) with δ replaced by δ_1 and so these maximum values of $pval_{\text{true}}(\bar{\mathbf{y}})$ may also be obtained from Table 3.4. Provided one accepts p-values as a measure of evidence against H_0, it follows from the decreasing relationship between max $pval_{\text{true}}(\bar{\mathbf{y}})$ and q in Table 3.4 that, for a given set of variables with given $\mathbf{\Delta}$, multivariate tests can again be potentially more misleading than univariate tests.

An alternative illustration of this effect is given in Table 3.5 which displays the nominal significance levels required to ensure that the actual significance level of T^2 is at most 5% (for the equal δ_i case) or that the maximum actual p-value is 5% for nominally significant values of T^2. To illustrate the use of this table, suppose we are interested in a subset of four coefficients of a regression coefficient vector $\boldsymbol{\beta}$. Suppose we estimate the 4×4 matrix $\mathbf{\Delta}$ in equation (3.31)

Table 3.5 Nominal levels (%) of T^2-test required to ensure actual level is less than 5%

Eigenvalue (δ)	Degrees of freedom (q) 1	2	3	4
1.0	5.0	5.0	5.0	5.0
1.5	1.6	1.1	0.9	0.7
2.0	0.6	0.3	0.2	<0.1
2.5	0.2	<0.1	<0.1	<0.1
3.0	<0.1	<0.1	<0.1	<0.1

using the appropriate **A** matrix (i.e. the $4 \times k$ matrix with each row containing a 1 corresponding to the coefficient of interest and with 0's elsewhere) and suppose the largest eigenvalue of the estimated matrix is 1.5. Then we know from Table 3.5 that, by fixing a nominal level of at most 1.6%, any univariate IID-based normal test that one of the four coefficients is equal to zero will have an actual significance level not greater than 5%. However, to ensure that the actual p-value is at most 5% for nominally significant values of the IID-based Wald test of the multivariate hypothesis that all four coefficients are zero, it is necessary to reduce the nominal level to 0.7%. Thus, in this sense, the multivariate test is more sensitive to the design.

As a simple numerical example, consider the 2×1 vector

$$\mathbf{y} = \begin{bmatrix} \text{log(expenditure on housing)} \\ \text{log(expenditure on services)} \end{bmatrix}.$$

Since $\chi_2^2(0.05) = 5.99$, the IID-based test of $H_0{:}\boldsymbol{\mu} = \boldsymbol{\mu}_0$ with approximate 5% nominal level is obtained by rejecting H_0 if:

$$n_o(\bar{\mathbf{y}} - \boldsymbol{\mu}_0)'\mathbf{V}_{yy}^{-1}(\bar{\mathbf{y}} - \boldsymbol{\mu}_0) > 5.99. \tag{3.34}$$

From the FES data we have

$$\mathbf{V}_{yy}/n_o = 10^{-5}\begin{bmatrix} 6.6 & 3.7 \\ 3.7 & 17.5 \end{bmatrix}$$

and, treating this as fixed, the critical region for $\bar{\mathbf{y}}$ defined by inequality (3.34) is displayed in Figure 3.2 as the exterior of the dotted ellipse.

The estimated actual covariance matrix of $\bar{\mathbf{y}}$ is

$$\hat{\boldsymbol{\Sigma}} = 10^{-5}\begin{bmatrix} 33.3 & 19.8 \\ 19.8 & 30.6 \end{bmatrix}$$

and a constant probability contour of $N_2(\boldsymbol{\mu}_0, \hat{\boldsymbol{\Sigma}})$, the (estimated) actual distribution of $\bar{\mathbf{y}}$, containing $\bar{\mathbf{y}}$ with 95% probability is given by

$$(\bar{\mathbf{y}} - \boldsymbol{\mu}_0)'\hat{\boldsymbol{\Sigma}}^{-1}(\bar{\mathbf{y}} - \boldsymbol{\mu}_0) = 5.99. \tag{3.35}$$

and is displayed as the larger of the two solid ellipses in Figure 3.2. The actual p-value of any point on this ellipse is 0.05, which is the same as the nominal p-value of any point on the dotted ellipse. The actual p-value for any point $\bar{\mathbf{y}}$ is found by obtaining the value of k for which

$$(\bar{\mathbf{y}} - \boldsymbol{\mu}_0)'\hat{\boldsymbol{\Sigma}}^{-1}(\bar{\mathbf{y}} - \boldsymbol{\mu}_0) = k, \tag{3.36}$$

and then calculating the probability that $\bar{\mathbf{y}}$ lies outside this ellipse, which from equation (3.33) is just $P(\chi_q^2 > k)$. It is not difficult to see that the points on the dotted ellipse with the largest actual p-values are points A and A′. The ellipse (equation 3.36) which intersects these two points and which determines the

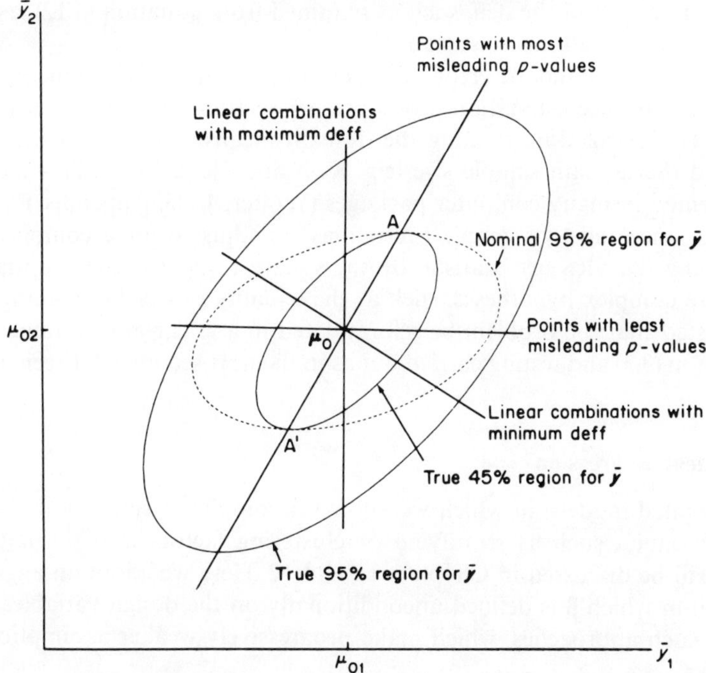

Figure 3.2 Probability contours for bivariate normal distribution of $\bar{\mathbf{y}}$ under IID assumptions (---) and under actual design (———).

actual p-value is displayed as the smaller solid ellipse. The actual p-values at points A and A′ turn out to be 0.55.

It may be shown that the vectors from $\boldsymbol{\mu}_0$ to the points A and A′ are right eigenvectors of $(\mathbf{V}_{yy}/n_o)\hat{\boldsymbol{\Sigma}}^{-1}$, corresponding to its minimum eigenvalue, which is equal to the reciprocal of the maximum eigenvalue of $\boldsymbol{\Delta}$. It is interesting to note that the linear combination of y_1 and y_2, defined by A and A′ to have the most misleading nominal p-value, is not the same as the linear combination of y_1 and y_2 which has the maximum deff, which is defined by the first right eigenvector of $\boldsymbol{\Delta}$ and is also depicted on Figure 3.2. Similarly the linear combinations with least misleading nominal p-values and smallest deff are also seen to be different. This illustrates how the theory of multivariate deffs is more complex than that of univariate deffs.

3.4 ALLOWING FOR COMPLEX DESIGNS

3.4.1 Domain Means

The simplest approach to allowing for complex designs in standard errors of domain means is to multiply the conventional SRS standard error by an estimate of the square root of the deff, such as obtained from equation (3.12), assuming a deff for the overall mean is available.

Otherwise the standard error may be estimated from first principles using one of the variance estimation procedures of Section 2.13. The most common method is linearization, treating the domain mean as a ratio of the domain total and the domain sample size (e.g. Kish and Hess, 1959). This method is implemented in many computer packages (Wolter, 1985, Appendix E).

Simple one- and two-sample t-tests may be adjusted for a complex design by dividing the SRS test statistic by the square root of the appropriate deff. For more complex hypotheses, such as the equality of $k > 2$ domain means, it is suggested that the problem be reformulated in a linear model framework as in equation (3.3) and testing carried out as in the next section (cf. Freeman *et al.*, 1977).

3.4.2 Linear Regression

Disaggregated models, in which we allow for complex designs by introducing design variables such as stratifying or clustering factors into the regression model, will be discussed in Chapters 11 and 12. Here we adopt an aggregated approach in which $\boldsymbol{\beta}$ is defined unconditionally on the design variables. Three possible such approaches, which make progressively weaker assumptions, are as follows.

1. Suppose that, unconditional on the design variables, the population values $(y_t, \mathbf{x}_t)$, $t \in U$, are identically distributed according to a model ξ in which y_t has a linear regression on $\mathbf{x}_t$:
$$E_\xi(y_t \mid \mathbf{x}_t) = \alpha + \mathbf{x}_t'\boldsymbol{\beta}. \tag{3.37}$$
2. Suppose again that the $(y_t, \mathbf{x}_t)$ are identically distributed, unconditional on the design variables, according to a model ξ and define α and $\boldsymbol{\beta}$ as the values of the coefficients a and $\mathbf{b}$ respectively of the linear predictor $\hat{y}_t = a + \mathbf{x}_t'\mathbf{b}$ with minimum mean-squared prediction error $E_\xi(y_t - \hat{y}_t)^2$.
3. Define α and $\boldsymbol{\beta}$ to be the values of the coefficients a and $\mathbf{b}$ of the linear predictor $\hat{y}_t$ with minimum mean-squared prediction error in the finite population $N_0^{-1}\Sigma_U(y_t - \hat{y}_t)^2$.

As we move from (1) to (2) to (3) the assumptions required to define $\boldsymbol{\beta}$ becomes weaker but the explanatory interpretation of $\boldsymbol{\beta}$ becomes more difficult. Thus in (1) the components of $\boldsymbol{\beta}$ retain their usual interpretation from equation (3.37) as the expected increase in y_t resulting from a unit increase in each individual $\mathbf{x}_t$ variable, whereas in (2) and (3) $\boldsymbol{\beta}$ is interpreted with respect to the best fitting linear predictor even though linear prediction will only be sensible if equation (3.37) is at least approximately correct. For the purpose of obtaining a consistent point estimator of $\boldsymbol{\beta}$, however, it is not necessary to choose between the approaches since in each case

$$\boldsymbol{\beta} = \mathrm{Var}(\mathbf{x}_t)^{-1}\,\mathrm{Cov}(\mathbf{x}_t, y_t) \tag{3.38}$$

where the covariances are defined with respect to the model in (1) and (2) and with respect to the finite population in (3). Property (3.38) follows for (1) by noting that

$$\begin{aligned}\mathrm{Cov}(\mathbf{x}_t, y_t) &= \mathrm{Cov}[\mathbf{x}_t, E(y_t|\mathbf{x}_t)]\\ &= \mathrm{Cov}(\mathbf{x}_t, \alpha + \mathbf{x}_t'\boldsymbol{\beta}) \quad \text{from equation (3.37)}\\ &= \mathrm{Var}(\mathbf{x}_t)\boldsymbol{\beta}\end{aligned}$$

and for (2) and (3) by differentiating the mean-squared prediction errors by a and $\mathbf{b}$. It follows that we may estimate $\boldsymbol{\beta}$ for each approach by consistently estimating the individual covariances in equation (3.38), for example using the OLS estimator $\hat{\boldsymbol{\beta}}$ in equation (3.15) under a self-weighting design, and $\hat{\boldsymbol{\beta}}$ will be consistent for each $\boldsymbol{\beta}$ of (1), (2) and (3) providing the respective assumptions of each approach hold. Of course, the aim of the initial model-building process should be to define variables y and $\mathbf{x}$ so that equation (3.37) holds as closely as possible.

In this section we shall only consider the OLS estimator $\hat{\boldsymbol{\beta}}$ and assume that it is consistent for $\boldsymbol{\beta}$ in equation (3.38) (with respect to either the true design or model). The question of consistency and estimators of $\boldsymbol{\beta}$ which allow for weighting or use of auxiliary information will be discussed in Chapter 7. We shall also ignore any finite population correction in variance estimation, since regression analysis is usually concerned with investigating causal mechanisms and hence model parameters.

Three methods for allowing for complex designs in the standard error of $\hat{\boldsymbol{\beta}}$ are:

1. To adjust SRS-based standard errors using an estimated deff;
2. To include extra terms in the fitted model to allow for the design;
3. To use a variance estimation technique as in Section 2.13.

For method (1), a heteroscedasticity-robust SRS standard error is recommended since, as noted in Section 3.3.4, heteroscedasticity can bias standard errors even more than complex designs. If a deff for the mean of the dependent variable is available, then multiplying the standard error of each coefficient by the square root of this deff will usually give a conservative (if over-conservative) standard error (see Section 3.3.4).

Method (2) is discussed by Holt and Scott (1981), Scott and Holt (1982), Wu *et al.* (1988), Pfeffermann and Nathan (1981), Pfeffermann and Smith (1985), Christensen (1984, 1987) and King and Evans (1986). This approach should be used with caution, however, since, as we saw in Section 3.3.4, the simple extension of the regression model to include additive terms representing clusters or strata may not lead to very robust estimators of standard errors.

To illustrate method (3), the linearization estimator of $\mathrm{var}(\hat{\beta})$ is derived for the case of one explanatory variable. Consider first the sample covariance V_{xy} which may be written in the notation of equations (2.34) and (2.35) as

$$V_{xy} = g(\bar{\mathbf{u}})$$

where

$$\bar{\mathbf{u}} = (\bar{u}_1, \bar{u}_2, \bar{u}_3)', \qquad g(\bar{\mathbf{u}}) = \bar{u}_1 - \bar{u}_2 \bar{u}_3,$$

$$\bar{u}_1 = \sum_s x_{hdct} y_{hdct}/n_o, \qquad \bar{u}_2 = \sum_s x_{hdct}/n_o = \bar{x}, \qquad \bar{u}_3 = \sum_s y_{hdct}/n_o = \bar{y}$$

and, for example, x_{hdct} is the value of x for the tth unit in the cth 2SU in the dth PSU in stratum h.

The statistic V_{xy} is an estimator of the population covariance:

$$\Omega_{xy} = g(\boldsymbol{\mu}),$$

where $\boldsymbol{\mu} = (\mu_1, \mu_2, \mu_3)'$ and μ_1, μ_2, μ_3 are the population means of xy, x and y respectively. As discussed in Chapter 2, $\mathrm{var}(V_{xy})$ may be approximated to first order by the variance of

$$\begin{aligned} g(\boldsymbol{\mu}) &+ \sum_{j=1}^{3} (\partial g(\boldsymbol{\mu})/\partial \mu_j)(\bar{u}_j - \mu_j) \\ &= \mu_1 - \mu_2\mu_3 + (\bar{u}_1 - \mu_1) - \mu_3(\bar{u}_2 - \mu_2) - \mu_2(\bar{u}_3 - \mu_3) \\ &= \sum_s (x_{hdct} - \mu_1)(y_{hdct} - \mu_2)/n_o. \end{aligned} \tag{3.39}$$

It follows that the linearization estimator of the variance of V_{xy}, $v_L(V_{xy})$, may be obtained from the usual variance formula for the mean $\bar{w}$ of the derived variable

$$w_{hdct} = [x_{hdct} - \bar{x}][y_{hdct} - \bar{y}]. \tag{3.40}$$

Hence the problem of estimating the variance of the sample second moment V_{xy} has been reduced to the problem of estimating the variance of the sample first moment $\bar{w}$. Extending this argument, the linearization estimator $v_L(\hat{\beta})$ of $\mathrm{var}(\hat{\beta})$ is identical to the linearization estimator of the variance of the ratio of means $\bar{w}_1/\bar{w}_2 (= \hat{\beta})$ of the derived variables:

$$w_{1hdct} = (x_{hdct} - \bar{x})(y_{hdct} - \bar{y}) \quad \text{and} \quad w_{2hdct} = (x_{hdct} - \bar{x})^2.$$

Thus using equations (2.38) and (2.40) we obtain

$$v_L(\hat{\beta}) = V_{xx}^{-2} \sum_h l_h/(l_h - 1) \sum_d (z_{hd} - \bar{z}_h)^2$$

where

$$\begin{aligned} z_{hd} &= \sum_c \sum_t z_{hdct}, \\ z_{hdct} &= n_o^{-1}(w_{1hdct} - \hat{\beta} w_{2hdct}) \\ &= n_o^{-1}(x_{hdct} - \bar{x})[y_{hdct} - \bar{y} - \hat{\beta}(x_{hdct} - \bar{x})] \end{aligned}$$

and in the case of SRS this reduces to the formula in (3.21). The corresponding expression for $\mathbf{V}_L(\hat{\boldsymbol{\beta}})$ for multiple regression and possibly unequal weighting is given by Fuller (1975). This estimator is implemented in several software packages such as SUPERCARP (Hidiroglou *et al.*, 1980), PC CARP (Fuller, 1986), SURREGR (Holt, 1977) and OSIRIS (Van Eck, 1979).

The main testing procedures used in conventional linear regression are t-tests of hypotheses $H_1:\beta_j=0$ concerning single coefficients β_j and F-tests of hypotheses $H_2:\boldsymbol{\beta}_q=\mathbf{0}$ concerning a set of q coefficients $\boldsymbol{\beta}_q$. Tests of H_1 may again be constructed by referring the ratio of the OLS coefficient $\hat{\beta}_j$ and its estimated standard error to a standard normal distribution. Tests of H_2 may be obtained by rejecting H_2 if the Wald test statistic $X^2_{\mathrm{W}}=\hat{\boldsymbol{\beta}}'_q\mathbf{V}_L(\hat{\boldsymbol{\beta}}_q)^{-1}\hat{\boldsymbol{\beta}}_q$ is in the upper tail of the χ^2_q distribution, where $\mathbf{V}_L(\hat{\boldsymbol{\beta}}_q)$ is the appropriate $q\times q$ submatrix of the estimated covariance matrix of $\hat{\boldsymbol{\beta}}$. If v, the d.f. used to estimate $\mathbf{V}_L(\hat{\boldsymbol{\beta}}_q)$, is low then a more refined approximation is to refer X^2_{W}/q to an F-distribution with (q, v) d.f. Shah *et al.* (1977) and Lemeshow and Stoddard (1984) simulate the performance of this procedure for stratified multistage sampling and show that it performs much better than the conventional OLS F-test.

3.4.3 Inference about Covariance Matrices

Many multivariate procedures involve inference about a covariance matrix $\boldsymbol{\Omega}$. One approach to inference, which extends naturally to complex designs, is based on the linearization variance estimation method. This approach, which is recommended as robust to departures from distributional assumptions by a number of authors in classical IID multivariate analysis (e.g. Browne, 1984; Fuller 1987b, Section 4.2), is now described.

Let $\mathbf{V}$ be a consistent estimator of the $p\times p$ covariance matrix $\boldsymbol{\Omega}$. For example, $\mathbf{V}$ might be the usual sample covariance matrix under a self-weighting design. Let $\bar{\mathbf{w}}=(V_{11}, V_{21}, V_{22}, V_{31}, V_{32}, V_{33},\ldots, V_{pp})'$ be the $\frac{1}{2}p(p+1)$ vector of distinct elements of $\mathbf{V}$, denoted by vech$(\mathbf{V})$, and suppose we may write

$$\bar{\mathbf{w}}=n_{\mathrm{o}}^{-1}\sum_s \mathbf{w}_t,$$

where $\mathbf{w}_t$ is a vector of cross-product terms of form (3.40), possibly weighted, say, for unequal selection probabilities. Then $\bar{\mathbf{w}}$ is consistent for vech$(\boldsymbol{\Omega})=\boldsymbol{\mu}$, say. Under IID assumptions, $\bar{\mathbf{w}}$ will generally be asymptotically normally distributed with mean $\boldsymbol{\mu}$ and applying the argument used in Section 3.4.2 for the OLS regression coefficient, the linearization estimator of the asymptotic covariance matrix of $\bar{\mathbf{w}}$ may be expressed as

$$\mathbf{V}_0=\sum_s(\mathbf{w}_t-\bar{\mathbf{w}})(\mathbf{w}_t-\bar{\mathbf{w}})'/[n_{\mathrm{o}}(n_{\mathrm{o}}-1)].$$

As an example of how this result may be applied, consider a linear hypothesis about $\boldsymbol{\Omega}$, which may be expressed as $H_0:\mathbf{A}\boldsymbol{\mu}=\mathbf{0}$, where $\mathbf{A}$ is a given $q\times\frac{1}{2}p(p+1)$

matrix of rank q. As an IID-based procedure for testing H_0, the Wald test statistic:

$$X^2_{W0} = (\mathbf{A}\bar{\mathbf{w}})'[\mathbf{A}\mathbf{V}_0\mathbf{A}']^{-1}\mathbf{A}\bar{\mathbf{w}}$$

may be referred to the upper tail of χ^2_q.

Under a complex design this procedure may be modified by replacing $\mathbf{V}_0$ by the linearization estimator $\mathbf{V}_L(\bar{\mathbf{w}})$ of $\text{Var}(\bar{\mathbf{w}})$ which allows for the design, and again this may be obtained simply from the usual formula (2.33) for the estimated covariance matrix of $\bar{\mathbf{w}}$, viewed as a statistic linear in the derived variables $\mathbf{w}_t$. This generates a modified Wald statistic X^2_W. An example of this approach is given by Pervaiz (1986) who extends Layard's (1972) procedure of this type for testing the equality of covariance matrices.

This approach assumes approximate normality of $\bar{\mathbf{w}}$, an approximation which can usually be improved by transformation, e.g. log transformation of variances and $\tanh^{-1}$ transformation of correlations (Layard, 1972, 1974). In moderate to large-scale surveys, however, evidence from Pervaiz (1986) suggests that deviations of the null distribution of X^2_W from χ^2_q are more likely to occur because v, the d.f. of $\mathbf{V}_L(\bar{\mathbf{w}})$ is low than because of the non-normality of $\bar{\mathbf{w}}$. The problem is that if p is at all large then $p(p+1)/2$ can be very large, often greater than v, and hence $\mathbf{V}_L(\bar{\mathbf{w}})$ may become very unstable and even non-singular. One alternative approximate approach when v is large compared with $p(p+1)/2$ is simply to correct X^2_{W0} for its first moment, i.e. to refer

$$X^2_{W0}q/\text{tr}[(\mathbf{A}\mathbf{V}_0\mathbf{A}')^{-1}\mathbf{A}\mathbf{V}_L(\bar{\mathbf{w}})\mathbf{A}']$$

to χ^2_q.

3.4.4 Parametric Modelling 1—Pseudo MLE

In this section we discuss an extension of the approach adopted for regression in Section 3.4.2 to general parametric models.

A conventional approach under SRS to fitting parametric models is via maximum likelihood estimation (MLE), where observations $\mathbf{y}_t$, $t \in s$, are assumed to be independent with probability density functions (p.d.f.) $f_t(\mathbf{y}_t; \boldsymbol{\theta})$ and the functions f_t are known but the $k \times 1$ parameter vector $\boldsymbol{\theta}$ is unknown. The ML estimator $\hat{\boldsymbol{\theta}}$ maximizes the log likelihood

$$l(\boldsymbol{\theta}) = \sum_s \log f_t(\mathbf{y}_t; \boldsymbol{\theta}),$$

and, in regular cases, solves the k likelihood equations

$$\dot{l}(\boldsymbol{\theta}) = \partial l(\boldsymbol{\theta})/\partial \boldsymbol{\theta} = \sum_s \mathbf{u}_t(\boldsymbol{\theta}) = \mathbf{0}, \tag{3.41}$$

where

$$\mathbf{u}_t(\boldsymbol{\theta}) = \partial \log f_t(\mathbf{y}_t; \boldsymbol{\theta})/\partial \boldsymbol{\theta}. \tag{3.42}$$

Now suppose that we wish to allow for a complex design but retain $\boldsymbol{\theta}$ as an aggregate parameter. Three possible approaches to the definition of $\boldsymbol{\theta}$, analagous to those given for regression in Section 3.4.2, are as follows:

1. Suppose that, unconditional on the design variables, the population values $\mathbf{y}_t$, $t \in U$, follow a model ξ in which the $\mathbf{y}_t$ remain independent with p.d.f. $f_t(\mathbf{y}_t; \boldsymbol{\theta})$.
2. Let $f_t(\mathbf{y}_t; \boldsymbol{\theta})$ be a nominal p.d.f. specified for $\mathbf{y}_t$ which is not necessarily correct and let $\boldsymbol{\theta}$ be the value (assumed unique) of $\tilde{\boldsymbol{\theta}}$ which maximizes
$$E_\xi N_o^{-1} \sum_U \log f_t(\mathbf{y}_t; \tilde{\boldsymbol{\theta}}), \qquad (3.43)$$
where ξ is the true model.
3. Let $f_t(\mathbf{y}_t; \boldsymbol{\theta})$ again be a nominal p.d.f. for $\mathbf{y}_t$, not necessarily correct, and let $\boldsymbol{\theta}$ be the value of $\tilde{\boldsymbol{\theta}}$ which maximizes
$$N_o^{-1} \sum_U \log f_t(\mathbf{y}_t; \tilde{\boldsymbol{\theta}}). \qquad (3.44)$$

As we move from (1) to (2) to (3) the assumptions required to define $\boldsymbol{\theta}$ become weaker but the interpretation of $\boldsymbol{\theta}$ becomes more difficult. In (1) $\boldsymbol{\theta}$ has its usual interpretation with respect to the true model. In (2) and (3) $\boldsymbol{\theta}$ is interpreted as defining that model $f_t(\mathbf{y}_t; \boldsymbol{\theta})$, amongst all models $f_t(\mathbf{y}_t; \tilde{\boldsymbol{\theta}})$ with varying values of $\tilde{\boldsymbol{\theta}}$, which provides the 'best' approximation, in the sense of equations (3.43) or (3.44), to the true model ξ or the population distribution of $\mathbf{y}_t$ respectively. Of course, the best approximation amongst a set of poor approximations is still a poor approximation, and so even for approaches (2) and (3) it is important in the model-building process to select f_t to be a reasonable approximation to reality.

As in Section 3.4.2 it is not necessary to choose between (1), (2) and (3) when selecting a point estimator of $\boldsymbol{\theta}$ since all three approaches share the property that $\boldsymbol{\theta}$ maximizes
$$E \log f_t(\mathbf{y}_t; \boldsymbol{\theta})$$
where in (1) E is the expectation with respect to ξ (see e.g. Silvey, 1970, p. 74) and in (2) and (3) E is defined as in equations (3.43) and (3.44). Alternatively, assuming sufficient regularity and after differentiation, all three approaches share the property that $\boldsymbol{\theta}$ solves the set of equations
$$\mathbf{T}(\boldsymbol{\theta}) = E\mathbf{u}_t(\boldsymbol{\theta}) = \mathbf{0}, \qquad (3.45)$$
where E is as above and $\mathbf{u}_t(\boldsymbol{\theta})$ is defined in equation (3.42). This may be viewed as a population version of the likelihood equations (3.41).

Example. Normal Linear Regression

Consider the linear regression model
$$y_t = \alpha + \mathbf{x}_t'\boldsymbol{\beta} + \varepsilon, \qquad \varepsilon \sim N(0, \sigma^2)$$

with fixed $\mathbf{x}_t$ so that $\boldsymbol{\theta} = (\alpha, \boldsymbol{\beta}', \sigma^2)'$ and

$$f_t(y_t; \boldsymbol{\theta}) = (2\pi\sigma^2)^{-1/2} \exp\{-(y_t - \alpha - \mathbf{x}_t'\boldsymbol{\beta})^2/2\sigma^2\}.$$

Then

$$\mathbf{u}_t(\boldsymbol{\theta}) = \begin{bmatrix} (y_t - \alpha - \mathbf{x}_t'\boldsymbol{\beta})/\sigma^2 \\ \mathbf{x}_t(y_t - \alpha - \mathbf{x}_t'\boldsymbol{\beta})/\sigma^2 \\ -1/(2\sigma^2) + (y_t - \alpha - \mathbf{x}_t'\boldsymbol{\beta})^2/2\sigma^4 \end{bmatrix}$$

and, for example, the solution $\boldsymbol{\theta}$ of the set of equations

$$N_o^{-1} \sum_{t \in U} \mathbf{u}_t(\boldsymbol{\theta}) = 0$$

is such that $\boldsymbol{\beta} = \boldsymbol{\Sigma}_{xx}^{-1} \boldsymbol{\Sigma}_{xy}$, where

$$\boldsymbol{\Sigma}_{xx} = N_o^{-1} \sum_U (\mathbf{x}_t - \boldsymbol{\mu}_x)(\mathbf{x}_t - \boldsymbol{\mu}_x)', \qquad \boldsymbol{\Sigma}_{xy} = N_o^{-1} \sum_U (\mathbf{x}_t - \boldsymbol{\mu}_x)(y_t - \mu_y),$$

$$\boldsymbol{\mu}_x = N_o^{-1} \sum_U \mathbf{x}_t, \qquad \mu_y = N_o^{-1} \sum_U y_t,$$

which corresponds to equation (3.38).

Having defined the target parameter $\boldsymbol{\theta}$ as the solution of equation (3.45) we need to choose a point estimator of $\boldsymbol{\theta}$. The full MLE would require an expression for the exact likelihood which may be very complicated and require many assumptions since it involves modelling the relation between $\mathbf{y}_t$ and the design variables. A much simpler approach is to replace the population mean $\mathbf{T}(\boldsymbol{\theta})$ of the $\mathbf{u}_t(\boldsymbol{\theta})$ in equation (3.45) (treating $\boldsymbol{\theta}$ as fixed and $\mathbf{u}_t(\boldsymbol{\theta})$ as a variable known for each unit t in the sample s) by a consistent estimator $\hat{\mathbf{T}}(\boldsymbol{\theta})$ and to define the *pseudo MLE* $\hat{\boldsymbol{\theta}}_{\text{PML}}$ as the solution of

$$\hat{\mathbf{T}}(\boldsymbol{\theta}) = \mathbf{0}. \tag{3.46}$$

The estimation of population means is a problem which has been extensively studied by survey statisticians (Cochran, 1977) and all the techniques that have been developed for that problem (such as ratio estimation, regression estimation, post-stratification, adjustment for non-response) may be used to construct $\hat{\mathbf{T}}(\boldsymbol{\theta})$.

For example, in the absence of auxiliary information, a standard design-based estimator is

$$\hat{\mathbf{T}}(\boldsymbol{\theta}) = \sum_s w_t \mathbf{u}_t(\boldsymbol{\theta}), \tag{3.47}$$

where w_t is inversely proportional to the probability of selecting unit t. It is not necessary, however, for pseudo MLEs to be derived from a design-based approach, even though most of the literature employing pseudo MLEs has been within this framework (e.g. Binder, 1983). From a model-based viewpoint, if the model f_t is correct and the sampling design does not depend on the $\mathbf{y}_t$ then the exact MLE of $\boldsymbol{\theta}$ is still obtained as the solution of equation (3.41), which is equivalent to pseudo MLE with constant weights w_t in equation (3.47). If the model f_t is not assumed to be correct for purposes of inference (even though it is used as a 'working model' to define the target parameter $\boldsymbol{\theta}$), then methods based upon weaker model assumptions might still be used. For example, if stratified sampling is employed and the $\mathbf{u}_t(\boldsymbol{\theta})$ are assumed to be IID within strata but not between strata then the pseudo MLE of form (3.47) with weights w_t equal to the inverse sampling fractions within strata would provide a natural model-based 'predictor' of $\boldsymbol{\theta}$ (cf. Skinner, 1983).

In general $\hat{\boldsymbol{\theta}}_{\text{PML}}$ is not the exact MLE, nor does it share the optimal asymptotic properties of ML such as efficiency and it is certainly not unique since several consistent estimators of $\mathbf{T}(\boldsymbol{\theta})$ may exist. However, $\hat{\boldsymbol{\theta}}_{\text{PML}}$ is robust in the sense that it is consistent for $\boldsymbol{\theta}$, defined with respect to at least one of the approaches (1)–(3), even if f_t is not the true p.d.f.

Now consider the estimation of the covariance matrix of $\hat{\boldsymbol{\theta}}_{\text{PML}}$. A standard consistent estimator of the covariance matrix of $\hat{\boldsymbol{\theta}}$ in conventional MLE is the inverse observed information matrix $\mathbf{I}(\hat{\boldsymbol{\theta}})^{-1}$, where

$$\mathbf{I}(\boldsymbol{\theta}) = -\,\partial \dot{l}(\boldsymbol{\theta})/\partial \boldsymbol{\theta}. \tag{3.48}$$

A more robust estimator of $\text{Var}(\hat{\boldsymbol{\theta}})$, which assumes independent observations but not that the f_t are necessarily the true p.d.f.s of $\mathbf{y}_t$, is the *linearization estimator* (Royall, 1986):

$$\mathbf{V}_{\text{L,SRS}}(\hat{\boldsymbol{\theta}}) = \mathbf{I}(\hat{\boldsymbol{\theta}})^{-1}\mathbf{V}_{\text{L,SRS}}[\dot{l}(\hat{\boldsymbol{\theta}})]\mathbf{I}(\hat{\boldsymbol{\theta}})^{-1},$$

where

$$\mathbf{V}_{\text{L,SRS}}[\dot{l}(\hat{\boldsymbol{\theta}})] = n_{\text{o}} \sum_s \mathbf{u}_t(\hat{\boldsymbol{\theta}})\mathbf{u}_t(\hat{\boldsymbol{\theta}})'.$$

This estimator may be extended naturally to estimate $\text{Var}(\hat{\boldsymbol{\theta}}_{\text{PML}})$, under a complex design. Suppose that $\hat{\mathbf{T}}(\hat{\boldsymbol{\theta}})$ takes the weighted form in equation (3.47), then the linearization estimator of $\text{Var}(\hat{\boldsymbol{\theta}}_{\text{PML}})$ is

$$\mathbf{V}_L(\hat{\boldsymbol{\theta}}_{\text{PML}}) = \mathbf{I}(\hat{\boldsymbol{\theta}}_{\text{PML}})^{-1}\mathbf{V}_L[\hat{\mathbf{T}}(\hat{\boldsymbol{\theta}}_{\text{PML}})]\mathbf{I}(\hat{\boldsymbol{\theta}}_{\text{PML}})^{-1}, \tag{3.49}$$

where

$$\mathbf{I}(\boldsymbol{\theta}) = -\,\partial \hat{\mathbf{T}}(\boldsymbol{\theta})/\partial \boldsymbol{\theta}$$

$$\mathbf{V}_L[\hat{\mathbf{T}}(\hat{\boldsymbol{\theta}}_{\text{PML}})] = \sum_{h=1}^{H} \frac{l_h}{(l_h - 1)} \sum_{d=1}^{l_h} (\mathbf{z}_{hd} - \bar{\mathbf{z}}_h)(\mathbf{z}_{hd} - \bar{\mathbf{z}}_h)'$$

and $\mathbf{z}_{hd}$ is the sum of the $w_t\mathbf{u}_t(\hat{\boldsymbol{\theta}}_{\text{PML}})$ across sample units t in PSU $d = 1, \ldots, l_h$ in stratum h as in equation (2.41). Again the consistency of $\mathbf{V}_L(\hat{\boldsymbol{\theta}}_{\text{PML}})$ does not depend on the assumption that f_t is the true p.d.f.

Example. Logistic Regression

A logistic regression model for a 0–1 dependent variable y_t and a vector $\mathbf{x}_t$ of independent variables is represented by the p.d.f.

$$f_t(y_t; \boldsymbol{\theta}) = \pi_t(\boldsymbol{\theta})^{y_t}[1 - \pi_t(\boldsymbol{\theta})]^{1 - y_t}$$

where

$$\pi_t(\boldsymbol{\theta}) = \Pr(y_t = 1) = 1 - [1 + \exp(\mathbf{x}_t'\boldsymbol{\theta})]^{-1}.$$

It follows from equation (3.42) that

$$\mathbf{u}_t(\boldsymbol{\theta}) = [y_t - \pi_t(\boldsymbol{\theta})]\mathbf{x}_t$$

and, if $\hat{\boldsymbol{\theta}}_{\text{PML}}$ is defined from equations (3.46) and (3.47)

$$\mathbf{I}(\boldsymbol{\theta}) = \sum_{t \in s} w_t \pi_t(\boldsymbol{\theta})[1 - \pi_t(\boldsymbol{\theta})]\mathbf{x}_t\mathbf{x}_t'.$$

The linearization estimator $\mathbf{V}_L(\hat{\boldsymbol{\theta}}_{\text{PML}})$ is then as given in Roberts *et al.* (1987, equation 2.4). This example is discussed further in Chapter 9. Further examples of the application of pseudo ML are given in Chapter 4 for contingency tables, in Binder (1983) for generalized linear models and in Chambless and Boyle (1985) for proportional hazards models.

Conventional likelihood-ratio tests about $\boldsymbol{\theta}$ are in general non-robust to departures both from the assumption that f_t is the correct p.d.f. and the assumption that the y_t are independent. Wald tests may be used instead. The hypothesis H_0:$\mathbf{A}(\boldsymbol{\theta}) = \mathbf{0}$, where $\mathbf{A}(.)$ is a given $q \times 1$ vector function, may be tested by referring

$$X_{\text{W}}^2 = \mathbf{A}(\hat{\boldsymbol{\theta}}_{\text{PML}})'[\dot{\mathbf{A}}(\hat{\boldsymbol{\theta}}_{\text{PML}})\mathbf{V}_L(\hat{\boldsymbol{\theta}}_{\text{PML}})\dot{\mathbf{A}}(\hat{\boldsymbol{\theta}}_{\text{PML}})']^{-1}\mathbf{A}(\hat{\boldsymbol{\theta}}_{\text{PML}}),$$

to the χ_q^2 distribution, where $\dot{\mathbf{A}}(\boldsymbol{\theta}) = \partial\mathbf{A}(\boldsymbol{\theta})/\partial\boldsymbol{\theta}$ is the $k \times q$ matrix of first derivatives of $\mathbf{A}(\boldsymbol{\theta})$. For linear hypotheses, where $\mathbf{A}(\boldsymbol{\theta}) = \mathbf{A}\boldsymbol{\theta}$ and $\mathbf{A}$ is a given $k \times q$ matrix of constants of rank q, $\dot{\mathbf{A}}(\boldsymbol{\theta})$ reduces to $\mathbf{A}$.

We note finally that the class of estimators defined by solutions of estimating equations of form (3.46) with $\hat{\mathbf{T}}$ as in equation (3.47) includes many other examples, such as non-linear least-squares estimators, robust M-estimators or quasi-likelihood estimators (Binder, 1983), and the covariance matrix of any estimator in this class may be estimated as in equation (3.49).

3.4.5 Parametric Modelling 2—Moment Structures and GLS

In the pseudo ML approach, a parametric model is specified by a family of distributions $f_t(.;\boldsymbol{\theta})$. An alternative approach, which does not require the specification of a distributional form, is to define a parametric model for a $k \times 1$ vector $\boldsymbol{\mu}$ of population moments

$$\boldsymbol{\mu} = \mathbf{A}(\boldsymbol{\theta}), \tag{3.50}$$

where $\mathbf{A}(.)$ is a given function and $\boldsymbol{\theta}$ is a $q \times 1$ vector of unknown parameters, $q \leqslant k$. An important special case is the linear model $\mathbf{A}(\boldsymbol{\theta}) = \mathbf{A}\boldsymbol{\theta}$ where $\mathbf{A}$ is a given $k \times q$ matrix of constants. Some examples of such moment-structure models follow. This approach assumes the availability of a consistent estimator $\mathbf{T}$ of $\boldsymbol{\mu}$ (usually the corresponding vector of sample moments, suitably weighted) and also a consistent estimator $\mathbf{V}$ of the covariance matrix $\text{Var}(\mathbf{T})$ of $\mathbf{T}$. Consistency here may refer either to a design-based or a model-based framework.

Example 1

Let $\boldsymbol{\mu} = (\mu_{11}, \ldots, \mu_{IJ})'$ be the $IJ \times 1$ vector of population domain means μ_{ij} of a variable Y within domains (i, j) defined by two discrete variables X_1 and X_2 with I and J categories respectively. Let $\mathbf{T} = (\bar{y}_{11}, \ldots, \bar{y}_{IJ})'$ be the corresponding vector of sample domain means, assumed consistent for $\boldsymbol{\mu}$ and let $\mathbf{V}$ be the usual linearization estimator of $\text{Var}(\mathbf{T})$. The additive linear model

$$\mu_{ij} = \mu + \alpha_i + \beta_j \qquad i = 1, \ldots, I, \quad j = 1, \ldots, J,$$

where $\alpha_I = \beta_J = 0$ may be represented as

$$\boldsymbol{\mu} = \mathbf{A}\boldsymbol{\theta}$$

where $\boldsymbol{\theta} = (\mu, \alpha_1, \ldots, \alpha_{I-1}, \beta_1, \ldots, \beta_{J-1})'$ and $\mathbf{A}$ is the appropriate $IJ \times (I + J - 1)$ matrix of 0's and 1's.

Example 2

Let $\boldsymbol{\mu} = (\mu_{11}, \mu_{12}, \mu_{21}, \mu_{22})'$, where $(\mu_{i1}\,\mu_{i2})'$ is the population mean vector of $(Y_1, Y_2)'$ in domain $i = 1, 2$. The model $\mu_{11} = \mu_{21}, \mu_{12} = \mu_{22}$ is discussed by Koch and Lemeshow (1972), in an example where Y_1 = height, Y_2 = weight and the two domains are Negro and White six-year-old male children. The model may be expressed as $\boldsymbol{\mu} = \mathbf{A}\boldsymbol{\theta}$, where

$$\mathbf{A} = \begin{pmatrix} 1 & 0 & 1 & 0 \\ 0 & 1 & 0 & 1 \end{pmatrix}' \qquad \boldsymbol{\theta} = \begin{pmatrix} \theta_1 \\ \theta_2 \end{pmatrix}.$$

Koch and Lemeshow take $\mathbf{T}$ as the corresponding vector of weighted sample means and $\mathbf{V}$ as the balanced half-sample estimator of $\text{Var}(\mathbf{T})$.

Example 3

Let $\boldsymbol{\mu} = (\Sigma_{11}, \Sigma_{21}, \Sigma_{22}, \Sigma_{31} \cdots \Sigma_{pp})'$ be the $\frac{1}{2}p(p+1) \times 1$ vector of population covariances between p variables. The *factor analysis* model

$$\Sigma_{ij} = \sum_{r=1}^{m} \lambda_{ir}\lambda_{jr} + \delta_{ij}\psi_i$$

where $m < p$ and δ_{ij} is the Kronecker δ, may also be represented in the form $\boldsymbol{\mu} = \mathbf{A}(\boldsymbol{\theta})$, $\boldsymbol{\theta} = (\lambda_{11}, \ldots, \lambda_{pm}, \psi_1, \ldots, \psi_p)'$, where now $\mathbf{A}(.)$ is a non-linear function (Fuller, 1987a). Consistent estimators $\mathbf{T}$ of $\boldsymbol{\mu}$ and $\mathbf{V}$ of $\text{Var}(\mathbf{T})$ may be obtained as in Section 3.4.3.

A *generalized least-squares* (GLS) point estimator $\hat{\boldsymbol{\theta}}_{\text{GLS}}$ of $\boldsymbol{\theta}$ in the model (3.50) is obtained as the value of $\boldsymbol{\theta}$ which minimizes

$$[\mathbf{T} - \mathbf{A}(\boldsymbol{\theta})]'\mathbf{V}^{-1}[\mathbf{T} - \mathbf{A}(\boldsymbol{\theta})]. \tag{3.51}$$

In particular, for the linear model $\mathbf{A}(\boldsymbol{\theta}) = \mathbf{A}\boldsymbol{\theta}$ we obtain the usual expression

$$\hat{\boldsymbol{\theta}}_{\text{GLS}} = (\mathbf{A}'\mathbf{V}^{-1}\mathbf{A})^{-1}\mathbf{A}'\mathbf{V}^{-1}\mathbf{T}.$$

For general non-linear models, numerical methods may be needed to minimize expression (3.51). A consistent estimator of $\text{Var}(\hat{\boldsymbol{\theta}}_{\text{GLS}})$, assuming equation (3.50) is correct, is (Fuller, 1984)

$$\mathbf{V}_L(\hat{\boldsymbol{\theta}}_{\text{GLS}}) = [\dot{\mathbf{A}}(\hat{\boldsymbol{\theta}}_{\text{GLS}})'\mathbf{V}^{-1}\dot{\mathbf{A}}(\hat{\boldsymbol{\theta}}_{\text{GLS}})]^{-1}$$

where $\dot{\mathbf{A}}(\boldsymbol{\theta}) = \partial\mathbf{A}(\boldsymbol{\theta})/\partial\boldsymbol{\theta}$ is the $k \times q$ matrix of first derivatives of $\mathbf{A}(\boldsymbol{\theta})$. For the linear model, $\dot{\mathbf{A}}(\boldsymbol{\theta})$ reduces to $\mathbf{A}$ and $\mathbf{V}_L(\hat{\boldsymbol{\theta}}_{\text{GLS}})$ reduces to the estimator given, for example, by Koch *et. al.* (1975). A goodness-of-fit test of the model (3.50) is obtained by referring the Wald test statistic

$$X^2_{\text{W}} = [\mathbf{T} - \mathbf{A}(\hat{\boldsymbol{\theta}}_{\text{GLS}})]'\mathbf{V}^{-1}[\mathbf{T} - \mathbf{A}(\hat{\boldsymbol{\theta}}_{\text{GLS}})]$$

to the χ^2_{k-q} distribution. Nested hypotheses may be tested similarly. For example, in the linear model $\boldsymbol{\mu} = \mathbf{A}\boldsymbol{\theta}$ the hypothesis $H_0{:}\mathbf{C}\boldsymbol{\theta} = \mathbf{0}$, where $\mathbf{C}$ is a given $r \times q$ matrix, may be tested by referring

$$\hat{\boldsymbol{\theta}}'\mathbf{C}'[\mathbf{C}(\mathbf{A}'\mathbf{V}^{-1}\mathbf{A})^{-1}\mathbf{C}']^{-1}\mathbf{C}\hat{\boldsymbol{\theta}}$$

to the χ^2_r distribution. Koch *et al.* (1975) give examples of the application of this procedure to fitting alternative models to data from the US National Center for Health Statistics. The implementation of this procedure in the computer package SAS is described in SAS/IML (1985, p. 61).

As in Section 3.4.3, problems may occur with these test procedures and indeed with the stability of $\hat{\boldsymbol{\theta}}_{\text{GLS}}$ itself if k is large compared with v, the d.f. upon which $\mathbf{V}$ is based, since $\mathbf{V}^{-1}$ may then be unstable. In this case, an approach which should not usually result in much loss of efficiency is to replace $\mathbf{V}$ in equation (3.51) by $\mathbf{V}_{\text{SRS}}$, the covariance matrix estimator under SRS assumptions which is usually based on many d.f., giving a point estimator $\hat{\boldsymbol{\theta}}_{\text{GLS,SRS}}$. A consistent estimator of $\text{Var}(\hat{\boldsymbol{\theta}}_{\text{GLS,SRS}})$ is now

$$\mathbf{V}_L(\hat{\boldsymbol{\theta}}_{\text{GLS,SRS}}) = (\tilde{\mathbf{A}}'\mathbf{V}_{\text{SRS}}^{-1}\tilde{\mathbf{A}})^{-1}\tilde{\mathbf{A}}'\boldsymbol{\Delta}\mathbf{V}_{\text{SRS}}^{-1}\tilde{\mathbf{A}}(\tilde{\mathbf{A}}'\mathbf{V}_{\text{SRS}}^{-1}\tilde{\mathbf{A}})^{-1},$$

where $\tilde{\mathbf{A}} = \dot{\mathbf{A}}(\hat{\boldsymbol{\theta}}_{\text{GLS,SRS}})$ is the $k \times q$ matrix of first derivatives evaluated at $\hat{\boldsymbol{\theta}}_{\text{GLS,SRS}}$, $\boldsymbol{\Delta} = \mathbf{V}_{\text{SRS}}^{-1}\mathbf{V}$ is the estimated design effects matrix and again $\tilde{\mathbf{A}} = \mathbf{A}$ for the linear model $\mathbf{A}(\boldsymbol{\theta}) = \mathbf{A}\boldsymbol{\theta}$. The test statistic

$$X^2_{\text{W,SRS}} = [\mathbf{T} - \mathbf{A}(\hat{\boldsymbol{\theta}}_{\text{GLS,SRS}})]'\mathbf{V}_{\text{SRS}}^{-1}[\mathbf{T} - \mathbf{A}(\hat{\boldsymbol{\theta}}_{\text{GLS,SRS}})]$$

no longer has an asymptotic chi-squared distribution, however, but is distributed as a weighted sum of $k - q$ independent χ^2_1 random variables where the weights are the non-zero eigenvalues of

$$\mathbf{H} = \mathbf{\Delta} - \mathbf{\Delta}\mathbf{V}_{\text{SRS}}^{-1}\tilde{\mathbf{A}}(\tilde{\mathbf{A}}'\mathbf{V}_{\text{SRS}}^{-1}\tilde{\mathbf{A}})^{-1}\tilde{\mathbf{A}}'.$$

As a one-moment approximation $(k-q)\, X^2_{\text{W,SRS}}/\text{tr}(\mathbf{H})$ may be referred to χ^2_{k-q} to test the goodness of fit of equation (3.50). Note that when $\mathbf{V}_{\text{SRS}} = \mathbf{V}$ we have $\text{tr}(\mathbf{H}) = k - q$ and so

$$X^2_{\text{W}} = X^2_{\text{W,SRS}} = (k-q)X^2_{\text{W,SRS}}/\text{tr}(\mathbf{H}).$$

CHAPTER 4

Chi-Squared Tests for Contingency Tables

J. N. K. Rao and D. R. Thomas

4.1 INTRODUCTION

Statistical methods for the analysis of cross-classified count data are used extensively by survey researchers. In particular, the Pearson chi-squared test of independence in a two-way contingency table is probably best known and most often used. Analyses of multi-way contingency tables are also now quite common, largely due to the development of hierarchical loglinear models and related logit models and associated methods for systematic testing of hypotheses, similar to analysis of variance for continuous data.

Standard computer packages used to implement these methods, for example GLIM (Payne, 1985), SPSS^{X} (1986, Chapters 20, 29, 30) and SAS (1985, Chapter 14), are based upon the assumption of multinomial sampling which is equivalent to the assumption of IID observations or the assumption of SRS, as discussed in Chapter 2. This chapter begins by examining the impact of complex designs on these IID-based methods. It is shown, for example, that clustering can have a substantial effect on significance levels of the standard Pearson chi-squared test X_{P}^2 and the standard likelihood-ratio chi-squared test X_{LR}^2, in the sense that the actual type I error rate can be much bigger than the nominal level thus leading to unnecessarily complex models.

Alternative test procedures, which take account of the complexity of the design, are considered in Sections 4.3–4.5. These consist of first- and second-order corrections to X_{P}^2 and X_{LR}^2, as introduced in Section 2.11, as well as Wald tests and their modified versions. The first-order corrections have the advantage that they only require knowledge of the design effects (deffs) for the individual cells and marginals of the contingency table, whereas the second-order corrections and Wald tests require knowledge of the full covariance matrix of the estimated cell proportions.

Standard errors of parameter estimates and 'smoothed' cell estimates and the use of residual analysis for detecting model deviations are considered in Sections 4.6 and 4.7. Finally in Section 4.8 the various methods are illustrated on some data from the Canada Health Survey (1978–79), a typical complex survey based on a multistage design involving stratification and cluster sampling. The commonly used tests of goodness of fit, homogeneity and independence in two-way tables, and independence in three-way contingency tables are studied in detail. Multi-way tables under nested loglinear models are also studied. The results given here can be used either in the design-based framework pertaining to finite population cell proportions or for model-based inference on cell probabilities in some underlying superpopulation model.

4.2 IMPACT OF DESIGN ON STANDARD TESTS

4.2.1 Simple Goodness of Fit

The impact of survey design on X_P^2 or X_{LR}^2 is best understood by considering first the special case of simple goodness of fit with I cells and associated finite population proportions or cell probabilities under a superpopulation model, denoted by μ_i, $i = 1, \ldots, I$ ($\mu_i \geqslant 0$, $\sum \mu_i = 1$). Denote the observed counts in a sample, s, of n_o ultimate units, drawn according to a specified sample design $p(s)$, by $n_1, \ldots, n_I$ ($\sum n_i = n_o$). The Pearson statistic for testing $H_0{:}\mu_i = \mu_{0i}$, $i = 1, \ldots, I$ is then given by

$$\begin{aligned}\tilde{X}_P^2 &= \sum_{i=1}^{I} (n_i - n_o\mu_{0i})^2/(n_o\mu_{0i}) \\ &= n_o \sum_{i=1}^{I} (\bar{y}_i - \mu_{0i})^2/\mu_{0i},\end{aligned} \tag{4.1}$$

where $\bar{y}_i = \sum_{t \in s} y_{it}/n_o = n_i/n_o$ and y_{it} is the 0–1 indicator variable for the ith category associated with the tth unit*. The statistic $\tilde{X}_P^2$ is used when $\bar{y}_i$ is unbiased for μ_i. In the design-based framework for example, this condition is satisfied for SRS or, more generally, for any self-weighting design with equal inclusion probabilities. For cases where $\bar{y}_i$ is biased for μ_i, $\tilde{X}_P^2$ is inappropriate and it is necessary to consider a more general statistic

$$X_P^2 = n_o \sum_{i=1}^{I} (\hat{\mu}_i - \mu_{0i})^2/\mu_{0i}, \tag{4.2}$$

where the estimator $\hat{\mu}_i$ is a consistent estimator of μ_i (see Chapter 1 and Part C). If $\hat{\mu}_i = \bar{y}_i$, then X_P^2 reduces to $\tilde{X}_P^2$. Published tables often report weighted-up counts $\hat{N}_i$, i.e. consistent estimates of population counts N_i, so that equation (4.2) can be readily implemented by setting $\hat{\mu}_i = \hat{N}_i/\hat{N}_o$ where $\hat{N}_o = \sum \hat{N}_i$. Note

*The test statistics considered here are also applicable to a specific subpopulation (or domain) if n_o is interpreted as the number of sample units falling in that domain (see the examples in Section 4.8)

that equation (4.1) should never be applied with the weighted-up counts $\hat{N}_i$ in place of the n_i because the resulting Pearson statistic would equal $(\hat{N}_o/n_o)X^2_P$, a meaningless quantity much bigger than X^2_P.

The likelihood ratio statistic for testing H_0 is given by

$$X^2_{LR} = 2n_o \sum_{i=1}^{I} \hat{\mu}_i \log(\hat{\mu}_i/\mu_{0i}). \tag{4.3}$$

Under multinomial sampling, it is well known that both X^2_P and X^2_{LR} are distributed asymptotically as χ^2_{I-1}, a chi-squared random variable with $I-1$ degrees of freedom (d.f.), when H_0 holds. It is therefore customary to reject H_0 when X^2_P or X^2_{LR} exceeds $\chi^2_{I-1}(\alpha)$, the upper α-point of χ^2_{I-1}, so that the type I error rate is approximately equal to α (say 0.05 or 0.01). Finite-sample results under multinomial sampling provide further support to the customary test procedure (Larntz, 1978).

For complex designs, the customary test procedure is not valid even asymptotically. Rao and Scott (1979, 1981) have, in fact, shown that both X^2_P and X^2_{LR} are distributed asymptotically as a weighted sum, $\delta_1 W_1 + \cdots + \delta_{I-1} W_{I-1}$ of $I-1$ independent χ^2_1 variables W_i. Here the weights δ_i are the eigenvalues of the design effects matrix $\mathbf{\Omega}_{0I}^{-1}\mathbf{\Sigma}_I$ (see Section 2.11), where $\mathbf{\Omega}_{0I}$ is the multinomial covariance matrix under H_0,

$$\mathbf{\Omega}_{0I} = n_o^{-1}[\text{diag}(\boldsymbol{\mu}_{0I}) - \boldsymbol{\mu}_{0I}\boldsymbol{\mu}'_{0I}],$$

$\mathbf{\Sigma}_I$ is the covariance matrix under the actual design (or model) $\mathbf{\Sigma}_I = \text{Var}(\hat{\boldsymbol{\mu}}_I)$ and

$$\hat{\boldsymbol{\mu}}_I = (\hat{\mu}_1, \ldots, \hat{\mu}_{I-1})', \qquad \boldsymbol{\mu}_{0I} = (\mu_{01}, \ldots, \mu_{0,I-1})', \qquad \delta_1 \geqslant \cdots \geqslant \delta_{I-1} > 0.$$

For the special case of $I = 2$ categories, the design effects matrix reduces to the ordinary deff, namely $\text{var}(\hat{\mu}_1)/[n_o^{-1}\mu_{01}(1-\mu_{01})]$. As in Section 2.11, the δ_i may be interpreted as 'generalized design effects' and, for example, δ_1 is the largest possible deff taken over all possible linear combinations of the $\hat{\mu}_i$'s.

In the special case of multinomial sampling, all $\delta_i = 1$ and the weighted sum of W_i's reduces to χ^2_{I-1}, the standard result. In other cases, the weighted sum could be substantially different from a χ^2_{I-1} variable. We now consider a few special designs to illustrate the nature of the weighted sum, $\sum \delta_i W_i$ (Rao and Scott, 1979, 1981).

1. *SRS without replacement*. In the case when the μ_i are finite population cell proportions, all $\delta_i = 1 - n_o/N_o$ for any $\boldsymbol{\mu}_{0I} = (\mu_{01}, \ldots, \mu_{0,I-1})'$ so that X^2_P is distributed asymptotically as $(1 - n_o/N_o)\chi^2_{I-1}$. This result implies that the Pearson statistic X^2_P is always asymptotically conservative in the sense that the actual type I error rate is less than the nominal level, α. A simple finite population correction to X^2_P, given by $(1 - n_o/N_o)^{-1}X^2_P$, ensures that the actual type I error rate is equal to α asymptotically, and the test is asymptotically more powerful than X^2_P. The first-order correction to X^2_P,

to be given in Section 4.3, reduces to this finite population correction in this special case.

If the finite population is regarded as a random sample from an infinite superpopulation with cell probabilities μ_i and the hypothesis $H_0:\mu_i = \mu_{0i}$, $i = 1,\ldots,I$ is specified with respect to these probabilities, then the Pearson statistic X_P^2 is asymptotically correct, i.e. is distributed asymptotically as χ_{I-1}^2 under H_0 and no finite population correction is necessary.

2. *Stratified random sampling* (*proportional allocation*). Suppose simple random samples are drawn independently with replacement from H strata. In this case all $\delta_i \leqslant 1$ for any $\boldsymbol{\mu}_0$ so that X_P^2 is again asymptotically conservative. Rao and Scott (1979, 1981), however, have shown that X_P^2 is well approximated by χ_{I-1}^2 if I is large and H relatively small. These results are valid for both design-based inference and model-based inference.
3. *Two-stage sampling*. Brier (1980) proposed a simple model for within-cluster dependence. The design involves two-stage sampling with multinomial sampling within each sampled cluster (or PSU), c, assuming equal subsample sizes $n_c = n$, $c = 1,\ldots,m$. Thus the observed within-cluster counts $\mathbf{n}_{cI} = (n_{c1},\ldots,n_{c,I-1})'$ follow a multinomial distribution with probabilities $\boldsymbol{\mu}_{cI} = (\mu_{c1},\ldots,\mu_{c,I-1})'$. The probabilities $\boldsymbol{\mu}_{cI}$ are then assumed to be sampled independently from a Dirichlet distribution. This clustering model leads to $\delta_i = \delta(>1)$ for all i and for any $\boldsymbol{\mu}_0$. The type I error rate under this clustering model, for nominal level α, is approximately given by $\Pr[\chi_{I-1}^2 > \delta^{-1}\chi_{I-1}^2(\alpha)]$ which increases with δ and can bc made arbitrarily large by increasing δ (see Table 3.4). Although Brier's model is fairly restrictive (since it implies the same deff for all individual cell estimates and all linear combinations of cell estimates), it clearly illustrates the effect of clustering on the type I error rate of standard tests based on X_P^2 or X_{LR}^2.

Some empirical evidence of the effect of complex designs is provided by Holt *et al.* (1980), who estimated the actual type I error rates of X_P^2 for several variables from two large-scale UK surveys, the General Household Survey (GHS) of 1971 and the British Election Study (BES) of 1974. As expected, these rates are much larger than the nominal level, α, for variables having high design effects. For example, in the GHS the estimated actual type I error rate, for $\alpha = 0.05$, ranged from 0.30 to 0.48 as the average cell design effect increased from 2.17 to 3.27. Thus, standard goodness-of-fit tests can be very misleading with complex survey data, in the sense that correct hypotheses about population proportions may be judged to be false too often.

4.2.2 Multi-way Contingency Tables

Using the notation of Section 4.2.1, the cells in a multi-way contingency table are numbered lexicographically as $i = 1,\ldots,I$ with corresponding finite population proportions or cell probabilities μ_i and their estimates $\hat{\mu}_i$. A loglinear model on the μ_i's may be written as

$$\log \boldsymbol{\mu} = \tilde{u}(\boldsymbol{\theta})\mathbf{1} + \mathbf{X}\boldsymbol{\theta}. \tag{4.4}$$

Here $\log \boldsymbol{\mu}$ is the I-vector of log probabilities, $\mathbf{X}$ is a known $I \times r$ matrix of full rank $r \leqslant I - 1$, such that $\mathbf{X}'\mathbf{1} = \mathbf{0}$, $\boldsymbol{\theta}$ is an r-vector of parameters, $\mathbf{1}$ is the I-vector of 1's, $\mathbf{0}$ is the null vector and $\tilde{u}(\boldsymbol{\theta})$ is the normalizing factor which ensures that $\sum \mu_i = 1$. If $r = I - 1$ in equation (4.4), the 'saturated' loglinear model is obtained.

As a simple illustration of the general loglinear model (4.4), consider a 2×2 table with cell probabilities $\mu_{ab}(a = 1, 2; b = 1, 2; I = 4)$. The saturated loglinear model may then be written, in notation similar to that of Bishop *et al.* (1975, p. 17), as

$$\log \mu_{ab} = \tilde{u} + u_{1(a)} + u_{2(b)} + u_{12(ab)}; \qquad a, b = 1, 2, \tag{4.5}$$

with $\sum_a u_{1(a)} = \sum_b u_{2(b)} = \sum_a u_{12(ab)} = \sum_b u_{12(ab)} = 0$. In the notation of equation (4.4), we can write $\boldsymbol{\theta} = (u_{1(1)}, u_{2(1)}, u_{12(11)})'$ and $\mathbf{X}$ as a 4×3 matrix of 1's and -1's:

$$\mathbf{X} = \begin{bmatrix} 1 & 1 & 1 \\ 1 & -1 & -1 \\ -1 & 1 & -1 \\ -1 & -1 & 1 \end{bmatrix}, \tag{4.6}$$

the rows of $\mathbf{X}$ corresponding to cells (11), (12), (21) and (22) respectively. Under the hypothesis of independence, we have $u_{12(11)} = 0$ so that the model is now given by equation (4.4) with $\boldsymbol{\theta} = (u_{1(1)}, u_{2(1)})'$ and $\mathbf{X}$ given by the first two columns of equation (4.6).

The maximum likelihood estimator (MLE) of $\boldsymbol{\mu}(\boldsymbol{\theta}) = [\mu_1(\boldsymbol{\theta}), \ldots, \mu_I(\boldsymbol{\theta})]'$ under multinomial sampling, denoted by $\boldsymbol{\mu}(\hat{\boldsymbol{\theta}})$, is obtained from the likelihood equations:

$$\mathbf{X}'\boldsymbol{\mu}(\hat{\boldsymbol{\theta}}) = \mathbf{X}'\bar{\mathbf{y}}, \tag{4.7}$$

where $\bar{\mathbf{y}} = (\bar{y}_1, \ldots, \bar{y}_I)'$. The method of iterative proportional fitting (IPF) gives the fitted proportions $\boldsymbol{\mu}(\hat{\boldsymbol{\theta}})$ from equation (4.7) without evaluating the MLE $\hat{\boldsymbol{\theta}}$ of $\boldsymbol{\theta}$ (Bishop *et al.*, 1975, p. 83). The Newton–Raphson method is often used as an alternative since $\hat{\boldsymbol{\theta}}$ is then obtained and the iteration has quadratic convergence unlike the IPF method.

Maximum likelihood estimation of $\boldsymbol{\mu}(\boldsymbol{\theta})$ for general designs cannot usually be implemented due to difficulties in defining appropriate likelihood functions. Hence it is customary to use a 'pseudo MLE' of $\boldsymbol{\mu}(\boldsymbol{\theta})$ obtained from equation (4.7) by substituting the survey estimator $\hat{\boldsymbol{\mu}} = (\hat{\boldsymbol{\mu}}_1, \ldots, \hat{\boldsymbol{\mu}}_I)'$ for $\bar{\mathbf{y}}$ (see Section 3.4.4). The pseudo MLE is consistent for $\boldsymbol{\mu}(\boldsymbol{\theta})$, provided $\hat{\boldsymbol{\mu}}$ is consistent. In our 2×2 example, $\mu_{ab}(\hat{\boldsymbol{\theta}}) = \hat{\mu}_{a.}\hat{\mu}_{.b}$, where $\{\hat{\mu}_{ab}\}$ are the survey estimators of $\{\mu_{ab}\}$ with marginal row and column totals $\{\hat{\mu}_{a.}\}$ and $\{\hat{\mu}_{.b}\}$ respectively.

Following equation (4.2), the general Pearson statistic for testing the goodness of fit of the model (4.4) with $r < I - 1$ is given by

$$X_{\mathrm{P}}^2 = n_{\mathrm{o}} \sum_{i=1}^{I} [\hat{\mu}_i - \mu_i(\hat{\boldsymbol{\theta}})]^2 / \mu_i(\hat{\boldsymbol{\theta}}). \tag{4.8}$$

Similarly, the likelihood ratio statistic may be written as

$$X_{\mathrm{LR}}^2 = 2n_{\mathrm{o}} \sum_{i=1}^{I} \hat{\mu}_i \log[\hat{\mu}_i / \mu_i(\hat{\boldsymbol{\theta}})]. \tag{4.9}$$

Rao and Scott (1984) have shown that X_{P}^2 (or X_{LR}^2) is distributed asymptotically as a weighted sum, $\delta_1 W_1 + \cdots + \delta_{I-r-1} W_{I-r-1}$, of $I - r - 1$ independent χ_1^2 variables W_i, ($\delta_1 \geqslant \cdots \geqslant \delta_{I-r-1} > 0$). This result is similar to the formula in Section 4.2.1 for the special case of simple goodness of fit, but the weights δ_i are now the eigenvalues of a more complex design effects matrix given by

$$\boldsymbol{\Delta} = (\mathbf{C}'\mathbf{D}(\boldsymbol{\mu})^{-1}\mathbf{C})^{-1}(\mathbf{C}'\mathbf{D}(\boldsymbol{\mu})^{-1}\boldsymbol{\Sigma}\mathbf{D}(\boldsymbol{\mu})^{-1}\mathbf{C}), \tag{4.10}$$

where $\mathbf{D}(\boldsymbol{\mu}) = \mathrm{diag}(\boldsymbol{\mu})$, the $I \times I$ diagonal matrix with diagonal elements $\mu_1, \ldots, \mu_I$, $\boldsymbol{\Sigma} = \mathrm{Var}(\hat{\boldsymbol{\mu}})$, the covariance matrix of $\hat{\boldsymbol{\mu}}$ under the actual complex design (or true model) and $\mathbf{C}$ is any $I \times (I - r - 1)$ full rank matrix such that $\mathbf{C}'\mathbf{X} = \mathbf{0}$ and $\mathbf{C}'\mathbf{1} = \mathbf{0}$. In particular, under the standard parametrization of a loglinear model, $\mathbf{C}$ may be chosen as the matrix complementing $\mathbf{X}$ to form the model matrix in the saturated case. For example, in the case of the hypothesis of independence in the 2×2 table, $\mathbf{C}$ is the last column of the saturated model matrix (4.6). The δ_i may again be interpreted as generalized deffs, δ_1 being the largest possible deff taken over all linear combinations of the elements of the vector $\mathbf{C}' \log \hat{\boldsymbol{\mu}}$.

In the case of multinomial sampling, the generalized deffs δ_i are all equal to 1 and hence the standard result that X_{P}^2 or X_{LR}^2 is distributed asymptotically as a χ_{I-r-1}^2 variable follows as a special case.

The previous results for special designs remain valid:

1. All $\delta_i = 1 - n_{\mathrm{o}}/N_{\mathrm{o}}$ for SRS without replacement;
2. All $\delta_i \leqslant 1$ for stratified random sampling with proportional allocation;
3. $\delta_i = \delta(> 1)$ for all i under Brier's model for two-stage cluster sampling.

Holt *et al.* (1980) found from their empirical study with UK surveys that, provided the variables cut across strata and clusters, the type I error rate inflation in X_{P}^2 for testing independence in a two-way table is much less severe in comparison to corresponding tests for simple goodness of fit and homogeneity in a two-way table. The estimated actual type I error rate, for a nominal rate of $\alpha = 0.05$, ranged from 0.04 to 0.13 for 36 tables from the BES, while the range for GHS variables is as follows: 0.02–0.09 for 27 tables, 0.11–0.28 for 8 tables and 0.62 for 1 table. Even though the effect of a complex design is less severe in this case, the performance of these tests remains very unsatisfactory in suggesting too often that association exists between pairs of variables which are in fact independent.

4.2.3 Nested Models

Writing $\mathbf{X\theta}$ as $\mathbf{X}_1\boldsymbol{\theta}_1 + \mathbf{X}_2\boldsymbol{\theta}_2$ we are often interested in testing the nested hypothesis H_0:$\boldsymbol{\theta}_2 = \mathbf{0}$ given the loglinear model (4.4) and $r < I - 1$, where $\mathbf{X}_1$ is $I \times r_1$ of rank r_1 and $\mathbf{X}_2$ is $I \times r_2$ of rank r_2 ($r_1 + r_2 = r$). The general Pearson statistic and the likelihood ratio statistic in this case are given by

$$X_P^2(2|1) = n_o \sum_{i=1}^{I} [\mu_i(\hat{\boldsymbol{\theta}}) - \mu_i(\hat{\hat{\boldsymbol{\theta}}}_1)]^2/\mu_i(\hat{\hat{\boldsymbol{\theta}}}_1) \tag{4.11}$$

and

$$X_{LR}^2(2|1) = 2n_o \sum_{i=1}^{I} \mu_i(\hat{\boldsymbol{\theta}}) \log[\mu_i(\hat{\boldsymbol{\theta}})/\mu_i(\hat{\hat{\boldsymbol{\theta}}}_1)]. \tag{4.12}$$

Here $\mu_i(\hat{\hat{\boldsymbol{\theta}}}_1)$ is the 'pseudo MLE' under the reduced model, obtained from the 'pseudo likelihood equations' $\mathbf{X}'_1\boldsymbol{\mu}(\hat{\hat{\boldsymbol{\theta}}}_1) = \mathbf{X}'_1\hat{\boldsymbol{\mu}}$. The likelihood ratio statistic $X_{LR}^2(2|1)$ is usually preferred to $X_P^2(2|1)$ due to its additive property

$$X_{LR}^2(2|1) + X_{LR}^2(1) = X_{LR}^2(2),$$

where

$$X_{LR}^2(1) = 2n_o \sum \hat{\mu}_i \log[\hat{\mu}_i/\mu_i(\hat{\boldsymbol{\theta}})]$$

and

$$X_{LR}^2(2) = 2n_o \sum \hat{\mu}_i \log[\hat{\mu}_i/\mu_i(\hat{\hat{\boldsymbol{\theta}}}_1)].$$

However, the two tests are asymptotically equivalent, under a general design.

Rao and Scott (1984) showed that $X_P^2(2|1)$ (or $X_{LR}^2(2|1)$) is distributed asymptotically as a weighted sum,

$$\delta_1(2|1)W_1 + \cdots + \delta_{r_2}(2|1)W_{r_2},$$

of r_2 independent χ_1^2 variables W_i,

$$\delta_1(2|1) \geqslant \cdots \geqslant \delta_{r_2}(2|1) > 0.$$

Here the weights $\delta_i(2|1)$ are the eigenvalues of the design effects matrix

$$\boldsymbol{\Delta}(2|1) = (\tilde{\mathbf{X}}'_2\boldsymbol{\Omega}\tilde{\mathbf{X}}_2)^{-1}(\tilde{\mathbf{X}}'_2\boldsymbol{\Sigma}\tilde{\mathbf{X}}_2),$$

where $\tilde{\mathbf{X}}_2 = [\mathbf{I} - \mathbf{X}_1(\mathbf{X}'_1\boldsymbol{\Omega}\mathbf{X}_1)^{-1}\mathbf{X}'_1\boldsymbol{\Omega}]\mathbf{X}_2$ and $\boldsymbol{\Omega} = \mathbf{n}_o^{-1}[\mathbf{D}(\boldsymbol{\mu}) - \boldsymbol{\mu}\boldsymbol{\mu}']$, the multinomial covariance matrix. The $\delta_i(2|1)$ may be interpreted as generalized deffs, with $\delta_1(2|1)$ as the largest possible deff taken over all linear combinations of the elements of the vector $\tilde{\mathbf{X}}'_2\hat{\boldsymbol{\mu}}$. In the case of multinomial sampling, the standard result that $X_P^2(2|1)$ or $X_{LR}^2(2|1)$ is distributed asymptotically as a $\chi_{r_2}^2$ variable is obtained. In the general case of complex designs, however, the actual type I error rate will usually differ from the nominal level α assumed when referring $X_P^2(2|1)$ or $X_{LR}^2(2|1)$ to the αth critical point of $\chi_{r_2}^2$.

An alternative test of H_0, used for example in GLIM (Payne, 1985, p. 107), refers the F-statistic

$$F = \frac{X^2_{\text{LR}}(2|1)/r_2}{X^2_{\text{LR}}(1)/(I-r-1)} \tag{4.13}$$

to the αth critical point of the F-distribution with r_2 and $I-r-1$ degrees of freedom. Under the condition that both sets of eigenvalues $\delta_1(2|1),\ldots,\delta_{r_2}(2|1)$ and all the eigenvalues of $\boldsymbol{\Delta}$ in equation (4.10) are equal, this test procedure remains valid under complex designs (Rao and Scott, 1987). This property is not possessed by the two previous tests based on $X^2_{\text{P}}(2|1)$ and $X^2_{\text{LR}}(2|1)$. Whilst the above condition is very restrictive and unlikely to hold exactly in practice, this property does suggest that the F-test may be relatively more robust to complex designs than the other two tests. Nevertheless, none of these tests can be regarded as entirely satisfactory for complex designs, since the type I error rate can, in principle, be grossly distorted in every case.

4.3 FIRST-ORDER CORRECTIONS TO STANDARD TESTS

The impact of survey design on X^2_{P} and X^2_{LR}, studied in Section 4.2, clearly demonstrates that standard tests without some adjustment for survey design can give quite erroneous results in practice. This section gives simple first-order corrections to X^2_{P} and X^2_{LR} that can be implemented from published tables if these provide cell deffs and the deffs for marginals in the contingency table. The basic idea is to note that for, say, X^2_{P} defined in equation (4.8), the true asymptotic null distribution of $X^2_{\text{P}}/\delta_.$ has the same first moment as χ^2_{I-r-1}, the limiting distribution of X^2_{P} under the standard multinomial assumption, where $\delta_. = \sum \delta_i/(I-r-1)$ is the mean generalized deff. A first-order correction refers

$$X^2_{\text{P}}(\hat{\delta}_.) = X^2_{\text{P}}/\hat{\delta}_. \quad \text{or} \quad X^2_{\text{LR}}(\hat{\delta}_.) = X^2_{\text{LR}}/\hat{\delta}_.$$

to χ^2_{I-r-1}, where $\hat{\delta}_.$ is a consistent estimator of $\delta_.$, i.e. H_0 is rejected if the corrected test statistic falls into the standard critical region. A consistent estimator of $\delta_.$ under model (4.4) is given by

$$(I-r-1)\hat{\delta}_. = \sum \hat{\delta}_i = \operatorname{tr}(\hat{\boldsymbol{\Delta}}) \tag{4.14}$$

where $\hat{\delta}_1,\ldots,\hat{\delta}_{I-r-1}$ are the eigenvalues of $\hat{\boldsymbol{\Delta}}$, the estimated design effects matrix obtained from equation (4.10) by replacing $\boldsymbol{\mu}$ by $\boldsymbol{\mu}(\hat{\boldsymbol{\theta}})$ and $\boldsymbol{\Sigma}$ by $\hat{\boldsymbol{\Sigma}}$, a consistent estimator of the covariance matrix of $\hat{\boldsymbol{\mu}}$ under the actual design (or model). Such estimators are discussed in Section 2.13. In fact, for most of the special cases described below, it is not necessary to have the full matrix $\hat{\boldsymbol{\Sigma}}$ but only variance estimates (or equivalently deffs) for individual cell estimates (i.e.

the diagonal elements of $\hat{\boldsymbol{\Sigma}}$) and for certain marginals. These results are now spelled out for the commonly used tests of simple goodness of fit, homogeneity and independence in a two-way table, and independence in a three-way table.

The test based on $X_{\rm P}^2(\hat{\delta}.)$ is asymptotically correct in the case of constant deffs, $\delta_i = \delta$ for all i, but in general it should perform well when the coefficient of variation of the δ_i is small. The second-order correction of Section 4.4 takes account of the variability in the generalized deffs, δ_i, unlike $X_{\rm P}^2(\hat{\delta}.)$.

4.3.1 Simple Goodness of Fit

In the case of a simple goodness of fit test, $(I-1)\hat{\delta}. = \mathrm{tr}(\boldsymbol{\Omega}_{oI}^{-1}\hat{\boldsymbol{\Sigma}}_I)$ reduces to

$$(I-1)\hat{\delta}. = \sum_{i=1}^{I} \frac{\hat{\mu}_i}{\mu_{oi}}(1-\hat{\mu}_i)\hat{D}_i, \tag{4.15}$$

where $\hat{D}_i$ is the estimated deff of the estimated cell proportion $\hat{\mu}_i$, viz.

$$\hat{D}_i = \widehat{\mathrm{var}}(\hat{\mu}_i)/[\hat{\mu}_i(1-\hat{\mu}_i)n_o^{-1}].$$

This deff does not depend on H_0 and is the one which would commonly be published along with the cell estimates. It follows from equation (4.15) that only estimated cell deffs, and not the full matrix $\hat{\boldsymbol{\Sigma}}_I$, are needed to implement the correction $X_{\rm P}^2(\hat{\delta}.)$ or $X_{\rm LR}^2(\hat{\delta}.)$.

A simulation study by Thomas and Rao (1987) indicated that the type I error performance of $X_{\rm P}^2(\hat{\delta}.)$ can be improved, when the degrees of freedom, ν, of $\hat{\boldsymbol{\Sigma}}$ are not large, by treating

$$FX_{\rm P}^2(\hat{\delta}.) = X_{\rm P}^2(\hat{\delta}.)/(I-1) \tag{4.16}$$

as an F-variable with d.f. $I-1$ and $\nu(I-1)$ respectively. As discussed in Section 2.14, ν may often be taken as the number of sampled clusters minus the number of strata. The correction $FX_{\rm P}^2(\hat{\delta}.)$ depends only on the estimated cell deffs, $\hat{D}_i$, as in the case of $X_{\rm P}^2(\hat{\delta}.)$, so that no additional information is required to implement this modification to $X_{\rm P}^2(\hat{\delta}.)$.

For the two UK surveys referred to in Holt *et al.* (1980), the first-order correction $X_{\rm P}^2(\hat{\delta}.)$ performed well in all cases in controlling the type I error rate which ranged from 0.05 to 0.07, for nominal $\alpha = 0.05$.

Another simple adjustment to $X_{\rm P}^2$ is given by

$$X_{\rm P}^2(\hat{D}.) = X_{\rm P}^2/\hat{D}., \tag{4.17}$$

where $\hat{D}. = \sum \hat{D}_i/I$ is the average cell deff (Fellegi, 1980). Note that $\hat{D}.$ is not in general the same as $\hat{\delta}.$, so that the first moment matching is not achieved by using equation (4.17). However, $X_{\rm P}^2(\hat{D}.)$ should be close to $X_{\rm P}^2(\hat{\delta}.)$ in most practical situations when $H_0\!:\mu_i = \mu_{0i}$, $i = 1, \ldots, I$ is tenable.

4.3.2 Two-way Tables

Test of independence

The general Pearson statistic (4.8) for testing $H_0:\mu_{ab} = \mu_{a\cdot}\mu_{\cdot b}$, $a = 1,\ldots,A$; $b = 1,\ldots,B$ in an $A \times B$ contingency table reduces to

$$X_P^2 = n_o \sum_{a=1}^{A} \sum_{b=1}^{B} (\hat{\mu}_{ab} - \hat{\mu}_{a\cdot}\hat{\mu}_{\cdot b})^2/(\hat{\mu}_{a\cdot}\hat{\mu}_{\cdot b}). \tag{4.18}$$

The formula for $\hat{\delta}_\cdot$ in equation (4.14) reduces to

$$(A-1)(B-1)\hat{\delta}_\cdot = \sum_a \sum_b \frac{\hat{\mu}_{ab}}{\hat{\mu}_{a\cdot}\hat{\mu}_{\cdot b}}(1-\hat{\mu}_{ab})\hat{D}_{ab} - \sum_a (1-\hat{\mu}_{a\cdot})\hat{D}_a(1)$$
$$- \sum_b (1-\hat{\mu}_{\cdot b})\hat{D}_b(2) \tag{4.19}$$

(Rao and Scott, 1982, 1984; Bedrick, 1983; Gross, 1984). Here

$$\hat{D}_{ab} = \widehat{\text{var}}\,(\hat{\mu}_{ab})/[\hat{\mu}_{ab}(1-\hat{\mu}_{ab})n_o^{-1}]$$

is the estimated design effect of $\hat{\mu}_{ab}$, $\hat{D}_a(1)$ and $\hat{D}_b(2)$ are the estimated deffs of row and column marginals $\hat{\mu}_{a\cdot}$ and $\hat{\mu}_{\cdot b}$ respectively. The result (4.19) suggests that published two-way tables should report the deffs of cells and of their marginals along with the cell estimates $\hat{\mu}_{ab}$. The corrected statistic $X_P^2(\hat{\delta}_\cdot)$ $(= X_P^2/\hat{\delta}_\cdot)$ or $X_{LR}^2(\hat{\delta}_\cdot)(= X_{LR}^2/\hat{\delta}_\cdot)$ is treated as a χ^2 variable with $(A-1)(B-1)$ d.f. under H_0.

An F-version $X_P^2(\hat{\delta}_\cdot)$ can be implemented along the lines of equation (4.16) but its properties have not been studied. Alternative corrections to X_P^2 include Fellegi's correction $X_P^2(\hat{D}_{\cdot\cdot}) = X_P^2/\hat{D}_{\cdot\cdot}$ and $X_P^2(\hat{D}_m) = X_P^2/\hat{D}_m$, where $\hat{D}_{\cdot\cdot}(= \sum_a\sum_b \hat{D}_{ab}/AB)$ is the average cell deff and $\hat{D}_m$ is the smaller of the two average marginal deffs, i.e. $\min(\sum_a \hat{D}_a(1)/A, \sum_b \hat{D}_b(2)/B)$. The correction $X_P^2(\hat{D}_m)$ is particularly useful when only marginal deffs are reported. However, the empirical results from the UK surveys indicated that both $X_P^2(\hat{D}_{\cdot\cdot})$ and $X_P^2(\hat{D}_m)$ give very conservative tests in general, resulting in a significant loss of power relative to $X_P^2(\hat{\delta}_\cdot)$.

Test of homogeneity

Suppose that the rows of the two-way table correspond to different populations and that sampling is done independently from each population. The population proportions in the ath population are denoted by $\mu_{b(a)}$ with corresponding survey estimates $\hat{\mu}_{b(a)}$ based on a sample of size n_{ao}, $a = 1,\ldots,A$; $b = 1,\ldots,B$, $(\sum_b \hat{\mu}_{b(a)} = 1$, $\sum_a n_{ao} = n_o)$. The homogeneity hypothesis is given by $H_0:\mu_{b(1)} = \cdots = \mu_{b(A)}$ for each b, which is equivalent to the independence hypothesis $\mu_{ab} = \mu_{a\cdot}\mu_{\cdot b}$ with $\mu_{ab} = (n_{ao}/n_o)\mu_{b(a)}$. The results for testing independence in a two-way table can therefore be used, but it is desirable to express the tests in terms of estimated population proportions $\hat{\mu}_{b(a)}$ and the associated estimated deffs

$$\hat{D}_{b(a)} = \widehat{\text{var}}(\hat{\mu}_{b(a)})/[\hat{\mu}_{b(a)}(1-\hat{\mu}_{b(a)})n_{ao}^{-1}]$$

since published tables usually report the latter quantities.

The general Pearson statistic (4.8) for testing homogeneity can be written as

$$X_P^2 = \sum_{a=1}^{A} n_{ao} \sum_{b=1}^{B} (\hat{\mu}_{b(a)} - \hat{\mu}_b)^2/\hat{\mu}_b, \tag{4.20}$$

with $\hat{\mu}_b = \sum_a n_{ao}\hat{\mu}_{b(a)}/n_o$. The formula for $\hat{\delta}_.$, as given by Scott and Rao (1981), is

$$(A-1)(B-1)\hat{\delta}_. = \sum_{a=1}^{A}\sum_{b=1}^{B} \frac{\hat{\mu}_{b(a)}}{\mu_b}(1-\hat{\mu}_{b(a)})(1-n_{ao}/n_o)\hat{D}_{b(a)}. \tag{4.21}$$

The correction $X_P^2/\hat{\delta}_.$ can therefore be implemented from published tables which report only the estimates $\hat{\mu}_{b(a)}$ and their associated deffs $D_{b(a)}$. The corrected test statistic is again treated as a χ^2 variable with $(A-1)(B-1)$ d.f. under H_0.

4.3.3 Three-way Tables

In a three-way table with cell probabilities μ_{abc} ($a = 1, \ldots, A$; $b = 1, \ldots, B$; $c = 1, \ldots, C$) one can specify four different types of hypothesis of independence in terms of the saturated loglinear model

$$\begin{aligned}\log \mu_{abc} = \tilde{u} + u_{1(a)} + u_{2(b)} + u_{3(c)} + u_{12(ab)} \\ + u_{13(ac)} + u_{23(bc)} + u_{123(abc)}\end{aligned} \tag{4.22}$$

where the u-parameters sum to zero when summed over any subscript a, b, c and $\tilde{u}$ is the normalizing factor.

The *hypothesis of complete independence* of variables 1, 2 and 3 is given by

$$\begin{aligned}H_0(1)\colon u_{12(ab)} = u_{13(ac)} = u_{23(bc)} = u_{123(abc)} = 0 \\ \Leftrightarrow \mu_{abc} = \mu_{a..}\mu_{.b.}\mu_{..c} \quad \text{for all } (a, b, c)\end{aligned} \tag{4.23}$$

where $\mu_{a..}$, $\mu_{.b.}$ and $\mu_{..c}$ are the one-way marginal probabilities corresponding to 1, 2 and 3 respectively. The pseudo MLE of μ_{abc} when $H_0(1)$ holds is obtained as $\hat{\mu}_{abc}(1) = \hat{\mu}_{a..}\hat{\mu}_{.b.}\hat{\mu}_{..c}$, where $\hat{\mu}_{abc}(1)$ is the survey estimator of μ_{abc} with one-way marginals $\hat{\mu}_{a..}$, $\hat{\mu}_{.b.}$ and $\hat{\mu}_{..c}$ respectively.

The *hypothesis of mutual independence*, 1 independent of (2, 3), is given by

$$\begin{aligned}H_0(2)\colon u_{12(ab)} = u_{13(ac)} = u_{123(abc)} = 0 \\ \Leftrightarrow \mu_{abc} = \mu_{a..}\mu_{.bc} \quad \text{for all } (a, b, c)\end{aligned} \tag{4.24}$$

where $\mu_{.bc}$ are the two-way marginal probabilities corresponding to variables (2, 3). The pseudo MLE of μ_{abc} when $H_0(2)$ holds is obtained as $\hat{\mu}_{abc}(2) = \hat{\mu}_{a..}\hat{\mu}_{.bc}$. The hypotheses 2 independent of (1, 3) and 3 independent of (1, 2) are analogous to equation (4.24).

The *hypothesis of conditional independence* of 1 and 2 given 3 is given by

$$\begin{aligned} H_0(3)&: u_{12(ab)} = u_{123(abc)} = 0 \\ &\Leftrightarrow \mu_{abc} = (\mu_{a\cdot c}\mu_{\cdot bc})/\mu_{\cdot\cdot c} \quad \text{for all } (a, b, c). \end{aligned} \tag{4.25}$$

The pseudo MLE of μ_{abc} when $H_0(3)$ holds is obtained as

$$\hat{\mu}_{abc}(3) = (\hat{\mu}_{a\cdot c}\hat{\mu}_{\cdot bc})/\hat{\mu}_{\cdot\cdot c}.$$

The hypothesis that 2 and 3 are conditionally independent given 1, and 1 and 3 conditionally independent given 2 are analogous to equation (4.25).

Finally, the *hypothesis of no three-factor interaction* is given by

$$H_0(4): u_{123(abc)} = 0 \quad \text{for all } (a, b, c) \tag{4.26}$$

but it cannot be expressed explicitly in terms of marginal probabilities, unlike the previous three hypotheses. The interpretation of $H_0(4)$ is that the association in a two-way table corresponding to a level of the third variable is a constant for all levels. The pseudo MLE $\hat{\mu}_{abc}(4)$ under $H_0(4)$ is obtained by solving the 'pseudo likelihood equations' $\mu_{ab\cdot} = \hat{\mu}_{ab\cdot}$, $\mu_{\cdot bc} = \hat{\mu}_{\cdot bc}$ and $\mu_{a\cdot c} = \hat{\mu}_{a\cdot c}$ iteratively for μ_{abc}, as noted in Section 4.2.

The general Pearson statistic (4.8) for testing $H_0(l)$, $l = 1, 2, 3, 4$ is given by

$$X^2_{\mathrm{P}l} = n_o \sum_{a=1}^{A} \sum_{b=1}^{B} \sum_{c=1}^{C} [\hat{\mu}_{abc} - \hat{\mu}_{abc}(l)]^2/\hat{\mu}_{abc}(l). \tag{4.27}$$

The likelihood ratio statistic $X^2_{\mathrm{LR}l}$ is similarly defined, using equation (4.9).

For the first three hypotheses $\hat{\delta}_.$ can be expressed in terms of deffs of cell estimates and of their marginals, similar to equation (4.19) for the two-way table. Denoting $\hat{\delta}_.$ by $\hat{\delta}_{.l}$ for $H_0(l)$, it follows from Rao and Scott (1984) that

$$\begin{aligned} (ABC - A - B - C + 2)\hat{\delta}_{.1} = {} & \sum_a\sum_b\sum_c \frac{\hat{\mu}_{abc}(1-\hat{\mu}_{abc})}{\hat{\mu}_{a\cdot\cdot}\hat{\mu}_{\cdot b\cdot}\hat{\mu}_{\cdot\cdot c}}\hat{D}_{abc} \\ & - \sum_a (1-\hat{\mu}_{a\cdot\cdot})\hat{D}_a(1) - \sum_b (1-\hat{\mu}_{\cdot b\cdot})\hat{D}_b(2) \\ & - \sum_c (1-\hat{\mu}_{\cdot\cdot c})\hat{D}_c(3) \end{aligned} \tag{4.28}$$

$$\begin{aligned} (A-1)(BC-1)\hat{\delta}_{.2} = {} & \sum_a\sum_b\sum_c \frac{\hat{\mu}_{abc}(1-\hat{\mu}_{abc})}{\hat{\mu}_{a\cdot\cdot}\hat{\mu}_{\cdot bc}}\hat{D}_{abc} - \sum_a (1-\hat{\mu}_{a\cdot\cdot})\hat{D}_a(1) \\ & - \sum_b\sum_c (1-\hat{\mu}_{\cdot bc})\hat{D}_{bc}(2,3) \end{aligned} \tag{4.29}$$

$$\begin{aligned} C(A-1)(B-1)\hat{\delta}_{.3} = {} & \sum_a\sum_b\sum_c \frac{\hat{\mu}_{abc}(1-\hat{\mu}_{abc})}{(\hat{\mu}_{a\cdot c}\hat{\mu}_{\cdot bc}/\hat{\mu}_{\cdot\cdot c})}\hat{D}_{abc} \\ & - \sum_a\sum_c (1-\hat{\mu}_{a\cdot c})\hat{D}_{ac}(1,3) - \sum_b\sum_c (1-\hat{\mu}_{\cdot bc})\hat{D}_{bc}(2,3) \\ & + \sum_c (1-\hat{\mu}_{\cdot\cdot c})\hat{D}_c(3) \end{aligned} \tag{4.30}$$

Here $\hat{D}_{abc} = \widehat{\text{var}}(\hat{\mu}_{abc})/[\hat{\mu}_{abc}(1-\hat{\mu}_{abc})n_o^{-1}]$ is the estimated deff of $\hat{\mu}_{abc}$, $\hat{D}_a(1)$ is the estimated deff of the row marginal $\hat{\mu}_{a..}$ and $\hat{D}_{bc}(2,3)$ is the estimated deff of the two-way marginal $\hat{\mu}_{.bc}$, etc. Thus the first-order corrections $X_{Pl}^2/\hat{\delta}_{.l}$, $l = 1, 2, 3$ can be implemented from published tables which report the deffs of cells and their two-way and one-way marginals, along with cell estimates $\hat{\mu}_{abc}$. These corrections are treated as χ^2 variables with degrees of freedom $ABC - A - B - C + 2$, $(A-1)(BC-1)$ and $C(A-1)(B-1)$ under $H_0(1)$, $H_0(2)$ and $H_0(3)$ respectively.

In the case of $H_0(4)$, the hypothesis of no three-factor interaction, $\hat{\delta}_{.4}$ cannot be expressed in terms of deffs of cell estimates and of their marginals. The reader is referred to Rao and Scott (1987), who suggest an approximation to $\hat{\delta}_{.4}$ depending only on these deffs and also discuss extensions of the above results to higher dimensional contingency tables.

4.3.4 Nested Models

First-order corrections to the test statistics in equations (4.11) and (4.12) are given by $X_P^2(2|1)/\hat{\delta}.(2|1)$ and $X_{LR}^2(2|1)/\hat{\delta}.(2|1)$, where

$$r_2\hat{\delta}.(2|1) = \text{tr}\,[(\mathbf{X}'\mathbf{\Omega}(\hat{\hat{\boldsymbol{\theta}}}_1)\mathbf{X})^{-1}(\mathbf{X}'\hat{\mathbf{\Sigma}}\mathbf{X})] - \text{tr}\,[(\mathbf{X}_1'\mathbf{\Omega}(\hat{\hat{\boldsymbol{\theta}}}_1)\mathbf{X}_1)^{-1}(\mathbf{X}_1'\hat{\mathbf{\Sigma}}\mathbf{X}_1)] \quad (4.31)$$

and $\mathbf{\Omega}(\tilde{\boldsymbol{\theta}}) = n_o^{-1}\{\mathbf{D}[\boldsymbol{\mu}(\tilde{\boldsymbol{\theta}})] - \boldsymbol{\mu}(\tilde{\boldsymbol{\theta}})\boldsymbol{\mu}(\tilde{\boldsymbol{\theta}})'\}$ where $\tilde{\boldsymbol{\theta}}$ equals $\hat{\boldsymbol{\theta}}$ or $\hat{\hat{\boldsymbol{\theta}}}_1$ as appropriate. Note that empirical work suggests that it is best to estimate $\mathbf{\Omega}(\boldsymbol{\theta})$ under H_0 by using $\mathbf{\Omega}(\hat{\hat{\boldsymbol{\theta}}}_1)$ throughout equation (4.31). This correction, in general, depends on the full estimated covariance matrix, $\hat{\mathbf{\Sigma}}$. However, if both $\boldsymbol{\mu}(\boldsymbol{\theta})$ and $\boldsymbol{\mu}(\boldsymbol{\theta}_1)$ can be expressed explicitly in terms of cell probabilities μ_i and their marginals, then $\hat{\delta}.(2|1)$ can be written in terms of deffs of cell estimates and of their marginals (Rao and Scott, 1984). For example, in a three-way table we can write $\hat{\delta}.(2|1)$, for testing the nested hypothesis that variable 1 is independent of variables (2, 3) given that 1 and 2 are conditionally independent given 3, as follows:

$$(A-1)(C-1)\hat{\delta}.(2|1) = (A-1)(BC-1)\hat{\delta}_{.2} - C(A-1)(B-1)\hat{\delta}_{.3}. \quad (4.32)$$

Here $\hat{\delta}_{.2}$ and $\hat{\delta}_{.3}$ are given by equations (4.29) and (4.30) respectively.

4.4 SECOND-ORDER CORRECTION TO X_P^2 OR X_{LR}^2

If an estimate of the full covariance matrix $\mathbf{\Sigma} = \text{Var}(\hat{\boldsymbol{\mu}})$ is available, a more accurate second-order correction to X_P^2 or X_{LR}^2 can be obtained, using the well-known Satterthwaite approximation to the distribution of a weighted sum of independent χ_1^2 variables. This correction takes account of the variability in the generalized design effects, δ_i, unlike the first-order corrections.

The second-order correction can be implemented by treating

$$X_P^2(\hat{\delta}_., \hat{a}) = \frac{X_P^2(\hat{\delta}.)}{1+\hat{a}^2} \quad \text{or} \quad X_{LR}^2(\hat{\delta}_., \hat{a}) = \frac{X_{LR}^2(\hat{\delta}.)}{1+\hat{a}^2} \quad (4.33)$$

as χ^2_ν, a chi-squared random variable with $\nu = (I - r - 1)/(1 + \hat{a}^2)$ d.f. Here $\hat{a}$ is the coefficient of variation of the eigenvalues, $\hat{\delta}_i$, of the estimated design effects matrix $\hat{\mathbf{\Delta}}$, i.e.

$$\hat{a}^2 = \left\{ \sum_{i=1}^{I-r-1} \hat{\delta}_i^2/(I - r - 1)\hat{\delta}_{\cdot}^2 \right\} - 1. \tag{4.34}$$

The individual eigenvalues $\hat{\delta}_i$ and hence $\hat{a}^2$ can be calculated easily on a high-speed computer, but it is also possible to calculate $\hat{a}$ directly from $\hat{\mathbf{\Delta}}$, without evaluating the $\hat{\delta}_i$, since

$$(I - r - 1)\hat{\delta}_{\cdot} = \operatorname{tr}\hat{\mathbf{\Delta}} \quad \text{and} \quad \sum \hat{\delta}_i^2 = \operatorname{tr}\hat{\mathbf{\Delta}}^2. \tag{4.35}$$

The second-order corrections control the type I error rate much better than the first-order corrections when the variation in the δ_i is substantial, as shown by Thomas and Rao (1987) on the basis of a goodness-of-fit Monte Carlo study. The first-order corrections tend to give somewhat inflated type I error rates when the variation in the δ_i is appreciable.

For nested models, a second-order correction can be implemented by treating

$$\frac{X_P^2(2|1)}{\hat{\delta}_{\cdot}(2|1)\{1 + \hat{a}(2|1)^2\}} \quad \text{or} \quad \frac{X_{LR}^2(2|1)}{\hat{\delta}_{\cdot}(2|1)\{1 + \hat{a}(2|1)^2\}}$$

as $\chi^2_{\nu(2|1)}$, a chi-square random variable with $\nu(2|1) = r_2/\{1 + \hat{a}(2|1)^2\}$ d.f. Here $\hat{a}(2|1)$ is the coefficient of variation of the eigenvalues $\hat{\delta}_i(2|1)$ of the estimated design effects matrix

$$(\tilde{\mathbf{X}}_2'\mathbf{\Omega}(\hat{\hat{\mathbf{\theta}}}_1)\tilde{\mathbf{X}}_2)^{-1}(\tilde{\mathbf{X}}_2'\hat{\mathbf{\Sigma}}\tilde{\mathbf{X}}_2) = \hat{\mathbf{\Delta}}(2|1) \quad \text{(say)}, \tag{4.36}$$

where $\tilde{\mathbf{X}}_2 = [\mathbf{I} - \mathbf{X}_1(\mathbf{X}_1'\mathbf{\Omega}(\hat{\hat{\mathbf{\theta}}}_1)\mathbf{X}_1)^{-1}\mathbf{X}_1'\mathbf{\Omega}(\hat{\hat{\mathbf{\theta}}}_1)]\mathbf{X}_2$. Again, $\hat{a}(2|1)$ can be calculated directly from $\hat{\mathbf{\Delta}}(2|1)$ as in equation (4.35) without evaluating the individual eigenvalues $\hat{\delta}_i(2|1)$.

4.5 WALD TESTS

Whenever an estimate of the full covariance matrix of $\hat{\boldsymbol{\mu}}$ is available, a Wald test of goodness of fit of the loglinear model (4.4) can also be constructed. The Wald test is implemented by treating

$$X_W^2 = (\mathbf{C}' \log \hat{\boldsymbol{\mu}})'[\mathbf{C}'\mathbf{D}(\hat{\boldsymbol{\mu}})^{-1}\hat{\mathbf{\Sigma}}\mathbf{D}(\hat{\boldsymbol{\mu}})^{-1}\mathbf{C}]^{-1}(\mathbf{C}' \log \hat{\boldsymbol{\mu}}) \tag{4.37}$$

as χ^2_{I-r-1}. Note that the Wald test requires that all estimated proportions $\hat{\mu}_i$ must be non-zero. The statistic X_W^2 is invariant to the choice of $\mathbf{C}$.

The Wald test is asymptotically correct, but it can be lead to inflated type I error rates in finite samples, as the degrees of freedom, ν, for estimating the covariance matrix $\mathbf{\Sigma}$ decreases and the number of cells in the table, I, increases, as shown by Thomas and Rao (1987) on the basis of a goodness-of-fit Monte Carlo study. Note that X_W^2 is not defined if $\nu < I - r - 1$.

If the degrees of freedom for estimating $\mathbf{\Sigma}$ (are not large relative to $I-r-1$, an improved test statistic can be obtained by treating

$$F_{\mathrm{W}} = \frac{(v-I+r+2)}{v(I-r-1)} X_{\mathrm{W}}^2 \tag{4.38}$$

as an F random variable with $I-r-1$ and $v-I+r+2$ d.f. The Monte Carlo study of Thomas and Rao (1987) has shown that F_{W} gives a better control of type I error rate than X_{W}^2, although it too tends to perform poorly as v approaches $I-r-1$.

A Wald test can also be constructed for a nested hypothesis $H{:}\boldsymbol{\theta}_2 = \mathbf{0}$ or a more general hypothesis $H{:}\mathbf{K}\boldsymbol{\theta} = \mathbf{k}$, where $\mathbf{K}$ is a $q \times r$ full rank matrix of known constants $(q < r)$ and $\mathbf{k}$ is a $q \times 1$ vector of constants. The Wald test is implemented by treating

$$X_{\mathrm{W}}^2(2|1) = (\mathbf{K}\hat{\boldsymbol{\theta}} - \mathbf{k})'[\mathbf{K}\,\widehat{\mathrm{Var}}(\hat{\boldsymbol{\theta}})\mathbf{K}']^{-1}(\mathbf{K}\hat{\boldsymbol{\theta}} - \mathbf{k}) \tag{4.39}$$

as χ_q^2, where $\widehat{\mathrm{Var}}(\hat{\boldsymbol{\theta}})$ is the estimated covariance matrix of $\hat{\boldsymbol{\theta}}$ given by equation (4.41) below. For the hypothesis $H{:}\boldsymbol{\theta}_2 = \mathbf{0}$, we substitute $\mathbf{K} = [\mathbf{0}, \mathbf{I}]$ and $\mathbf{k} = \mathbf{0}$ in equation (4.39), where $\mathbf{0}$ is an $r_2 \times r$ matrix of zeros and $\mathbf{I}$ is the $r_2 \times r_2$ identity matrix $(q = r_2)$.

4.6 SMOOTHED ESTIMATES

The tests given in the previous sections enable us to fit a parsimonious loglinear model to the data at hand. Such a fitted model helps our understanding of the relationship of the variables under study. When we fit a loglinear model, the resulting fitted cell estimates, $\boldsymbol{\mu}(\hat{\boldsymbol{\theta}})$ provide a smoothed description of the data.

The uncertainty in the smoothed estimates, $\boldsymbol{\mu}(\hat{\boldsymbol{\theta}})$, may be measured through the estimated asymptotic covariance matrix given by

$$\widehat{\mathrm{Var}}\{\boldsymbol{\mu}(\hat{\boldsymbol{\theta}})\} = n_{\mathrm{o}}^2 \boldsymbol{\Omega}(\hat{\boldsymbol{\theta}})\mathbf{X}\,\widehat{\mathrm{Var}}(\hat{\boldsymbol{\theta}})\mathbf{X}'\boldsymbol{\Omega}(\hat{\boldsymbol{\theta}}), \tag{4.40}$$

where

$$\widehat{\mathrm{Var}}(\hat{\boldsymbol{\theta}}) = n_{\mathrm{o}}^{-2}(\mathbf{X}'\boldsymbol{\Omega}(\hat{\boldsymbol{\theta}})\mathbf{X})^{-1}(\mathbf{X}'\hat{\boldsymbol{\Sigma}}\mathbf{X})(\mathbf{X}'\boldsymbol{\Omega}(\hat{\boldsymbol{\theta}})\mathbf{X})^{-1}. \tag{4.41}$$

The diagonal elements of equation (4.40) are the squared standard errors of the smoothed cell estimates, $\mu_i(\hat{\boldsymbol{\theta}})$, $i = 1, \ldots, I$.

4.7 RESIDUAL ANALYSIS

Whenever a test rejects the hypothesis of goodness of fit of a loglinear model of the form (4.4), residual analysis can provide an understanding of the nature of deviations from the hypothesis. We can use the standardized residuals

$$r_i = \{\hat{\mu}_i - \mu_i(\hat{\boldsymbol{\theta}})\}/\text{s.e.}\{\hat{\mu}_i - \mu_i(\hat{\boldsymbol{\theta}})\}$$

to detect deviations from the hypothesis, using the fact that the r_i's are approximately $N(0, 1)$ under the hypothesis. The squared standard errors of the residuals, $\hat{\mu}_i - \mu_i(\hat{\boldsymbol{\theta}})$, under the given survey design, are given by the diagonal elements of the estimated asymptotic covariance matrix of $\mathbf{r} = \hat{\boldsymbol{\mu}} - \boldsymbol{\mu}(\hat{\boldsymbol{\theta}})$:

$$\widehat{\text{Var}}(\mathbf{r}) = [\mathbf{I} - \boldsymbol{\Omega}(\hat{\boldsymbol{\theta}})\mathbf{X}(\mathbf{X}'\boldsymbol{\Omega}(\hat{\boldsymbol{\theta}})\mathbf{X})^{-1}\mathbf{X}']\hat{\boldsymbol{\Sigma}}[\mathbf{I} - \boldsymbol{\Omega}(\hat{\boldsymbol{\theta}})\mathbf{X}(\mathbf{X}'\boldsymbol{\Omega}(\hat{\boldsymbol{\theta}})\mathbf{X})^{-1}\mathbf{X}']. \quad (4.42)$$

Under multinomial sampling, $\widehat{\text{Var}}(\mathbf{r})$ reduces to

$$\boldsymbol{\Omega}(\hat{\boldsymbol{\theta}}) - \boldsymbol{\Omega}(\hat{\boldsymbol{\theta}})\mathbf{X}(\mathbf{X}'\boldsymbol{\Omega}(\hat{\boldsymbol{\theta}})\mathbf{X})^{-1}\mathbf{X}'\boldsymbol{\Omega}(\hat{\boldsymbol{\theta}}),$$

and the standard methods use the corresponding standardized residuals r_i^* to detect deviations from the hypothesis of goodness of fit. Since $r_i = r_i^*/\widehat{\text{deff}}\{\hat{\mu}_i - \mu_i(\hat{\boldsymbol{\theta}})\}$, it is clear that the use of the r_i^*, instead of the r_i, could lead to erroneous interpretations of the deviations from the hypothesis if some of the deffs of the residuals are large.

The formula (4.42) for $\widehat{\text{Var}}(\boldsymbol{r})$ is applicable for general loglinear models and multi-way tables. In the special case of simple goodness of fit, $\mu_i = \mu_{0i}$, $i = 1, \ldots, I$, the standardized residuals are simply given by

$$r_i = (\hat{\mu}_i - \mu_{0i})/\{\widehat{\text{var}}(\hat{\mu}_i)\}^{1/2} = r_i^*/\hat{D}_i^{1/2}, \quad (4.43)$$

where r_i^* are the standardized residuals under the assumption of multinomial sampling and $\hat{D}_i$ is the estimated deff of $\hat{\mu}_i$.

In the case of a test of homogeneity of proportions across A regions the standardized residuals, $r_{b(a)}$, may be expressed in terms of estimated population proportions, $\hat{\mu}_{b(a)}$, and the associated estimated deffs, $\hat{D}_{b(a)}$, as follows:

$$r_{b(a)} = (\hat{\mu}_{b(a)} - \hat{\mu}_b)/\{\widehat{\text{var}}(\hat{\mu}_{b(a)} - \hat{\mu}_b)\}^{1/2}, \quad (4.44)$$

where

$$\widehat{\text{var}}(\hat{\mu}_{b(a)} - \hat{\mu}_b) = \frac{1}{n_o^2}\hat{\mu}_b(1 - \hat{\mu}_b)\left\{\frac{n_o(n_o - n_{ao})}{n_{ao}}\hat{D}_{b(a)} + \sum_{a=1}^{A} n_{ao}\hat{D}_{b(a)}\right\} \quad (4.45)$$

4.8 EXAMPLES

The following examples are drawn from the Canada Health Survey (CHS), a multi-stage cluster sample designed to gather a wide range of data on the health status of Canadians. One of the primary objectives of the CHS was to relate health status to a variety of health risk factors, over time. In order to provide results at both the provincial and national levels, the annual sample size was set at 12 000 households. The CHS was conducted for a twelve-month period, during 1978–79, but was prematurely terminated as a result of government budget cuts. Further information on the design of the CHS, together with a detailed description of corresponding variance estimation methods, is given by Hidiroglou and Rao (1987a, 1987b).

4.8.1 Example 1: Goodness of Fit

When assessing a health risk factor it is useful to compare the age distribution of persons exposed to the risk factor to the age distribution of the population as a whole. Table 4.1 gives the age distribution of current smokers, based on CHS data, and compares this to the census age distribution. The CHS survey proportions, $\hat{\mu}_i$, based on weighted-up counts, are shown in Table 4.1 together with the census proportions, μ_{0i}, for the seven age categories studied. The weighted-up counts incorporate post-stratification (or ratio weight) adjustments designed to match the CHS age–sex distributions to the projected census distributions, at the provincial level. Also shown in Table 4.1 are the cell design effects, $\hat{D}_i$, and corresponding standardized residuals, given in this case by $r_i = r_i^*/\hat{D}_i^{1/2}$ (see Sections 4.3.1 and 4.7). The domain size for Table 4.1, i.e. the number of respondents classed as current smokers, is $n_o = 7889$.

First- and Second-order Corrections

Using equations (4.2) and (4.3), and the data of Table 4.1, values of the unadjusted Pearson and likelihood ratio statistics for testing the goodness-of-fit hypothesis, $H_0\colon \mu_i = \mu_{0i}; i = 1, \ldots, 7$, are easily computed as $X_P^2 = 308.5$ and $X_{LR}^2 = 338.1$ respectively. From equation (4.15), the value of $\hat{\delta}.$ becomes 1.34, yielding first-order corrections $X_P^2(\hat{\delta}.) = 231.1$ and $X_{LR}^2(\hat{\delta}.) = 253.2$. Though the reductions in X_P^2 and X_{LR}^2 are considerable, it is clear that when $X_P^2(\hat{\delta}.)$ and $X_{LR}^2(\hat{\delta}.)$ are referred to $\chi_6^2(0.05) = 12.59$, both still provide strong evidence against H_0. In other words, on the basis of the estimated survey proportions, we can conclude that the age distribution of smokers differs from that of the population as a whole. The nature of these differences is revealed by the age category residuals shown in Table 4.1. Compared to the census age distribution, smokers are over-represented in the 20–54-year age range, and increasingly under-represented from age 55 years and up.

Since an estimate, $\hat{\boldsymbol{\Sigma}}$, of the covariance matrix of the survey proportions, $\hat{\mu}_i$, $i = 1, \ldots, 7$, is available for the CHS data, second-order corrections to the standard tests can be obtained as described in Section 4. In practice, second-order corrections would be routinely used whenever an estimate $\hat{\boldsymbol{\Sigma}}$ is available, given their superior performance (Thomas and Rao, 1987). For a simple goodness-of-fit (g.o.f.) hypothesis, the estimated design effects matrix (see Section 4.2.1) is given by

$$\hat{\boldsymbol{\Delta}}(\text{g.o.f.}) = \boldsymbol{\Omega}_{0I}^{-1}\hat{\boldsymbol{\Sigma}}_I, \tag{4.46}$$

where $\boldsymbol{\Omega}_{0I}$ is the 6×6 multinomial covariance matrix under H_0, based on the first $I - 1 = 6$ census proportions, and $\hat{\boldsymbol{\Sigma}}$ is the corresponding estimated covariance matrix of the first six survey proportions. The second-order corrections are given in equation (4.33) as functions of $\hat{a}^2$, the squared coefficient

Table 4.1 A comparison of the age distribution of current smokers to the projected census age distribution

Ages ranges	Census proportion	CHS proportion	Design effects	Standardized residuals
15–19	0.133	0.118	1.39	−3.5
20–24	0.127	0.159	1.81	5.9
25–34	0.218	0.235	1.67	2.7
35–44	0.152	0.171	1.11	4.2
45–54	0.140	0.152	1.22	2.7
55–64	0.115	0.101	1.33	−3.7
65+	0.115	0.064	0.90	−19.5

Note: $n_o = 7889$.

of variation of the eigenvalues of $\hat{\mathbf{\Delta}}$(g.o.f.). The latter are, in decreasing order, 2.52, 2.13, 1.11, 0.98, 0.84 and 0.42, so that with the aid of equation (4.34) we get $\hat{a}^2 = 0.31$. Note that direct evaluation of the eigenvalues can be avoided if necessary by means of equations (4.35). Evaluation of equation (4.33) yields the second-order corrected statistics $X_P^2(\hat{\delta}_., \hat{a}) = 176.7$, and $X_{LR}^2(\hat{\delta}_., \hat{a}) = 193.7$, on 4.6 fractional d.f. Interpolation from standard tables gives $\chi_{4.6}^2(0.05) = 10.44$, so that both second-order corrected tests are again highly significant, as expected.

Under the assumption that the second-order Pearson test $X_P^2(\hat{\delta}_., \hat{a})$ achieves a type I level approximately equal to the nominal 5% test level, the actual test level of the uncorrected test X_P^2, and the first-order corrected test $X_P^2(\hat{\delta}_.)$ can be estimated. The results are 17.2% and 6.9% respectively, (with similar results for the tests based on the likelihood ratio) showing clearly that the effect of the complex CHS design on goodness-of-fit tests is potentially serious, and that the first-order correction can reduce the inflation of the uncorrected tests to an acceptable level.

A Wald Test

A Wald statistic for testing the simple goodness-of-fit hypothesis H_0 is given by

$$X_W^2(\text{g.o.f.}) = (\hat{\boldsymbol{\mu}}_I - \boldsymbol{\mu}_{0I})'\hat{\boldsymbol{\Sigma}}^{-1}(\hat{\boldsymbol{\mu}}_I - \boldsymbol{\mu}_{0I}), \tag{4.47}$$

where $\hat{\boldsymbol{\mu}}_I$ and $\boldsymbol{\mu}_{0I}$ are vectors of sample and census proportions of length $I-1$, having elements $\hat{\mu}_i$, μ_{0i}, $i = 1, \ldots, 6$ respectively. Using the data of Table 4.1, and an estimate of the covariance matrix (available from the authors), the Wald statistic can be evaluated as $X_W^2(\text{g.o.f}) = 618.0$, on 6 d.f., again highly significant. As noted in Section 4.5, Wald statistics can sometimes result in inflated test error rates, a deficiency that is clearly illustrated in the present example. Though the Wald test is asymptotically exact, its value in this example is approximately twice that of the unadjusted Pearson and likelihood ratio statistics, which we know to be asymptotically inflated. The degree of inflation of the Wald statistic in this case can best be illustrated by comparison with the second-order correction to the Pearson test, modified so that it can be referred to the upper 5% point of χ_6^2, rather than $\chi_{4.6}^2$ (see Hidiroglou and Rao, 1987a). Compared to the Wald value of 618.0, we get

$$X_P^2(\hat{\delta}_., \hat{a}, 0.05) = X_P^2(\hat{\delta}_., \hat{a})[\chi_6^2(0.05)/\chi_{4.6}^2(0.05)] = 213.1,$$

a ratio of nearly three to one.

4.8.2 Example 2: Test of Independence

The CHS dataset is of particular interest to health researchers because it includes objective physical measurements of health status for a subset of the respondents. A variety of physical measures, including a sub-maximal exercise test of cardio-respiratory fitness appropriate for the respondent's age and sex, were administered by registered nurses to selected individuals from a subsample of 4200 households. On the basis of these measures individuals were classified into several predefined health status categories. For example, on the basis of the exercise fitness test, respondents were categorized into three fitness levels namely: *recommended*, *minimally acceptable* and *unacceptable*. Together with self-reports on various health risk factors such as smoking, these objective measures provide a rare opportunity for assessing relationships between health status and health risk. As an example, we consider the relationship between smoking and fitness. The required data are presented in Table 4.2, in the form of weighted-up proportions, $\hat{\mu}_{ab}$, and cell design effects, $\hat{D}_{ab}$, for a cross-tabulation of fitness level by three categories of smoking behaviour namely: *current smoker*, *occasional smoker* and *never smoked*. Former regular and occasional smokers were omitted from this table in order to provide a clearly ordinal smoking variable. Also, respondents who were screened out of the fitness test by a preliminary 'physical

Table 4.2 Survey proportions and design effects (in brackets) for a cross-tabulation of smoking status by fitness level

	Fitness level		
Smoking status	Recommended	Minimally acceptable	Unacceptable/ screened out
Current smoker	0.220 (3.50)	0.150 (4.59)	0.170 (1.50)
Occasional smoker	0.023 (3.45)	0.010 (1.07)	0.011 (1.09)
Never smoked	0.203 (3.49)	0.099 (2.07)	0.114 (1.51)

Note: $n_o = 2505$.

activity readiness questionnaire' have been included in the unacceptable fitness category. The domain size is $n_0 = 2505$ for the data of Table 4.2, considerably less than for the previous example owing to the reduced size of the sample receiving the physical tests.

First-order Corrections

A test of independence based on the unadjusted Pearson statistic X_P^2 given in equation (4.18) provides an initial assessment of the relationship between smoking behaviour and fitness. With cell proportions $\hat{\mu}_{ab}$ from Table 4.2 and corresponding marginal proportions (the $\hat{\mu}_{a\cdot}$'s and $\hat{\mu}_{\cdot b}$'s) given by (0.540, 0.044, 0.416) and (0.446, 0.259, 0.295) respectively, the Pearson statistic X_P^2 becomes 18.1 on 4 d.f. In this case, the unadjusted likelihood ratio and Pearson test statistics agree to four significant digits.

The first-order correction to X_P^2 is obtained by evaluating $\hat{\delta}_{\cdot}$ of equation (4.19) which requires marginal design effects $\hat{D}_a(1)$ and $\hat{D}_b(2)$ corresponding to the above marginal proportions. For smoking and fitness, these are given by (1.44, 2.32, 2.44) and (4.69, 5.96, 1.71) respectively. Hence $\hat{\delta}_{\cdot} = 1.69$ so that the first-order corrected statistics become $X_P^2(\hat{\delta}_{\cdot}) = X_{LR}^2(\hat{\delta}_{\cdot}) = 10.72$, significant at the 5% level ($\chi_4^2(0.05) = 9.49$)).

Second-order Corrections

As described in Section 4.4, second-order corrections to standard tests take into account variation among the generalized design effects (the $\hat{\delta}_i$'s) using the square of the coefficient of variation $\hat{a}^2$. For a loglinear model, $\hat{a}^2$ can be evaluated from the estimated design effects matrix $\hat{\mathbf{\Delta}}$ using equations (4.34) and (4.35). The matrix $\hat{\mathbf{\Delta}}$ is defined in terms of the fitted cell proportions under the model in question, the estimated covariance matrix of the survey proportions and a contrast matrix **C** which is orthogonal to the model matrix **X** (see Section 4.2.2). For the 3×3 independence model, the unsaturated model matrix **X** is of dimension 9×4. If we construct it following the strategy outlined in Section 4.2.2, based on an obvious extension of equation (4.5) to a 3×3 table, **X** will have the form

$$\mathbf{X}' = \begin{bmatrix} 1 & 1 & 1 & 0 & 0 & 0 & -1 & -1 & -1 \\ 0 & 0 & 0 & 1 & 1 & 1 & -1 & -1 & -1 \\ 1 & 0 & -1 & 1 & 0 & -1 & 1 & 0 & -1 \\ 0 & 1 & -1 & 0 & 1 & -1 & 0 & 1 & -1 \end{bmatrix}, \tag{4.48}$$

where the first two columns of **X** (rows of **X**′) correspond to the loglinear model u-terms $u_{1(1)}$ and $u_{1(2)}$ that represent the smoking main effect, and the third and fourth columns correspond to the u-terms $u_{2(1)}$ and $u_{2(2)}$ that represent the fitness main effect. As noted in the text following equation (4.10), **C** can be chosen as the matrix complementing **X** to form the saturated loglinear model. The form of **C** is thus directly related to the structure of the interaction terms in the loglinear model. For the present example, instead of deriving **C** from the interaction terms of equation (4.5), it is convenient to use a **C** matrix corresponding to interactions amongst linear and quadratic contrasts of the smoking and fitness categories (each scored 1, 2, 3 on an interval scale), given by

$$\mathbf{C}' = \begin{bmatrix} 1 & 0 & -1 & 0 & 0 & 0 & -1 & 0 & 1 \\ 1 & -2 & 1 & 0 & 0 & 0 & -1 & 2 & -1 \\ 1 & 0 & -1 & -2 & 0 & 2 & 1 & 0 & -1 \\ 1 & -2 & 1 & -2 & 4 & -2 & 1 & -2 & 1 \end{bmatrix}. \tag{4.49}$$

It should be noted that all tests and estimates are invariant to the particular choice of **C** provided that it is orthogonal to **X**. However, compared to a choice of **C** based on the interaction terms of equation (4.5), the above choice will prove to be more convenient when we later discuss an extension of the independence model. With **C** as defined in equation (4.49), and with fitted cell proportions under the independence model given by

$$\boldsymbol{\mu}(\hat{\boldsymbol{\theta}}) = (0.241 \quad 0.139 \quad 0.159 \quad 0.020 \quad 0.011 \quad 0.013 \quad 0.186 \quad 0.108 \quad 0.123)', \tag{4.50}$$

where the elements of $\boldsymbol{\mu}(\hat{\boldsymbol{\theta}})$ are given in lexicographic order by $\mu_{ab}(\hat{\boldsymbol{\theta}}) = \hat{\mu}_{a\cdot}\hat{\mu}_{\cdot b}$, $a, b = 1, 2, 3$, the eigenvalues of $\hat{\boldsymbol{\Delta}}$ become 2.74, 1.84, 1.51 and 0.66. From equation (4.34), $\hat{a}^2 = 0.195$, which leads to $X^2_{\mathrm{P}}(\hat{\delta}_{\cdot}, \hat{a}) = X^2_{\mathrm{LR}}(\hat{\delta}_{\cdot}, \hat{a}) = 8.97$ on 3.35 adjusted d.f. Compared to an interpolated chi-squared critical value of 8.40, both second-order corrected tests are just significant at the 5% level, which suggests that the independence model does not provide an entirely adequate fit to the data of Table 4.1.

Residual Analysis

As noted in Section 4.7, a residual analysis should help to provide an understanding of the nature, and the extent, of any deviations from the hypothesis of independence. Routine evaluation of the estimated covariance matrix of the deviations $\hat{\mu}_i - \mu_i(\hat{\boldsymbol{\theta}})$, $i = 1, \ldots, 9$, yields standardized residuals

$$\mathbf{r} = (-3.4 \quad 1.7 \quad 1.8 \quad 1.0 \quad -1.1 \quad -0.6 \quad 2.7 \quad -1.4 \quad -1.5)' \tag{4.51}$$

where the elements r_i, $i = 1, \ldots, 9$, of the vector of standardized residuals **r** are defined in Section 4.7. The elements r_i are again given in lexicographic order, i.e. by rows of the smoking by fitness table. Using the Bonferroni multiple comparison method (Miller, 1981), the nine residuals (which have standard normal distributions asymptotically) can be assessed by reference to the upper $(2\frac{1}{2}/9)\%$ point of a standard normal distribution, namely $z_{\text{crit}} = 2.77$. By comparison with z_{crit}, the first element of **r** clearly demonstrates some lack of fit of the independence model. Furthermore, the seventh element of $\mathbf{r}$ is sufficiently close to z_{crit} to warrant considering it, also, as a possible model deviation. The first element (which is negative) corresponds to respondents at the recommended level of fitness who are current smokers; the seventh element (which is positive) corresponds to respondents at the recommended level of fitness who have never smoked. In summary, therefore, it appears that the independence model overestimates the proportion of current smokers who are at the recommended fitness level, and correspondingly underestimates the proportion of those at the recommended fitness level who have never smoked.

A Model of Uniform Association

The next step is to seek a more complex, but still unsaturated, model that adequately describes the relationship between smoking behaviour and fitness detected by the above analysis. In the interests of efficiency, this model should account for the ordinal nature of both smoking and fitness variables.

Agresti (1984, p. 76) describes a loglinear model for a two-way table in which both classifications are ordinal. The model includes all main effects, i.e. terms representing row and column effects, but includes only one interaction term, namely that corresponding to the interaction of linear contrasts on the two ordinal variables. In our example, this linear-by-linear interaction model can be written as

$$\log \mu_{ab} = \tilde{u} + u_{1(a)} + u_{2(b)} + \beta(v_a - \bar{v})(w_b - \bar{w}), \tag{4.52}$$

where v_a, $a = 1, 2, 3$ and w_b, $b = 1, 2, 3$ are known scores that can be assigned to the levels of the smoking and fitness variables respectively. The u-terms are again subject to side conditions $\sum_a u_{1(a)} = \sum_b u_{2(b)} = 0$. If equally spaced integer scores $v_a = a$, $w_b = b$, $a, b = 1, 2, 3$ are assigned to the smoking and fitness variables, then the resulting model corresponds to Goodman's (1979) model of 'uniform association'. With the parameter vector $\boldsymbol{\theta}$ chosen as

$$\boldsymbol{\theta} = (u_{1(1)}, u_{1(2)}, u_{2(1)}, u_{2(2)}, \beta)', \tag{4.53}$$

a model of the general form (4.4) is obtained. It turns out that the design matrix corresponding to this uniform association model can be obtained by adjoining the first column of the $\mathbf{C}$ matrix defined in equation (4.49) to the model matrix $\mathbf{X}$ for the independence model, given by equation (4.48). The result is the 9×5 design matrix given by

$$\mathbf{X}'_{\text{UA}} = \begin{bmatrix} 1 & 1 & 1 & 0 & 0 & 0 & -1 & -1 & -1 \\ 0 & 0 & 0 & 1 & 1 & 1 & -1 & -1 & -1 \\ 1 & 0 & -1 & 1 & 0 & -1 & 1 & 0 & -1 \\ 0 & 1 & -1 & 0 & 1 & -1 & 0 & 1 & -1 \\ 1 & 0 & -1 & 0 & 0 & 0 & -1 & 0 & 1 \end{bmatrix}. \tag{4.54}$$

Pseudo MLEs of the model parameter vector $\boldsymbol{\theta}$, corresponding to a solution of equation (4.7) with $\bar{\mathbf{y}}$ replaced by the survey proportions, $\hat{\boldsymbol{\mu}}$, of Table 4.2, were obtained using the Newton–Raphson method as

$$\hat{\boldsymbol{\theta}} = (0.93 \quad -1.58 \quad 0.33 \quad -0.22 \quad -0.09)'. \tag{4.55}$$

Corresponding standard errors, calculated as the square roots of the diagonal elements of equation (4.41), were found to be

$$\widehat{\text{s.e.}}(\hat{\boldsymbol{\theta}}) = 10^{-1}(0.42 \quad 1.00 \quad 0.61 \quad 0.72 \quad 0.30)'. \tag{4.56}$$

With the aid of equation (4.4), pseudo MLEs of the cell proportions $\boldsymbol{\mu}(\hat{\boldsymbol{\theta}})$ corresponding to the uniform association model become

$$\boldsymbol{\mu}(\hat{\boldsymbol{\theta}}) = (0.225 \quad 0.141 \quad 0.173 \quad 0.020 \quad 0.011 \quad 0.013 \quad 0.201 \quad 0.106 \quad 0.109)'. \tag{4.57}$$

Estimated standard errors of these fitted values were obtained from equations (4.40) and (4.41) as

$$\widehat{\text{s.e.}}[\boldsymbol{\mu}(\hat{\boldsymbol{\theta}})] = 10^{-1}(0.14 \quad 0.11 \quad 0.07 \quad 0.03 \quad 0.01 \quad 0.02 \quad 0.15 \quad 0.11 \quad 0.07)'. \tag{4.58}$$

Again, all vector elements are given in lexicographic order.

Uncorrected Pearson and likelihood ratio statistics for testing the fit of the uniform association model can be obtained by substituting the fitted cell estimates into equations (4.8) and (4.9). Thus, $X_{\text{P}}^2 = 4.55$ and $X_{\text{LR}}^2 = 5.55$, both on 3 d.f. Neither is significant at the traditional 5% level ($\chi_3^2(0.05) = 7.81$). For clustered data sets like that from the CHS, both first- and second-order corrections usually reduce test significance. Thus no correction need be applied unless a statistic, in its uncorrected form, provides evidence against the null hypothesis. The present analysis could therefore be terminated at this point with the conclusion that the uniform association model fits the smoking by fitness data well. To exhibit the general procedure, however, both first- and second-order corrections will be evaluated for the uniform association model. Results will be given only for corrected versions of the Pearson statistic, since both uncorrected Pearson and likelihood ratio statistics are again very close. Evaluation of the design effects matrix $\hat{\boldsymbol{\Delta}}$ requires a $\mathbf{C}$ matrix that is orthogonal to the design matrix $\mathbf{X}_{\text{UA}}$ given by equation (4.54). For the uniform association model, which includes only the linear × linear interaction term, an appropriate matrix, denoted $\mathbf{C}_{\text{UA}}$, consists of the last three columns of the matrix in equation (4.49). Thus

$$\mathbf{C}_{\text{UA}}' = \begin{bmatrix} 1 & -2 & 1 & 0 & 0 & 0 & -1 & 2 & -1 \\ 1 & 0 & -1 & -2 & 0 & 2 & 1 & 0 & -1 \\ 1 & -2 & 1 & -2 & 4 & -2 & 1 & -2 & 1 \end{bmatrix} \tag{4.59}$$

It is easily verified that $\mathbf{X}_{\text{UA}}'\mathbf{C}_{\text{UA}} = \mathbf{0}$ as required.

Given $\mathbf{C}_{UA}$ and the fitted cell proportions (4.57) corresponding to the uniform association model, together with the estimated covariance matrix of survey proportions, $\hat{\mathbf{\Sigma}}$, the design effects matrix (4.10) was estimated as

$$\hat{\mathbf{\Delta}}_{UA} = \begin{bmatrix} 2.000 & -0.845 & -0.763 \\ -0.154 & 2.342 & 0.592 \\ -0.034 & 0.243 & 0.769 \end{bmatrix}, \tag{4.60}$$

which has eigenvalues 2.66, 1.77 and 0.68. The first-order corrected Pearson statistic based on the mean of these eigenvalues ($\hat{\delta}_{.} = 1.70$) is $X^2_P(\hat{\delta}_{.}) = 3.27$ on 3 d.f. Of course, given knowledge of $\hat{\mathbf{\Sigma}}$, the second-order correction would always be used in preference to the first-order correction. With $\hat{a}^2$ computed as 0.226, the second-order correction becomes $X^2_P(\hat{\delta}_{.}, \hat{a}) = 2.67$ on 2.4 fractional d.f. As predicted, neither the first- nor the second-order corrected test is significant at the 5% level.

Approximate First-order Corrections

The uniform association model does not yield closed form expressions for the MLEs $\boldsymbol{\mu}(\hat{\boldsymbol{\theta}})$, so there is no formula giving $\hat{\delta}_{.}$ as a function of the cell and marginal proportions and their design effects, as there is in the case of independence in a two-way table (see equation (4.19) and in the case of the various independence hypotheses in a three-way table (see equations (4.28)–(4.30)). As described in Rao and Scott (1987), however, a 'nearly conservative' approximation, $\tilde{\delta}_{.}$, can be obtained from the 'closest' model that does yield closed form estimates. For uniform association this will be the model of independence, for which $\hat{\delta}_{.} = 1.69$. The Rao and Scott (1987) approximation is obtained by multiplying $\hat{\delta}_{.}$ by the ratio of the degrees of freedom of the independence model (d.f. = 4) to the degrees of freedom of the uniform association model (d.f. = 3). This yields $\tilde{\delta}_{.} = (4/3) \times 1.69 = 2.25$, which is conservative. The corresponding corrected Pearson statistic becomes $X^2_P(\tilde{\delta}_{.}) = 2.47$.

Wald Tests

Wald statistics for testing the goodness of fit of both independence and uniform association models can be obtained using equation (4.37). For the independence model, the quantities $\mathbf{C}'$ and $\hat{\boldsymbol{\mu}}$ specified in equation (4.37) are given by equations (4.49) and (4.50) respectively; for the uniform association model these quantities are given by equations (4.59) and (4.57) respectively. Then, for the independence model, $X^2_W = 11.82$ on 4 d.f., while for the uniform association model $X^2_W = 3.59$ on 3 d.f. The Wald tests thus reject the hypothesis of independence ($\chi^2_4(0.05) = 9.49$), and fail to reject the hypothesis of uniform association ($\chi^2_3(0.05) = 7.81$), test outcomes that are consistent with the earlier conclusions based on corrected chi-square statistics. For this two-way table analysis, there is no evidence of undue inflation of the Wald test statistics.

The Nested Model Approach

Had we first established the fit of the loglinear uniform association model, we could have tested the contribution of the linear by linear interaction term using the nested model approach of Section 4.2.3. Conditional on the fit of the uniform association model, this approach results in a more powerful test of independence than those described above, as noted by Agresti (1984, p. 82). With reference to Section 4.2.3, let $\mathbf{X}_1$ represent the first four columns of $\mathbf{X}_{\text{UA}}$ and let $\mathbf{X}_2$ represent the fifth column of $\mathbf{X}_{\text{UA}}$ corresponding to the linear-by-linear interaction. Thus the reduced model of Section 4.2.3, with design matrix $\mathbf{X}_1$ and pseudo MLEs $\mu_i(\hat{\hat{\boldsymbol{\theta}}}_1)$, $i=1,\ldots,9$, here corresponds to the independence model. Unadjusted statistics for testing $H_{\text{N}}{:}\theta_2=\beta=0$ are then given by the Pearson and likelihood ratio statistics (4.11) and (4.12). With pseudo MLEs of the fitted cell proportions under the independence model given by equation (4.50), and pseudo MLEs for the full uniform association model given by equation (4.57), equations (4.11) and (4.12) yield $X^2_{\text{P}}(2|1)=12.51$ and $X^2_{\text{LR}}(2|1)=12.55$ respectively, each on 1 d.f. The latter can also be obtained by subtraction of the previously quoted likelihood ratio statistics for testing the fit of the independence and uniform association models. As expected, both the above statistics are significant at the 5% level. First- and second-order corrections can be obtained from an estimate of the design effect matrix of Section 4.2.3, which can be expressed as

$$\hat{\boldsymbol{\Delta}}_{\text{N}}=(\tilde{\mathbf{X}}'_2\hat{\hat{\boldsymbol{\Omega}}}\tilde{\mathbf{X}}_2)^{-1}(\tilde{\mathbf{X}}'_2\hat{\boldsymbol{\Sigma}}\tilde{\mathbf{X}}_2), \tag{4.61}$$

where $\hat{\hat{\boldsymbol{\Omega}}}$ is the multinomial covariance matrix based on the reduced model estimates $\boldsymbol{\mu}(\hat{\hat{\boldsymbol{\theta}}}_1)$ and where $\tilde{\mathbf{X}}'_2$ of Section 4.2.3 is also defined in terms of $\hat{\hat{\boldsymbol{\Omega}}}$. In the present example, since $\tilde{\mathbf{X}}_2$ is a column vector, $\hat{\boldsymbol{\Delta}}_{\text{N}}$ is a scalar equal to 1.58. Thus $\hat{\delta}_{.}=1.58$ and $\hat{a}^2=0$ so that the first- and second-order corrected Pearson tests (identical in this case) are given by $X^2_{\text{P}}(\hat{\delta}_{.})=X^2_{\text{P}}(\hat{\delta}_{.},\hat{a})=7.94$ on 1 d.f. The corresponding test value for the likelihood ratio statistic is 8.03. Both are significant compared to the 5% critical value given by $\chi^2_1(0.05)=3.84$.

Equation (4.39) provides a general form of a Wald statistic for testing nested hypotheses. A Wald test of $H_{\text{N}}{:}\beta=0$ can be obtained from equation (4.39) by setting $\mathbf{K}=(0,0,0,0,1)'$ and $\mathbf{k}=(0,0,0,0,0)'$, which yields $X^2_{\text{W}}(2|1)=8.03$ on 1 d.f. In this single degree of freedom case, the Wald statistic and the corrected Pearson and likelihood ratio statistics are virtually identical.

Smoothed Estimates

The above analysis has established that the loglinear model with uniform association fits the smoking by fitness data well. Further confirmation of this is given by the residual analysis (not described in detail for lack of space) which shows that the largest (in absolute magnitude) of nine residuals is only 1.7. The fitted values of equation (4.57) can thus be regarded as smoothed estimates of the cell proportions, as discussed in Section 4.6. Corresponding standard errors of these smoothed estimates are then given by equation (4.58).

4.9 SUMMARY

This chapter has described the impact of survey design on the chi-squared tests commonly used in the analysis of contingency table data. Tests based on SRS assumptions have been shown to yield significance levels that differ from the nominal test level, even in large samples. In particular, for designs involving clustering, actual significance levels can greatly exceed the nominal level. First- and second-order corrections to standard chi-squared tests have been described. First-order corrections can be implemented given information on design effects, or estimated variances, for cells and specific marginals of the table in question, together with weighted estimates of the cell proportions. The more accurate second-order corrections, on the other hand, require an estimate of the full covariance matrix of cell proportions, which is not always readily available to the analyst. Details of these test procedures have been presented for several problems, namely simple goodness of fit, tests of independence on two-way tables, test of homogeneity of proportions, the analysis of three-way and multi-way contingency tables and the testing of nested hypotheses. Techniques for residual analysis have also been discussed, along with the generation of smoothed estimates of cell proportions, under a suitable model.

In addition to first- and second-order corrected tests, corresponding Wald procedures have been described in detail. These tests, which again require an estimate of the full covariance matrix of estimated cell proportions, can be useful when the ratio of the degrees of freedom of the hypothesis to the degrees of freedom available for variance estimation is small. When this is not the case, Wald procedures can yield seriously inflated test statistics. For lack of space, we have not been able to describe the jack knifed chi-square methods developed by Fay (1979, 1985), which have been shown to behave well (Thomas and Rao, 1987; Fay, 1987), and to be competitive with the second-order corrected tests developed by Rao and Scott.

Examples are included of the application of corrected chi-squared tests and Wald tests to goodness-of-fit and two-way table problems drawn from the CHS. Besides a standard test of independence, the two-way table example illustrates the testing of a model of uniform association, which yields no closed form pseudo maximum likelihood estimates. Tests of nested hypotheses, analysis of residuals and the generation of smoothed estimates are also illustrated. Due to space limitations, we have not included examples for three-way tables and instead refer the reader to the examples in Hidiroglou and Rao (1987b) and Rao and Thomas (1988).

CHAPTER 5

Measures of Association for Contingency Tables

E. A. Molina C.

5.1 INTRODUCTION

In Chapter 4 methods were described for testing the hypothesis of independence in a two-way contingency table. In moderate and large-scale surveys, such tests frequently lead to the rejection of this hypothesis, in which case it is natural to attempt to summarize the nature and strength of the dependence between the cross-classifying variables using a measure of association. Such measures are widely used in exploratory analysis, especially when there is a need to screen out many cross-classifications at minimal cost, as often happens in the analysis of large-scale surveys. Measures may also be used as a guide for the exploration of models, or to complement the results of fitting models, mainly when no single structural model that adequately describes the data can be found. This chapter is restricted to two-way classifications. It is possible, however, to define measures for many-way classifications.

Suppose that a population classified in an $A \times B$ contingency table is represented by the set of population cell proportions $\{\mu_{ab}\}$, where $\sum\sum \mu_{ab} = 1$, $(a = 1, \ldots, A; b = 1, \ldots, B)$. In other words, μ_{ab} is the proportion of population units which fall into category a on the first variable and category b on the second variable. Let $\mu_{a.}$ and $\mu_{.b}$ denote the corresponding marginal proportions. We denote by $\boldsymbol{\mu} = (\mu_{11}, \ldots, \mu_{AB})'$ the column vector obtained when the rows of the table are adjoined to form a single column, and by $\xi = \xi(\boldsymbol{\mu})$ a general measure of the association between the rows and columns of the table.

Example 1 (General Household Survey of the UK)

The data in Table 5.1 are taken from the 1971 General Household Survey of the United Kingdom. This survey employs a stratified three-stage self-weighting design and results in a sample of more than 13 000 households. Table 5.1 is

Table 5.1 Proportions μ_{ab} of households in Region 1 classified by tenure and type of accommodation

	Type of accommodation (b)			
Tenure (a)	1. Terraced	2. Semi-detached	3. Detached	$\mu_{a.}$
1. Rented	0.270	0.150	0.052	0.472
2. Bought	0.131	0.074	0.065	0.270
3. Buying	0.115	0.045	0.098	0.258
$\mu_{.b}$	0.516	0.269	0.215	1.000

based upon a sample of 2267 households from a region that we have called Region 1 (East and West Midlands, East Anglia). Table 5.1 shows estimated proportions μ_{ab} cross-classified according to tenure and type of accommodation of the household. This is a 3×3 table of sample rather than population proportions, but it will be useful to fix ideas.

In our notation, tenure is the row-classifying variable, with three categories labelled $a = 1$ for rented, $a = 2$ for bought, $a = 3$ for buying; the row-marginal proportions $\{\mu_{a.}\}$ are $\mu_{1.} = 0.472$, $\mu_{2.} = 0.270$, $\mu_{3.} = 0.258$. Similarly, the column-classifying variable, type of accommodation, has three categories labelled $b = 1$ for terraced; $b = 2$ for semi-detached, $b = 3$ for detached; the column-marginal proportions $\{\mu_{.b}\}$ are $\mu_{.1} = 0.516$, $\mu_{.2} = 0.269$, $\mu_{.3} = 0.215$.

Most measures of association may be classified into one of the following groups:

1. Measures based on Pearson's chi-square statistic;
2. Measures based in probabilistic models of predictive activity;
3. Measures based on the cross-ratios of the tables.

5.1.1 Measures Based on Pearson's Chi-square Statistic

The measures in this group basically attempt to standardize Pearson's chi-square statistic, in order to eliminate the dependence of the statistic on the number of cross-classified units. The earliest measures in this group were defined under continuity and normality assumptions which are unacceptable for the treatment of general categorical variables. Their later development as measures of pure contingency, i.e. of departure from the hypothesis of independence $\mu_{ab} = \mu_{a.}\mu_{.b}$, does not appear to be particularly useful because of the lack of meaning of the numbers so obtained, a point that is stressed by Goodman and Kruskal (1954). These coefficients derive their popularity mainly from tradition and from the simplicity of their application. We do not recommend the use of these statistics as measures of association for categorical variables and do not consider them again, beyond the following example.

Example 2 (Contingency Coefficient)

Let N_o denote the size of the population. The *contingency coefficient*, denoted by ξ_P, was defined by Pearson (1904) as

$$\xi_P = [X_P^2/(N_o + X_P^2)]^{1/2},$$

where

$$X_P^2 = N_o \sum_a \sum_b (\mu_{ab} - \mu_{a.}\mu_{.b})^2/(\mu_{a.}\mu_{.b})$$

is the population version of Pearson's chi-square statistic given in equation (4.18). The value of ξ_P for Table 5.1 is 0.264. It is difficult to attach a meaning to this figure other than lack of independence. This coefficient was defined in an attempt to norm X_P^2, so that it could work as a discrete analogue of the correlation coefficient. However, the coefficient does not necessarily have an upper limit of 1. To remedy this, several variations of this measure were proposed. Neither ξ_P nor its variations have a clear meaning, as opposed to the correlation coefficient, which measures linearity. They are measures of squared departure from independence; however, to measure the degree of association it is necessary to have some hypothesis as to the nature of the departure from independence to be measured. These measures are marginal-sensitive, a fact that discourages their use for comparative purposes, since they only measure departures from a particular table: the independence table for the given marginals (Altham, 1970a; Molina, 1982).

5.1.2 Measures Based on Probabilistic Models of Predictive Activity

The use of probabilistic models of predictive activity to define measures of association was suggested by Goodman and Kruskal (1954) to obtain coefficients with an operational interpretation. The measures are simple functions of probabilities defined to attack very specific situations, placing the problem of the measurement of association in the determination of a clear hypothesis about the kind of association under observation. Some of the coefficients in this family arise from considering the ability of one variable to predict another and are sometimes called *proportional reduction of error* (PRE) measures. Most of the measures discussed by Goodman and Kruskal can be written as $\zeta = v/\Delta$, where v and Δ are functions of the proportions μ_{ab}. We say that such measures have a *ratio structure*.

Example 3 (A Measure of Proportional Prediction)

Goodman and Kruskal (1954, Section 9) discussed the following model of activity: an individual is chosen at random from the population. We may guess the individual's row-category a either (1) using only the row marginal proportions $\mu_{a.}$ or (2) using the conditional proportions $\mu_{ab}/\mu_{.b}$, given that the individual belongs to column-category b. The relative decrease in the proportion of incorrect predictions as we go from case (1) to case (2) is

$$\tau_{a|b} = \left(\sum_a \sum_b \mu^2_{ab}/\mu_{.b} - \sum_a \mu^2_{a.} \right) \Big/ \left(1 - \sum_a \mu^2_{a.} \right) . \tag{5.1}$$

This is a PRE measure that takes values between 0 and 1. It is 0 if and only if there is independence, and 1 if knowledge of the column-categories completely determines the row-categories. It is indeterminate if and only if all the row marginal proportions but one are 0. The coefficient may also be interpreted as a measure of the proportion of variation in the row-categories attributable to the column-categories, i.e. a qualitative analogue of the coefficient of determination (Light and Margolin, 1971).

Example 4 (Numerical Example)

Consider the data in Table 5.1 and suppose that there is interest in knowing whether the type of accommodation helps in predicting the tenure of the household in Region 1. Applying formula (5.1) we obtain a value of $\tau_{a|b} = 0.041$. The conclusion is that a reduction in error of prediction of tenure of approximately 4.1% is obtained by knowing the household's type of accommodation. Thus, there is association in the predictive sense measured by the coefficient; however, the obtained reduction in error is too low to have much practical significance.

5.1.3 Measures Based on the Cross-ratios of the Tables

The measure described in the previous section has the advantage of a clear interpretation for a single table. However, when interest lies in comparing association in two or more populations, Yule (1912), Edwards (1963), Plackett (1965) and Mosteller (1968), amongst others, have argued that measures of association should be functions of the cross-product ratios of the tables alone, i.e. the $(A-1)(B-1)$ population parameters

$$\alpha_{ab}(\boldsymbol{\mu}) = (\mu_{ab}\mu_{AB})/(\mu_{aB}\mu_{Ab}), \qquad a = 1, \ldots, A-1; b = 1, \ldots, B-1.$$

This requirement is not satisfied by the chi-square-based measures, nor by most of Goodman and Kruskal's measures. Altham (1970a) notes that it is equivalent to basing measures upon contrasts of log proportions and argues that it is sensible to look at sums of squares of contrasts as measures of association.

To illustrate these ideas, consider a 3×3 contingency table with proportions $\boldsymbol{\mu}$. The logarithm of the first cross-ratio can be written as

$$\begin{aligned} \log \alpha_{11} &= \log \mu_{11} - \log \mu_{13} - \log \mu_{31} + \log \mu_{33} \\ &= (1, 0, -1, 0, 0, 0, -1, 0, 1)'(\log \mu_{11}, \log \mu_{12}, \ldots, \log \mu_{33}) = \mathbf{c}' \log \boldsymbol{\mu}, \end{aligned}$$

say, where $\mathbf{c}'\log\boldsymbol{\mu}$ is a contrast of log proportions. In general, given any vector $\mathbf{c} = (c_{11}, \ldots, c_{AB})'$ such that $\sum_a c_{ab} = \sum_b c_{ab} = 0$, we call the product $\mathbf{c}'\log\boldsymbol{\mu}$ a *contrast of log proportions* of the table. Suppose that we have a second table with population proportions $\boldsymbol{\eta} = (\eta_{11}, \ldots, \eta_{AB})'$, and consider the vector of log differences $\boldsymbol{\theta} = \log\boldsymbol{\mu} - \log\boldsymbol{\eta}$. Then

$$\mathbf{c}'\boldsymbol{\theta} = \mathbf{c}'\log\boldsymbol{\mu} - \mathbf{c}'\log\boldsymbol{\eta} = \sum\sum c_{ab}\log(\mu_{ab}/\eta_{ab})$$

is a contrast of the log ratios of $\boldsymbol{\mu}$ and $\boldsymbol{\eta}$. Altham (1970a) proposes selecting a set of $q = (A-1)(B-1)$ linearly independent contrasts $\{\mathbf{c}'_1\boldsymbol{\theta}, \ldots, \mathbf{c}'_q\boldsymbol{\theta}\}$ and then using the square root of the sum of squared contrasts

$$\xi(\boldsymbol{\mu}, \boldsymbol{\eta}) = \left\{\sum_j (\mathbf{c}'_j\boldsymbol{\theta})^2\right\}^{1/2} \tag{5.2}$$

to measure the distance between the pattern of association in $\boldsymbol{\mu}$ and the pattern of association in $\boldsymbol{\eta}$. The measure is zero if and only if all the cross-product ratios of both tables are the same.

This approach may also be used to define a measure of association for a single table $\boldsymbol{\mu}$ by taking the distance between the pattern of association in $\boldsymbol{\mu}$ and the pattern of association in an *independence table* $\mathbf{i} = (i_{11}, \ldots, i_{AB})'$ in which $i_{ab} = i_{a.}i_{.b}$. For, in the latter case, all the cross-product ratios of the $\mathbf{i}$ table are 1 and all the contrasts of log proportions $\mathbf{c}'\log\mathbf{i}$ are 0. Hence a measure of association for $\boldsymbol{\mu}$ is

$$\xi(\boldsymbol{\mu}, \mathbf{i}) = \left\{\sum_{j=1}^{q} (\mathbf{c}'_j\log\boldsymbol{\mu})^2\right\}^{1/2}. \tag{5.3}$$

This is a measure of the association of the rows and columns of $\boldsymbol{\mu}$ alone. The measure is zero if and only if the table is independent. The measure is undefined if the population proportion in any of the cells of the table is zero. Such structural zeros can be treated by selecting an appropriate set of contrasts (Altham, 1970b).

It is convenient to introduce a more compact notation in matrix terms. Let $\boldsymbol{\lambda}(\boldsymbol{\theta}) = (\mathbf{c}'_1\boldsymbol{\theta}, \mathbf{c}'_2\boldsymbol{\theta}, \ldots, \mathbf{c}'_q\boldsymbol{\theta})'$ and let $\mathbf{C}$ be the $AB \times q$ matrix with columns given by the vectors $\mathbf{c}_1, \ldots, \mathbf{c}_q$. With this notation, $\boldsymbol{\lambda}(\boldsymbol{\theta}) = \mathbf{C}'\boldsymbol{\theta}$ and the measure (5.2) becomes

$$\xi(\boldsymbol{\mu}, \boldsymbol{\eta}) = \{\boldsymbol{\lambda}(\boldsymbol{\theta})'\boldsymbol{\lambda}(\boldsymbol{\theta})\}^{1/2} = \{(\mathbf{C}'\boldsymbol{\theta})'(\mathbf{C}'\boldsymbol{\theta})\}^{1/2} = \{\boldsymbol{\theta}'\mathbf{C}\mathbf{C}'\boldsymbol{\theta}\}^{1/2}. \tag{5.4}$$

Similarly, in the case of a single table, equation (5.3) becomes

$$\xi(\boldsymbol{\mu}, \mathbf{i}) = \{\log\boldsymbol{\mu}'\mathbf{C}\mathbf{C}'\log\boldsymbol{\mu}\}^{1/2}. \tag{5.5}$$

Example 5 (A Measure Based on the Cross-ratios)

For a general $A \times B$ table, one set of linearly independent contrasts is given by the $q=(A-1)(B-1)$ parameters

$$\log \mu_{ab} - \log \mu_{aB} + \log \mu_{AB} - \log \mu_{Ab}, \qquad a=1,\ldots,A-1; b=1,\ldots,B-1. \quad (5.6)$$

We call these the *canonical contrasts*. Thus the parametric quantity

$$\xi_0(\boldsymbol{\mu},\mathbf{i}) = \left\{ \sum_a \sum_b (\log \mu_{ab} - \log \mu_{aB} + \log \mu_{AB} - \log \mu_{Ab})^2 \right\}^{1/2} \quad (5.7)$$

is a measure of the association between the rows and columns of the table. For the data in Table 5.1, the contrasts (5.6) are:

$$\log \mu_{11} - \log \mu_{13} + \log \mu_{33} - \log \mu_{31} = 1.470;$$
$$\log \mu_{12} - \log \mu_{13} + \log \mu_{33} - \log \mu_{32} = 1.823;$$
$$\log \mu_{21} - \log \mu_{23} + \log \mu_{33} - \log \mu_{31} = 0.534;$$
$$\log \mu_{22} - \log \mu_{23} + \log \mu_{33} - \log \mu_{32} = 0.894.$$

Thus, $\xi_0(\boldsymbol{\mu},\mathbf{i}) = \{(1.470)^2 + (1.823)^2 + (0.534)^2 + (0.894)^2\}^{1/2} = 2.563$. We conclude that there is association between tenure and type of accommodation in Region 1. However, we cannot say anything about the direction or the nature of this association, nor about its strength. To say something about the strength of this association we need another table in order to make comparisons.

We may obtain the contrasts (5.6) by means of the products $\mathbf{c}' \log \boldsymbol{\mu}$, where $\mathbf{c} = (c_{11},\ldots,c_{AB})'$ has 1's in the coordinates (a,b) and (A,B), -1's in the coordinates (a,B) and (A,b) and 0's in the rest of the coordinates. Let $\mathbf{C}_0$ denote the $AB \times q$ matrix with columns given by these vectors. We call this matrix the *canonical matrix*. Thus, for 3×3 contingency tables the transpose of the canonical matrix is

$$\mathbf{C}_0' = \begin{bmatrix} 1 & 0 & -1 & 0 & 0 & 0 & -1 & 0 & 1 \\ 0 & 1 & -1 & 0 & 0 & 0 & 0 & -1 & 1 \\ 0 & 0 & 0 & 1 & 0 & -1 & -1 & 0 & 1 \\ 0 & 0 & 0 & 0 & 1 & -1 & 0 & -1 & 1 \end{bmatrix}.$$

With this notation, $\xi_0(\boldsymbol{\mu},\mathbf{i}) = \{\log \boldsymbol{\mu}' \mathbf{C}_0 \mathbf{C}_0' \log \boldsymbol{\mu}\}^{1/2}$. Note that for Table 5.1 $\mathbf{C}_0' \log \boldsymbol{\mu} = (1.470,\ 1.823,\ 0.534,\ 0.894)'$.

Example 6 (Comparison of Two Tables)

The data in Table 5.2 are also taken from the 1971 General Household Survey of the UK and are based on a sample of 2847 households from a region that we have called Region 2 (North, North West, Yorkshire). As in Table 5.1, sample proportions are cross-classified according to tenure and type of accommodation. Suppose that we wish to compare the association between tenure and type of accommodation in Regions 1 and 2. Again we ignore the sampling scheme and

Table 5.2 Proportions η_{ab} of households in Region 2 classified by tenure and type of accommodation

	Type of accommodation (b)			
Tenure, (a)	Terraced	Semi-detached	Detached	$\eta_{a.}$
Rented	0.243	0.179	0.22	0.444
Bought	0.100	0.085	0.050	0.235
Buying	0.098	0.175	0.048	0.321
$\eta_{.b}$	0.441	0.439	0.120	1.000

assume that we have a table of population proportions. Let $\boldsymbol{\eta}$ denote the vector of proportions in this table. The vector of canonical contrasts for this table is

$$\mathbf{C}_0' \log \boldsymbol{\eta} = (1.718, 0.831, 0.007, -0.747)'.$$

The canonical contrasts of log differences are

$$\begin{aligned}\boldsymbol{\lambda}_0(\boldsymbol{\theta}) &= \mathbf{C}_0'\boldsymbol{\theta} = \mathbf{C}_0' \log \boldsymbol{\mu} - \mathbf{C}_0' \log \boldsymbol{\eta} \\ &= (1.470 - 1.718, 1.823 - 0.831, 0.534 - 0.007, 0.894 + 0.747)' \\ &= (-0.248, 0.992, 0.527, 1.641)'.\end{aligned}$$

Thus, the parameter

$$\xi_0(\boldsymbol{\mu}, \boldsymbol{\eta}) = \{\boldsymbol{\theta}'\mathbf{C}_0\mathbf{C}_0'\boldsymbol{\theta}\}^{1/2},$$

measures the distance between the patterns of association in the two tables. The computed values is

$$\xi_0(\boldsymbol{\mu}, \boldsymbol{\eta}) = \{(-0.248)^2 + (0.992)^2 + (0.527)^2 + (1.641)^2\}^{1/2} = 2.004.$$

We conclude that the association between tenure and type of accommodation differs when we go from Region 1 to Region 2. The association in Region 2 is

$$\xi_0(\boldsymbol{\eta}, \mathbf{i}) = \{\log \boldsymbol{\eta}'\mathbf{C}_0\mathbf{C}_0' \log \boldsymbol{\eta}\}^{1/2} = \{(1.718)^2 + \cdots + (-0.747)^2\}^{1/2} = 2.049.$$

Thus, the association between tenure and type of accommodation in Region 1 is judged to be higher than in Region 2.

The measures based on the cross-ratios are marginal free measures which concentrate on the inherent association between the cross-classified variables. The cross-ratios of a contingency table have acquired importance in the modern analysis of cross-classified data. However, arguments in favour of the cross-ratios are based on symmetry considerations. Thus, it does not seem possible to define asymmetric or predictive measures in terms of the cross-ratios of the table alone.

There is no single measure of association that can be preferred to the others in every situation. A measure of association is a drastic reduction of the data that should be used with care. A general account of the subject is available in the series of papers by Goodman and Kruskal (1954, 1959, 1963, 1972). A review of the comparison of associations and measures based on cross-ratios is given by Molina (1982, Chapters 2 and 3).

5.2 LARGE SAMPLE INFERENCE UNDER MULTINOMIAL SAMPLING

Suppose that a sample of size n_o is drawn from a finite population under simple random sampling (SRS) with replacement and that the sample is cross-classified after sampling. Denote the observed cell counts in the sample by $n_{11},\ldots,n_{AB}, \sum n_{ab} = n_o$. As in Chapter 4, the distribution of $(n_{11},\ldots,n_{AB})$ is multinomial and the distribution of the vector $\bar{\mathbf{y}}$ of sample cell proportions $\bar{y}_{ab} = n_{ab}/n_o$, $a = 1,\ldots,A; b = 1,\ldots,B$ is such that as $n_o \to \infty$,

$$n_o^{1/2}(\bar{\mathbf{y}} - \boldsymbol{\mu}) \sim \mathrm{N}(\mathbf{0}, \boldsymbol{\Omega}) \tag{5.8}$$

the multivariate normal distribution with mean vector $\mathbf{0}$ and covariance matrix $\boldsymbol{\Omega} = D(\boldsymbol{\mu}) - \boldsymbol{\mu}\boldsymbol{\mu}'$, where $D(\boldsymbol{\mu}) = \mathrm{diag}(\boldsymbol{\mu})$. A consistent estimator of $\boldsymbol{\Omega}$ is $\hat{\boldsymbol{\Omega}} = D(\bar{\mathbf{y}}) - \bar{\mathbf{y}}\bar{\mathbf{y}}'$.

5.2.1 Inference for Measures with a Ratio Structure

A consistent estimator $\hat{\zeta} = \hat{v}/\hat{\Delta}$ of the measure ζ is obtained by replacing each μ_{ab} in the definition of v and Δ by $\bar{y}_{ab}$. The estimator $\hat{\zeta}$ is a non-linear function of $\bar{\mathbf{y}}$ and its variance may estimated by any of the non-parametric methods discussed in Section 2.13. In particular, the linearization variance estimator takes the form

$$v_0 = n_o^{-1}\hat{\Delta}^{-4}\hat{\boldsymbol{\phi}}'\hat{\boldsymbol{\Omega}}\hat{\boldsymbol{\phi}}, \tag{5.9}$$

where $\hat{\boldsymbol{\phi}} = (\hat{\phi}_{11},\ldots,\hat{\phi}_{AB})'$, $\hat{\phi}_{ab} = \hat{v}'_{ab}\hat{\Delta} - \hat{v}\hat{\Delta}'_{ab}$, and $\hat{v}'_{ab}$, $\hat{\Delta}'_{ab}$ denote the estimates of the partial derivatives with respect to $\bar{y}_{ab}$. For the PRE measure $\tau_{a|b}$ this reduces to

$$\hat{\boldsymbol{\phi}}_{ab} = -2\left(\sum_a\sum_b \bar{y}_{ab}^2/\bar{y}_{.b}^2 - \sum_a \bar{y}_{a.}^2\right)\bar{y}_{a.} + \Delta\left\{2y_{ab}/\bar{y}_{.b} - \sum_a (\bar{y}_{ab}/\bar{y}_{.b})^2\right\}.$$

Similar formulae for other measures are available in Goodman and Kruskal (1963, 1972) and in Agresti (1984, Section 10.3).

To find an approximate confidence interval for $\tau_{a|b}$ we refer the statistic

$$(\hat{\tau}_{a|b} - \tau_{a|b})/v_0^{1/2} \tag{5.10}$$

to the $N(0, 1)$ distribution. Let $Z_{\alpha/2}$ be the upper $100(\alpha/2)\%$ point for the standard normal distribution. Then $(\hat{\tau}_{a|b} - Z_{\alpha/2}v_0^{1/2},\ \hat{\tau}_{a|b} + Z_{\alpha/2}v_0^{1/2})$ is an approximate confidence interval at the $1 - \alpha$ level of confidence. The measure $\tau_{a|b}$ takes values between 0 and 1 and so if the interval happens to go beyond those bounds, such inadmissible values should be excluded. We may test the null hypothesis that $\tau_{a|b} = \tau_{a|b0}$ at approximately the α level of significance by rejecting the null hypothesis when the value $\tau_{a|b0}$ lies outside this confidence interval. Care should be taken when $\tau_{a|b0}$ is a boundary value, since in that case the asymptotic distributions may degenerate. This discussion generalizes to any measure with a ratio structure.

***Example** 7* (General Household Survey of the UK)

As mentioned earlier, Table 5.1 is a table of sample rather than population proportions. Again, we ignore the actual complex design employed and assume that the sample was obtained by SRS with replacement. Here $n_o = 2267$. In Example 4 we obtained the value $\hat{\tau}_{a|b} = 0.041$. Substitution of the cell proportions $\bar{y}_{ab}$ into equation (5.9) gives the estimator $v_0 = 0.136/n_o$ and the estimated standard error of $\hat{\tau}_{a|b}$ is thus $v_0^{1/2} = 0.008$. A 95% asymptotic confidence interval for $\tau_{a|b}$ is given by $\hat{\tau}_{a|b} \pm 1.96 v_0^{1/2}$, or (0.026, 0.056). We are thus 95% confident that the reduction in error due to knowledge of the household's type of accommodation lies between 2.6 and 5.6%.

5.2.2 Measures Based on Contrasts of Log Ratios

For simplicity we analyse first the association in a single table. Given the vector of contrasts $\mathbf{C}' \log \boldsymbol{\mu}$, it follows from a suitable multivariate version of the δ-method theorem (e.g. Bishop *et al.*, 1975, Section 14.6.3) that the estimator $\mathbf{C}' \log \bar{\mathbf{y}}$ follows approximately a multivariate normal distribution with mean vector $\mathbf{C}' \log \boldsymbol{\mu}$ and covariance matrix $n_o^{-1} \mathbf{C}' D(\boldsymbol{\mu})^{-1} \mathbf{C}$. It is shown in Molina and Smith (1986) that we can find a matrix $\mathbf{C}_1$ of contrast coefficients such that the statistic

$$\xi_1(\bar{\mathbf{y}}, \mathbf{i})^2 = \log \bar{\mathbf{y}}' \mathbf{C}_1 \mathbf{C}_1' \log \bar{\mathbf{y}} \tag{5.11}$$

is asymptotically distributed as a non-central chi-squared random variable with $q = (A-1)(B-1)$ degrees of freedom (d.f.) and non-centrality parameter $\xi_1(\boldsymbol{\mu}, \mathbf{i})^2 = \log \boldsymbol{\mu}' \mathbf{C}_1 \mathbf{C}_1' \log \boldsymbol{\mu}$. The matrix $\mathbf{C}_1$ must satisfy the SRS condition

$$n_o^{-1} \mathbf{C}_1' D(\boldsymbol{\mu})^{-1} \mathbf{C}_1 = \mathbf{I}_q, \tag{5.12}$$

that the asymptotic covariance matrix of $\mathbf{C}_1 \log \bar{\mathbf{y}}$ equals $\mathbf{I}_q$, the identity matrix of order q. A FORTRAN program for computing such a matrix $\mathbf{C}_1$ is available from the author. In practice, $D(\boldsymbol{\mu})$ in equation (5.12) is replaced by $D(\bar{\mathbf{y}})$, which is a consistent estimator of $D(\boldsymbol{\mu})$.

To test the hypothesis of no association, $H_0 : \xi_1(\boldsymbol{\mu}, \mathbf{i}) = 0$, against the alternative $\xi_1(\boldsymbol{\mu}, \mathbf{i}) > 0$, we refer the statistic $\xi_1(\bar{\mathbf{y}}, \mathbf{i})^2$ to the ordinary chi-square distribution on q d.f. The test is appropriate since the non-centrality parameter is zero if and only if the table is independent. When the table exhibits some degree of association we may find an approximate $100(1-\alpha)\%$ confidence interval for $\xi_1(\boldsymbol{\mu}, \mathbf{i})^2$ following the procedures suggested in Johnson and Pearson (1969): let $\chi^2_m(U, \alpha)$ denote the lower $\alpha\%$ point for the non-central chi-square distribution on m d.f. and non-centrality parameter U. Then, $(\chi^2_m(U; \alpha/2), \chi^2_m(U; 1-\alpha/2))$, $U = \xi_1(\bar{\mathbf{y}}, \mathbf{i})^2$, is an approximate $100(1-\alpha)\%$ confidence interval for $\xi_1(\boldsymbol{\mu}, \mathbf{i})^2$. In order to find the percentage points of the distribution, inverse interpolation may be used in tables for the non-central distribution. However, we remark that the parameter $\xi_1(\boldsymbol{\mu}, \mathbf{i})$ is designed to make comparisons. Thus, the number, by itself, is of little use other than as an alternative test of independence. The relevant quantity to be estimated is the measure $\xi(\boldsymbol{\mu}, \boldsymbol{\eta})$.

5.2.3 Comparison of Two Contingency Tables

Suppose that a second sample of size n_o^* is drawn independently from a second population classified by the same variables used for table $\boldsymbol{\mu}$, and denote by $\boldsymbol{\eta}$ the new vector of population cell proportions. Let $\bar{\mathbf{y}}^*$ denote the corresponding vector of sampled cell proportions. Then, under SRS, $\bar{\mathbf{y}}^*$ is asymptotically $N(\boldsymbol{\eta}, n_o^{*-1}\boldsymbol{\Omega}(\boldsymbol{\eta}))$, where $\boldsymbol{\Omega}(\boldsymbol{\eta}) = D(\boldsymbol{\eta}) - \boldsymbol{\eta}\boldsymbol{\eta}'$. Given the measure $\tau_{a|b}$ we might examine the sampling difference $\tau_{a|b}(\bar{\mathbf{y}}) - \tau_{a|b}(\bar{\mathbf{y}}^*)$ by referring it to the normal distribution with zero mean and variance $v_0 + v_0^*$, where v_0^* is the version of equation (5.9) corresponding to the second sample. However, this comparison may be meaningless, as the same value of $\tau_{a|b}$ may be achieved in many ways (there are $q = (A-1)(B-1)$ d.f. in an $A \times B$ table; a PRE measure accounts for a single d.f.). Thus, when interest lies in knowing whether both populations achieve the association in a similar way, the procedure should be complemented with a comparison of the cross-ratios of the tables.

Let $\boldsymbol{\theta} = \log \boldsymbol{\mu} - \log \boldsymbol{\eta}$, $\hat{\boldsymbol{\theta}} = \log \bar{\mathbf{y}} - \log \bar{\mathbf{y}}^*$, and consider the vectors of contrasts $\boldsymbol{\lambda}_2 = \mathbf{C}_2'\boldsymbol{\theta}$, $\hat{\boldsymbol{\lambda}}_2 = \mathbf{C}_2'\hat{\boldsymbol{\theta}}$, where the matrix $\mathbf{C}_2$ satisfies the condition

$$\mathbf{C}_2'[n_o^{-1}D(\bar{\mathbf{y}})^{-1} + n_o^{*-1}D(\bar{\mathbf{y}}^*)^{-1}]\mathbf{C}_2 = \mathbf{I}_q. \tag{5.13}$$

It follows (Molina and Smith, 1986) that the statistic

$$\{\xi_2(\bar{\mathbf{y}}, \bar{\mathbf{y}}^*)\}^2 = \hat{\boldsymbol{\lambda}}_2'\hat{\boldsymbol{\lambda}}_2 \tag{5.14}$$

is asymptotically distributed as a non-central chi-squared random variable with non-centrality parameter $\{\boldsymbol{\lambda}_2'\boldsymbol{\lambda}_2\} = \{\xi_2(\boldsymbol{\mu}, \boldsymbol{\eta})\}^2$. Note that the parameter $\xi_2(\boldsymbol{\mu}, \boldsymbol{\eta})$ is of the type (5.4), and measures the distance between the associations in the tables. A FORTRAN program for computing $\mathbf{C}_2$ is available from the author. We can test the null hypothesis $H_0: \xi(\boldsymbol{\mu}, \boldsymbol{\eta}) = 0$ by referring the statistic (5.14) to the ordinary chi-square distribution on q d.f. If H_0 is true, all cross-ratios are the same in both tables.

Example 8 (Comparison of Two 3×3 Tables)

The numbers in Table 5.3 are fictitious proportions designed to illustrate the ideas discussed. The tables have been standardized to make their structure

Table 5.3 Proportions for two populations cross-classified by variables a and b

	Population 1					Population 2				
	b_1	b_2	b_3	b_4	Total	b_1	b_2	b_3	b_4	Total
a_1	0.20	0.02	0.02	0.01	0.25	0.20	0.02	0.02	0.01	0.25
a_2	0.02	0.20	0.01	0.02	0.25	0.01	0.02	0.02	0.20	0.25
a_3	0.01	0.02	0.20	0.02	0.25	0.02	0.20	0.01	0.02	0.25
a_4	0.02	0.01	0.02	0.20	0.25	0.02	0.01	0.20	0.02	0.25
Total	0.25	0.25	0.25	0.25	1.0	0.25	0.25	0.25	0.25	1.0

clearer, so that the marginal populations are the same. To fix ideas, suppose that the tables correspond to the populations of two countries, variable b being the occupation of fathers and variable a the occupations of sons, say. Note that the cross-classification corresponding to population 1 has a definite diagonal pattern, whereas that of population 2 is more erratic. This is an asymmetric situation, and the interest lies in knowing if we can predict the occupation of sons from the occupation of fathers. As an example, we consider the PRE measure $\tau_{a|b}$. We assume that the proportions in the tables were obtained under multinomial sampling. The estimator $\tau_{a|b}(\bar{\mathbf{y}})$ in each table is the same and equal to 0.539. Thus, the test statistic

$$[\tau_{a|b}(\bar{\mathbf{y}}) - \tau_{a|b}(\bar{\mathbf{y}}^*)]/(v_0 + v_0^*)^{1/2}$$

equals zero, and we can conclude that a reduction in error of approximately 53.9% is obtained when we predict the occupation of sons from the occupation of fathers in both cases. However, it does not tell us whether both populations behave similarly, and clearly they do not. This is not surprising since the measure is not designed to that end. To discuss whether the occupational mobility in both countries is similar, we apply the measure $\xi_2(\boldsymbol{\mu}, \boldsymbol{\eta})$. To test the null hypothesis that the association is the same in both countries, $H_0: \xi_2(\boldsymbol{\mu}, \boldsymbol{\eta}) = 0$, we refer the statistic (5.14) to the ordinary chi-square distribution on $(4-1)(4-1) = 9$ d.f. The calculated value is 56.354, which is highly significant for a test with size 0.05. We conclude that the behaviour of the populations towards occupational mobility is different in each country.

5.3 EFFECTS OF IGNORING THE SAMPLING SCHEME

As discussed in Chapter 4, the underlying assumption of multinomial sampling is often not tenable with survey data due to clustering and stratification. In this section we consider the consequences of departures from this assumption on the procedures discussed in Section 5.2. We assume that the sample proportions $\bar{\mathbf{y}}$ and $\bar{\mathbf{y}}^*$ are replaced by consistent estimators $\hat{\boldsymbol{\mu}}$ and $\hat{\boldsymbol{\eta}}$ but that, otherwise, no allowance is made for the complex design.

Suppose that a sample of size n_o is drawn from a finite population under a complex design and that the sample is cross-classified after sampling, the sampling method being employed only for economic, administrative or efficiency reasons. That is, we concentrate on the aggregate association between the cross-classified variables. We assume that an estimator $\hat{\boldsymbol{\mu}}$ of $\boldsymbol{\mu}$ is such that, as $n_o \to \infty$,

$$n_o^{1/2}(\hat{\boldsymbol{\mu}} - \boldsymbol{\mu}) \sim N(\mathbf{0}, \boldsymbol{\Sigma}). \tag{5.15}$$

In general $\boldsymbol{\Sigma}$ will be a non-negative definite matrix determined by the complex design.

5.3.1 Measures with a Ratio Structure

We consider first the measure ζ and its estimator $\hat{\zeta}$ obtained by substituting $\hat{\boldsymbol{\mu}}$ for $\boldsymbol{\mu}$ in the definition of ζ. This is a consistent estimator of ζ but v_0 is no longer a consistent estimator of the variance of $\hat{\zeta}$. The design effect of $\hat{\zeta}$ and v_0 is defined, as in Chapter 2, as

$$\text{deff}(\hat{\zeta}, v_0) = \sigma_y^2 / \sigma_0^2, \tag{5.16}$$

where $\sigma_y^2 = \text{var}(\hat{\zeta})$ is the true variance of $\hat{\zeta}$ and $\sigma_0^2 = E(v_0)$ is the true expected value of v_0.

Example 9 (Stratified SRS with Replacement)

Consider H strata and assume SRS with replacement within each stratum. Let $\boldsymbol{\mu}_h$ be the vector of conditional probabilities that a population unit falls into cells (a, b), given that the unit is in stratum h. Let W_h and n_h respectively denote the proportion of the population units and the number of sample units in stratum h and let $w_h = n_h/n_o$, $n_o = \sum_h n_h$. Then, the estimator $\bar{\mathbf{y}}$, the vector of observed proportions, satisfies equation (5.15) with

$$\boldsymbol{\Sigma} = \sum_h \{u_h^{-1} W_h^2 (D(\boldsymbol{\mu}_h) - \boldsymbol{\mu}_h \boldsymbol{\mu}_h')\},$$

where u_h denotes the asymptotic value of w_h. Molina and Smith (1988) have shown that in large samples if the allocation is proportional σ_y^2 reduces to $\sigma_y^2 = \sigma_0^2 - \sigma_b^2$, where the parameter

$$\sigma_b^2 = n_o^{-1} \Delta^{-4} \sum_h W_h \{\boldsymbol{\phi}'(\boldsymbol{\mu}_h - \boldsymbol{\mu})\}^2$$

represents the 'between strata' variation. Thus

$$\text{deff}(\hat{\tau}_{a|b}, v_0) = 1 - \sigma_b^2/\sigma_0^2 \leqslant 1,$$

and the SRS procedure is conservative. However, the more the strata differ in their population proportions, the larger σ_b^2 becomes, and the SRS procedure may become too conservative. If the allocation is not proportional v_0 may underestimate σ_y^2.

Example 10 (Altham's Model for Two-stage Sampling)
Consider M primary sampling units (PSUs) with N_c secondary units in the cth PSU. A two-stage sample of m subsamples is drawn with sizes n_c, $\sum_{c=1}^m n_c = n_o, n_c \leqslant N_c$. Define the random variable Z_{abct} by $Z_{abct} = 1$ if the tth population element of the cth primary unit is in cell (a, b) of the classification, 0 otherwise. Assume that the Z_{abct} in different PSUs are independent and that

$$\varepsilon(Z_{abct}) = \mu_{ab}, \qquad \varepsilon\{(Z_{abct} - \mu_{ab})(Z_{a'b'ct'} - \mu_{a'b'})\} = \psi_{aba'b'}, \qquad t \neq t' \quad (5.17)$$

where ε denotes expectation with respect to the model (Altham, 1976; Rao and Scott, 1981). Then $\bar{\mathbf{y}}$, the vector of sample proportions of secondary units in the cross-classification, is a model-unbiased estimator of $\boldsymbol{\mu}$ and, conditional on the realized sample sizes n_c, $\boldsymbol{\Sigma} = \boldsymbol{\Omega} + (\tilde{n} - 1)\boldsymbol{\Psi}$, and $\sigma_y^2 = \sigma_0^2 + (\tilde{n} - 1)\sigma_\psi^2$, where $\tilde{n} = \sum_{c=1}^m n_c^2/n_o$, $\sigma_\psi^2 = \Delta^{-4}(\boldsymbol{\phi}'\boldsymbol{\Psi}\boldsymbol{\phi})$, and $\boldsymbol{\Psi}$ is the $(AB) \times (AB)$ matrix with entries given by the model-covariances in equation (5.17). Therefore, for $\sigma_0^2 \neq 0$, $\text{deff}(\hat{\tau}_{a|b}, v_0) = 1 + (\tilde{n} - 1)\sigma_\psi^2/\sigma_0^2$, and it follows that ignoring the clustering may lead to underestimation of σ_y^2.

5.3.2 Measures Based on Contrasts of Log Ratios

For measures based on coantrasts of log ratios, we first note that assuming multinomial sampling amounts to selecting the contrasts $\boldsymbol{\lambda}_1 = \mathbf{C}_1' \log \boldsymbol{\mu}$, where the matrix $\mathbf{C}_1$ satisfies equation (5.12). Under general complex designs the matrix $\mathbf{C}_1$ will not always reduce the dispersion matrix of $\mathbf{C}_1 \log \hat{\boldsymbol{\mu}}$ to the identity, that is, selection of the same contrasts as for multinomial sampling may alter the large-sample distribution of the estimators. It is shown in Molina and Smith (1986) that under a general sampling scheme the dispersion matrix of $\mathbf{C}_1 \log \hat{\boldsymbol{\mu}}$ becomes $\boldsymbol{\Gamma}_1 = n_o^{-1}\mathbf{C}_1' D(\boldsymbol{\mu})^{-1}\boldsymbol{\Sigma} D(\boldsymbol{\mu})^{-1}\mathbf{C}_1$. Since the matrix $\boldsymbol{\Gamma}_1$ is positive definite there exists a non-singular, lower triangular matrix $\mathbf{L}$ such that $\boldsymbol{\Gamma}_1 = \mathbf{LL}'$. Let $\gamma_1 \geqslant \gamma_2 \geqslant \cdots \geqslant \gamma_q$ denote the eigenvalues of $\boldsymbol{\Gamma}_1$ and let $\mathbf{E}$ denote the corresponding matrix of eigenvectors. It follows that the measure of association $\xi_1(\hat{\boldsymbol{\mu}}, \mathbf{i})^2$ can be written as the linear combination

$$\sum_i \gamma_i Z_i^2, \quad (5.18)$$

where $Z_1, \ldots, Z_q$ are asymptotically independent $N(z_i, 1)$ random variables with means given by the vector $\mathbf{z} = (\mathbf{LE})^{-1}\mathbf{C}_1 \log \boldsymbol{\mu}$ (Molina and Smith, 1986). This shows that the effect of ignoring the design is to change the large sample distribution to a mixture of non-central chi-square random variables on 1 d.f. with weights given by the eigenvalues of $\boldsymbol{\Gamma}_1$. Thus, if the size of the eigenvalues departs sufficiently from unity, ignoring the sampling scheme may lead to very misleading results. The eigenvalues may be viewed as generalized design effects as in Section 2.11. Molina and Smith (1986) show that γ_1 represents the largest possible design effect over all possible linear combinations for the estimated contrasts. Similarly, γ_i is the largest possible design effect over all the linear combinations of the estimated contrasts which are orthogonal to the first $i-1$ eigenvectors of $\mathbf{\Gamma}_1$.

Example 11 (Two-Stage Sampling)

Consider M PSUs with N_c secondary units in the cth PSU. From the PSUs, m are selected with probability proportional to N_c with replacement. Subsamples each of size n are drawn by SRS with replacement independently from each selected PSU, so that $n_o = mn$. Let W_c denote the proportion of the population in the cth primary unit and let $\mu_{ab(c)}$ denote the probability for a population secondary unit of falling in cell (a, b) of the classification, given that the unit is in the cth PSU. Then the probability for a secondary unit of falling in cell (a, b) is $\mu_{ab} = \sum_c W_c \mu_{ab(c)}$. Let

$$\boldsymbol{\mu}_c = (\mu_{11(c)}, \ldots, \mu_{AB(c)})'.$$

Then the vector of sampled proportions $\bar{\mathbf{y}}$ is such that

$$E(\bar{\mathbf{y}}) = \boldsymbol{\mu}, \qquad \mathbf{\Sigma} = \mathbf{\Omega} + (n-1)\sum_c W_c(\boldsymbol{\mu}_c - \boldsymbol{\mu})(\boldsymbol{\mu}_c - \boldsymbol{\mu})'.$$

If the actual sampling scheme is ignored and a matrix $\mathbf{C}_1$ is selected assuming SRS, the resulting statistic, $\xi_1(\bar{\mathbf{y}}, \mathbf{i})^2$ will be distributed as the mixture (5.18), with the γ's being the eigenvalues of

$$\mathbf{\Gamma}_1 = \mathbf{I}_q + (n-1)\mathbf{\Psi}, \qquad \mathbf{\Psi} = n_o^{-1}\sum_c W_c(\mathbf{C}_1' D(\boldsymbol{\mu})^{-1}\boldsymbol{\mu}_c\boldsymbol{\mu}_c' D(\boldsymbol{\mu})^{-1}\mathbf{C}_1).$$

Let $\alpha_1 \geqslant \alpha_2 \geqslant \cdots \geqslant \alpha_q$, denote the eigenvalues of $\mathbf{\Psi}$. Then $\gamma_i = 1 + (n-1)\alpha_i \geqslant 1$ $i = 1, \ldots, q$. Therefore, if the α_i are at all large, we may underestimate the actual significance level of SRS-based tests.

5.4 INFERENCE PROCEDURES FOR COMPLEX SURVEYS

Suppose that we have an estimator $\hat{\boldsymbol{\mu}}$ satisfying equation (5.15). Initially we assume that a consistent estimator $\hat{\mathbf{\Sigma}}$ of $\mathbf{\Sigma}$ is available. We later consider approximate methods which do not require the full covariance matrix.

5.4.1 General Large-sample Inference

The linearization estimator v_L of the variance of the measure $\hat{\zeta}$ under a complex design is simply obtained by replacing $\hat{\mathbf{\Omega}}$ in equation (5.9) by $\hat{\mathbf{\Sigma}}$ and each $\bar{y}_{ab}$ by $\hat{\mu}_{ab}$ in the equations for $\hat{\boldsymbol{\phi}}$ and $\hat{\Delta}$. Inference then proceeds as in the multinomial case with v_L replacing v_0.

For measures based on contrasts of log ratios Molina and Smith (1986) have shown that we can find a matrix $\mathbf{C}_{1y}$ of contrast coefficients such that the statistic

$$\xi_{1y}(\hat{\boldsymbol{\mu}}, \mathbf{i})^2 = \log \hat{\boldsymbol{\mu}}' \, \mathbf{C}_{1y} \mathbf{C}'_{1y} \log \hat{\boldsymbol{\mu}} \tag{5.19}$$

is asymptotically distributed as a non-central chi-squared random variable with $(A-1)(B-1)$ d.f. and non-centrality parameter $\xi_{1y}(\boldsymbol{\mu}, \mathbf{i})^2$. The matrix $\mathbf{C}_{1y}$ must satisfy the condition

$$\mathbf{C}'_{1y} \boldsymbol{\Psi} \mathbf{C}_{1y} = \mathbf{I}_q, \tag{5.20}$$

where $\mathbf{I}_q$ denotes the identity matrix of order q, and $\boldsymbol{\Psi} = n_o^{-1} D(\boldsymbol{\mu})^{-1} \boldsymbol{\Sigma} D(\boldsymbol{\mu})^{-1}$. A FORTAN program for computing $\mathbf{C}_{1y}$ is available from the author.

To test the hypothesis of no association, $H_0{:}\xi_{1y}(\boldsymbol{\mu}, \mathbf{i}) = 0$ we refer the statistic $\xi_{1y}(\boldsymbol{\mu}, \mathbf{i})^2$ to the ordinary chi-square distribution with q d.f. The test is appropriate, for the non-centrality parameter is zero if and only if the table is independent. Thus, we obtain an alternative test of indepedence for complex designs.

Suppose that a second sample of size n_o^* is drawn independently from a second population classified by the same variables used for table $\boldsymbol{\mu}$, and denote by $\boldsymbol{\eta}$ the new vector of population cell proportions. We assume that there is an estimator $\hat{\boldsymbol{\eta}}$ which is asymptotically $N(\boldsymbol{\eta}, n_o^{*-1} \boldsymbol{\Sigma}^*)$, where $\boldsymbol{\Sigma}^*$ is determined by the sampling scheme.

Let $\boldsymbol{\theta} = \log \boldsymbol{\mu} - \log \boldsymbol{\eta}$, $\hat{\boldsymbol{\theta}} = \log \hat{\boldsymbol{\mu}} - \log \hat{\boldsymbol{\eta}}$, and consider the vectors of contrasts $\boldsymbol{\lambda}_{2y} = \mathbf{C}'_{2y} \boldsymbol{\theta}, \hat{\boldsymbol{\lambda}}_{2y} = \mathbf{C}'_{2y} \hat{\boldsymbol{\theta}}$, where the matrix $\mathbf{C}_{2y}$ satisfies the condition

$$\mathbf{C}'_{2y} [n_o^{-1} D(\boldsymbol{\mu})^{-1} \boldsymbol{\Sigma} D(\boldsymbol{\mu})^{-1} + n_o^{*-1} D(\boldsymbol{\eta})^{-1} \boldsymbol{\Sigma}^* D(\boldsymbol{\eta})^{-1}] \mathbf{C}_{2y} = \mathbf{I}_q. \tag{5.21}$$

It follows (Molina and Smith, 1986) that the statistic

$$\xi_{2y}(\hat{\boldsymbol{\mu}}, \hat{\boldsymbol{\eta}})^2 = \hat{\boldsymbol{\lambda}}'_{2y} \hat{\boldsymbol{\lambda}}_{2y}$$

is asymptotically distributed as a non-central chi-squared random variable with non-centrality parameter $\boldsymbol{\lambda}'_{2y} \boldsymbol{\lambda}_{2y} = \xi_{2y}(\boldsymbol{\mu}, \boldsymbol{\eta})^2$. The parameter $\xi_{2y}(\boldsymbol{\mu}, \boldsymbol{\eta})$ measures the distance among the associations in the tables and we can test the null hypothesis $H_0{:}\xi_{2y}(\boldsymbol{\mu}, \boldsymbol{\eta}) = 0$, by referring the statistic to the ordinary chi-square distribution on $(A-1)(B-1)$ d.f. If H_0 is true, the cross-ratios of both tables are the same, and we ensure that the $(A-1)(B-1)$ aspects of the association measured by the cross-ratios are the same in both tables. We remark that we can compare two samples drawn under two different designs. This is important since the tables may come from two regions or countries with differences that make it difficult to apply the same sample design. The procedure to compute the matrix $\mathbf{C}_{2y}$ is the same as in the case of a single table with input covariance matrix given by

$$n_o^{-1} D(\hat{\boldsymbol{\mu}})^{-1} \hat{\boldsymbol{\Sigma}} D(\hat{\boldsymbol{\mu}})^{-1} + n_o^{*-1} D(\hat{\boldsymbol{\eta}})^{-1} \hat{\boldsymbol{\Sigma}}^* D(\hat{\boldsymbol{\eta}})^{-1} \tag{5.22}$$

5.4.2 Bounds and Corrections for the Statistics

In practice an estimator of $\boldsymbol{\Sigma}$ is often not available. At best, some information about the variances, or the design effects, is at hand. Then it is of interest to obtain bounds and corrections for the statistics based directly on these estimators. If γ_1, γ_q or a bound for them can be specified we can find bounds for the distribution of the measures, since in that case

$$\gamma_q \sum Z_i^2 \leqslant \xi_1(\hat{\boldsymbol{\mu}}, \mathbf{i})^2 = \sum \gamma_i Z_i^2 \leqslant \gamma_1 \sum Z_i^2,$$

where $\sum Z_i^2$ is asymptotically $\chi_q^2(\mathbf{z}'\mathbf{z})$, the non-central chi-square distribution with q d.f. and non-centrality parameter $\mathbf{z}'\mathbf{z}$. Under the hypothesis of independence, $\mathbf{z}'\mathbf{z} = 0$, and $\xi_1(\hat{\boldsymbol{\mu}}, \mathbf{i})^2/\gamma_1$ provides us with a conservative test for complex designs. For example, Molina and Smith (1986) show that, under stratified sampling and proportional allocation, $\gamma_1 \leqslant 1$ so that the SRS test procedure based upon $\xi_1(\hat{\boldsymbol{\mu}}, \mathbf{i})^2$ provides us with a conservative test of independence. In the case of two tables, a conservative test for the comparison of associations is also obtained. If the allocation is not proportional, the SRS tests need not be conservative.

As an alternative correction, Molina and Smith (1986) argue that $\hat{\delta} = \operatorname{tr}[D(\boldsymbol{\mu})^{-1}\hat{\boldsymbol{\Sigma}}]/(AB - 1)$ should be close to the average of the γ_i's and thus that the corrected statistic $\xi_1(\hat{\boldsymbol{\mu}}, \mathbf{i})^2/\hat{\delta}$ may be teated approximately as a non-central chi-square random variable. Note that

$$\operatorname{tr}(D(\boldsymbol{\mu})^{-1}\boldsymbol{\Sigma}) = \sum\sum \operatorname{var}(\hat{\mu}_{ab})\mu_{ab}^{-1} = \sum\sum(1 - \mu_{ab})d_{ab},$$

where $d_{ab} = \operatorname{var}(\hat{\mu}_{ab})/\{\mu_{ab}(1 - \mu_{ab})\}$ is the design effect for $\hat{\mu}_{ab}$ and so this correction may be applied only with estimates of the d_{ab} and not of the full covariance matrix $\hat{\boldsymbol{\Sigma}}$.

Example 12 (General Household Survey of the UK)

As mentioned in previous examples, the data in Tables 5.1 and 5.2 were obtained using a complex sampling design. Since the design is self-weighting, the vectors of observed proportions and their respective dispersion matrices are used to estimate $\xi_{2y}(\boldsymbol{\mu}, \boldsymbol{\eta})$ as outlined in Section 5.4.1 (the values of $\hat{\boldsymbol{\Sigma}}$ and $\hat{\boldsymbol{\Sigma}}^*$ are given in Molina and Smith, 1986). The calculated value of $\xi_{2y}(\hat{\boldsymbol{\mu}}, \hat{\boldsymbol{\eta}})^2 = 77.77$, which is highly significant for a chi-squared random variable on 4 d.f., leading to the rejection of the hypothesis that the association between tenure and type of accommodation is the same for both tables. Suppose, however, that estimators of the full dispersion matrices are not available and the statistics $\xi_2(\bar{\mathbf{y}}, \bar{\mathbf{y}}^*)^2$ are computed under SRS assumptions, that is, the measure in equation (5.14) is used. The computed value is $\xi_2(\bar{\mathbf{y}}, \bar{\mathbf{y}}^*)^2 = 87.00$. The difference between the values of the two test statistics is to be expected, since the type of housing is closely related to the geographical areas which formed the clusters. Therefore, SRS assumptions are expected to underestimate the variances, leading to an overestimation of the measures. The statistic $\xi_2(\bar{\mathbf{y}}, \bar{\mathbf{y}}^*)^2$ is asymptotically distributed as the mixture equation (5.18) with weights given by the eigenvalues of the matrix

$$\mathbf{C}_2'[n_o^{-1}D(\boldsymbol{\mu})^{-1}\boldsymbol{\Sigma}D(\boldsymbol{\mu})^{-1} + n_o^{*-1}D(\boldsymbol{\eta})^{-1}\boldsymbol{\Sigma}^*D(\boldsymbol{\eta})^{-1}]\mathbf{C}_2.$$

The estimated weights are 2.365, 1.839, 1.241 and 0.731. The average value is $\bar{\gamma} = 1.544$ and a corrected statistic is $\xi_2(\bar{\mathbf{y}}, \bar{\mathbf{y}}^*)^2 = 56.35$, which in this case is conservative. An alternative correction is given by $\xi_2(\bar{\mathbf{y}}, \bar{\mathbf{y}}^*)^2/\hat{\delta}$, where

$$\hat{\delta} = \mathrm{tr}(D(\bar{\mathbf{y}})^{-1}\hat{\boldsymbol{\Sigma}} + D(\bar{\mathbf{y}})^{*-1}\hat{\boldsymbol{\Sigma}}^*)/(2AB-2).$$

The values are $\hat{\delta} = 1.645$, $\xi_2(\bar{\mathbf{y}}, \bar{\mathbf{y}}^*)^2/\hat{\delta} = 52.89$, which also gives a conservative correction. However, note that there can be large differences between the corrections and the original values of the statistics.

Example 13 (Example 7 continued)

Since a consistent estimator of the covariance matrix under the actual design, $\hat{\boldsymbol{\Sigma}}$, is available, a primary analysis of the predictive potential of the variable 'type of accommodation' is possible. Since the design is self-weighting, the sampled proportions from Region 1 satisfy condition (5.15), and $\tau_{a|b}(\bar{\mathbf{y}})$ is a consistent estimator of $\tau_{a|b}$. Hence the value computed under multinomial assumptions suffice. In Example 7 we obtained an estimated value of 0.041. To assess the significance of this value we need an estimate of the asymptotic variance. This can be obtained by the the method outlined in Section 5.4.1 from which we obtain a value $v_L = 0.285/n_o$, which, when compared with the SRS-based estimate of $v_0 = 0.136/n_o$, gives an estimated design effect of $0.285/0.136 = 2.096$. The estimated standard error of $\tau_{a|b}(\bar{\mathbf{y}})$ is now $v_L^{1/2} = 0.011$. A 95% confidence interval for $\tau_{a|b}$ is $\tau_{a|b}(\bar{\mathbf{y}}) \pm 1.96 v_L^{1/2}$ or (0.019, 0.063). This may be compared with the narrower confidence interval (0.026, 0.056), obtained under SRS assumptions.

In order to apply the results developed in the previous sections we need only estimators satisfying equation (5.15). When the design is self-weighting the observed proportions satisfy this requirement. Otherwise a table of derived estimators of the probabilities should be used. The results are valid when the sample sizes are large, which is usually the case in large-scale surveys. For measures based on log ratios, it may be necessary to use the usual continuity corrections if some very small frequencies occur in some cells of the tables.

PART B

Aggregated Analysis: Point Estimation and Bias

CHAPTER 6

Introduction to Part B

T. M. F. Smith

6.1 INTRODUCTION

In Part A we implicitly assumed that the estimator $\hat{\theta}$ of the aggregate target parameter θ of interest is approximately unbiased under both the true model and the misspecified model. In this case the misspecification effect (meff) defined in equation (2.2) by

$$\text{meff}(\hat{\theta}, v_0) = \text{var}_{\text{true}}(\hat{\theta})/E_{\text{true}}(v_0), \tag{6.1}$$

where v_0 is an estimator of the variance of $\hat{\theta}$ under the misspecified model and the subscript 'true' denotes the correctly specified model, is an adequate summary of the effect of selection. We have normally taken the misspecified model to be that of IID observations since that is the default option in most statistical software packages. Typically for clustered data, $\text{meff}(\hat{\theta}, v_0) > 1$, and this can have a serious impact on inferences if the IID model is wrongly assumed to hold. The nature and extent of these misspecification errors have been extensively explored in Part A.

In this part we continue our examination of the problem of estimating a well-defined aggregate target parameter. Now, however, the effect of selection is to introduce bias into the procedures. The meff in equation (6.1) does not take into account bias and so is no longer an adequate summary of the effects of misspecification. Attention will be given to methods for adjusting estimators for the bias, taking account of any auxiliary information available.

An example of a selection bias occurs in education. In the USA entrants to higher education are selected on the basis of their SAT (Scholastic Aptitude Test) scores (in England and Wales the A level score is used) and a student enters a given university if their score, z say, is greater than some threshold value z_0. If y is the score on a subsequent test, such as a final-year examination, then it has frequently been observed that the correlation, ρ^*, between y and z in the selected population is low and this is sometimes interpreted as meaning that the test score z is a poor selection variable. If interest centres on the use of z as a selection variable then the parameter of interest is ρ, the correlation between z and y before selection, not ρ^*. It is well known (see e.g. Lord and Novick, 1968) that under certain circumstances ρ^* is an attenuated version of ρ if the selection scheme is the truncation scheme in which only units for which $z > z_0$ are selected. Hence ρ^* will understimate the true correlation ρ, and ignoring this form of selection leads to bias in the estimation of the target parameter ρ.

In Chapter 7 some examples of bias effects due to selection are reviewed. In Chapter 8 a bias correction due initially to Pearson (1903), see Section 6.4, will be studied in detail and applied to analytic studies based on covariance matrices, such as regression analysis and principal component analysis. The Pearson adjustment is based on strong assumptions of linearity and homoscedasticity and so we also examine the robustness of this procedure, and of some alternative procedures, to departures from these assumptions. The alternatives are considered in Section 6.5, and in Chapters 7 and 8.

In Chapter 9 we consider the important special case of selection on the dependent variable. This selection scheme is widely used in retrospective medical studies where the disease outcome is the dependent variable and samples are drawn independently from cases (with the disease) and from controls (without the disease). This extreme form of selection would cause bias in linear regression models, but remarkably for logistic regression models there is no bias in the estimation of the regression coefficients of the independent variables; all the bias is concentrated on the constant term. The logistic regression analysis of case-control studies is the main content of Chapter 9.

6.2 THE BIAS EFFECT OF STRATIFICATION

In this section we show how even a simple selection scheme such as stratification can lead to bias. Strata can be formed prior to selection only if we have some auxiliary information. Following the notation in Chapter 1, Section 6.2 let $\mathbf{z}$ denote a $q \times 1$ vector of design variables and let $\mathbf{z}_U$ denote the $N_o \times q$ matrix of known values of $\mathbf{z}$ for all the population units. Similarly let $\mathbf{y}$ denote the $p \times 1$ vector of survey variables and $\mathbf{y}_U$ denote the $N_o \times p$ matrix of population values of $\mathbf{y}$. We let $\mathbf{Z}_U$ and $\mathbf{Y}_U$ denote the random variables for which $\mathbf{z}_U$ and $\mathbf{y}_U$ are the realized values.

We can distinguish two forms of stratification. The first is into natural population groupings of substantive interest, for example geographical region, type of hospital, male or female. In this case $\mathbf{z}$ would be a vector of indicator variables and $\mathbf{z}_U$ would be the matrix of 0's and 1's corresponding to the actual stratum membership of the population units. The second type of stratification is based on the surveyor's knowledge of a quantitative variable $\mathbf{z}$. Here strata are formed by design, with large values of $\mathbf{z}$ in one stratum and small values in another. A design grouping is not of substantive interest, but is carried out by the surveyor in order to improve the precision of estimation.

Whenever the stratification variable **z** is related to the survey variable **y** there is the possibility of bias if selection is ignored. In extreme cases the stratification variable can be a dependent survey variable of interest such as blood pressure in a retrospective study of the effect of diet on blood pressure or income in an income maintenance experiment. Further examples are given in Chapter 7.

An Example (Surveys of Aircraft Noise)

Surveys have been taken in the vicinity of most major airports to measure the annoyance due to aircraft noise. The surveys around London Heathrow had two objectives. The first was to describe the annoyance around Heathrow; the second was to use the results from Heathrow to predict the likely annoyance around possible sites for a third London airport. Geographical strata were formed using historical records of the loudness of the aircraft and of the number of flights over the area. An initial sample allocated proportional to population density was supplemented by an additional allocation to noise and number strata with few sample points. So the final allocation was not a proportional allocation.

Formally the N_o units in the target population were grouped into H strata of sizes $N_1, \ldots, N_H$, $\sum N_h = N_o$. The survey variables, **y**, were measures of annoyance, current values of noise and number and a host of other environmental, psychological and attitudinal variables. Let $\mathbf{y}_{ht}$ denote the value of **y** for unit t in stratum h. If we consider the stratification to be a natural grouping, which ignores the prior information on noise and number, then we can model the individual values using a simple one-way ANOVA model

$$y_{ht} = \mu_h + e_{ht}, \qquad h = 1, \ldots, H; t = 1, \ldots, N_h. \tag{6.2}$$

with $E_{\text{true}}(e_{ht}) = 0$, $\text{var}_{\text{true}}(e_{ht}) = \sigma_h^2$, $\text{cov}_{\text{true}}(e_{ht}, e_{h't'}) = 0$, for $h, t \neq h', t'$, where 'true' denotes the model (6.2).

The first problem in the analysis of survey data is to define the target parameter. One parameter of wide interest is the weighted mean

$$\mu^* = \sum w_h^* \mu_h. \tag{6.3}$$

If $w_h^* = W_h = N_h/N_o$ then μ^* is the mean of populations like the given finite population. If $w_h^* = 1$ for $h = 1$, and 0 otherwise, then we are estimating the first stratum mean which may be a domain of interest. In general μ^* represents the mean for some target population of interest to the analyst.

For the first descriptive purpose the finite population mean $\bar{Y}_U = \sum W_h \bar{Y}_h$ would be relevant, while the individual stratum means $\bar{Y}_h$ would also be useful for policy decisions. Since the stratum sizes N_h are large the model-based parameter $\bar{\mu} = \sum W_h \mu_h$ differs from $\bar{Y}_U$ only by a term of $O_p(N_o^{-1/2})$. Similarly $\mu_h = \bar{Y}_h + O_p(N_h^{-1/2})$. So $\bar{\mu}$ and μ_h, $h = 1, \ldots, H$, are chosen as the target parameters. Arguably, inferences are not simply required for the exact time at which the survey was taken, so the superpopulation mean $\bar{\mu}$, which represents populations like the one sampled but not identical to it, is more appropriate than $\bar{Y}$. For the second objective a target parameter of interest would be $\mu^* = \sum w_h^* \mu_h$, where the w_h^* are weights relevant to the proposed site, such as the population proportions in certain regions around the site.

Suppose the objective is to estimate $\mu^* = \sum w_h^* \mu_h$ and we mistakenly employ the SRS mean

$$\bar{y}_s = \frac{1}{n_o} \sum_h \sum_{t \in s} y_{ht} = \sum w_h \bar{y}_h,$$

where $w_h = n_h/n_o$. Then $E_{\text{true}}(\bar{y}_s) = \sum w_h \mu_h \neq \sum w_h^* \mu_h$, unless by chance $w_h^* = n_h/n_o$. The bias effect of this misspecification is

$$B(\bar{y}_s, \mu^*) = \sum (w_h - w_h^*) \mu_h,$$

and depends only on the wrongly chosen weights. We should note that this bias would exist even if the sample design was SRS and the target parameter was $\bar{\mu}$. If the population has natural strata which are ignored by the surveyor in the design then although under SRS

$$E_{\text{SRS}}(\bar{y}_s) = \bar{Y} \simeq \bar{\mu}; \qquad E_{\text{true}}(\bar{y}_s) = \sum w_h \mu_h \neq \bar{\mu}$$

under model (6.2). Thus inferences appropriate to descriptive surveys will not necessarily be appropriate for analytic surveys even when the target and descriptive parameters are closely related, such as $\bar{\mu}$ and $\bar{Y}$. This point is also discussed in Chapter 12.

If we know the weights w_h^* then the bias in $\bar{y}_s$ can be removed using the bias-corrected estimator

$$\bar{y}^* = \bar{y}_s - \hat{B}(\bar{y}_s, \mu^*),$$

where $\hat{B}(\bar{y}_s, \mu^*) = \sum (w_h - w_h^*) \bar{y}_h$, estimates the bias. Now clearly

$$\bar{y}^* = \sum w_h^* \bar{y}_h, \tag{6.4}$$

and so the bias-corrected estimator is the natural post-stratified estimator obtained by using the correct group weights, w_h^*.

In a model-based approach post-stratification reflects the population structure and removes a bias which exists when a model which does not use this structure is employed. In randomization inference this bias does not exist even when a SRS gives an obviously unbalanced sample (Holt and Smith, 1979), which we feel is an unsatisfactory feature of this type of inference.

The variance estimated under IID assumptions is also biased; this bias being the analogue of the meff in equation (6.1). If $\bar{y}_s$ is the IID point estimator then $v_0 = \sum\sum (y_{ht} - \bar{y}_s)^2 / [n_o(n_o - 1)]$ is the IID estimator of $\text{var}(\bar{y}_s)$. Under the model (6.2)

$$E_{\text{true}}(v_0) = \sum \sigma_h^2 w_h / n_o + \sum w_h (\mu_h - \bar{\mu}_s)^2 / (n_o - 1),$$

where $\bar{\mu}_s = \sum w_h \mu_h$, and

$$\text{var}_{\text{true}}(\bar{y}_s) = \sum \sigma_h^2 w_h / n_o.$$

Thus

$$\text{meff}(\bar{y}_s, v_0) = \frac{\sum \sigma_h^2 w_h / n_o}{\sum \sigma_h^2 w_h / n_o + \sum w_h (\mu_h - \bar{\mu}_s)^2 / (n_o - 1)} < 1.$$

Hence failing to take into account stratification effects causes bias in both the estimator of the mean and its variance. The meff, however, only measures the effect on the variance, and fails to take into account the bias effect on the mean. It is thus inadequate for measuring the total effects of misspecification in the presence of bias.

We could also form a bias-corrected estimator of the variance. The bias in v_0 as an estimator of $\text{var}_{\text{true}}(\bar{y}_s)$ is $\sum w_h(\mu_h - \bar{\mu}_s)^2/(n_o - 1)$, and so a first-order bias-corrected estimator is

$$\begin{aligned} v_0 - \sum w_h (\bar{y}_h - \bar{y}_s)^2 / (n_o - 1) &= \sum \frac{(n_h - 1)}{n_o (n_o - 1)} v_h \\ &\simeq \sum w_h v_h / n_o, \end{aligned}$$

if $(n_h - 1)/(n_o - 1) \simeq w_h$, where $v_h = \sum_t (y_{ht} - \bar{y}_h)^2 / (n_h - 1)$. Bias-corrected covariance matrices are considered further in Section 6.4 and in Chapter 8.

6.3 THE BIAS EFFECT OF AN AUXILIARY DESIGN VARIABLE

In the Heathrow surveys the design variables were in fact quantitative measures of loudness and number of aircraft at sites comprising clusters of houses. Thus in employing an ANOVA model for the survey variables we were not using the full information in the known design values $\mathbf{z}_U$. An alternative approach would be to fit some form of regression model relating the outcome variable, annoyance and the survey values of loudness and number, to the design variable $\mathbf{z}_U$. The non-proportional allocation for the stratification design was an attempt to select the sample points so that the parameters of the regression of annoyance on the actual values of noise and number (the noise and number index) could be estimated efficiently. Stratification, ignoring the actual values of $\mathbf{z}_U$, could be viewed as a robust alternative to the regression model.

We now consider the bias effect under a simple regression model. Suppose that we know the values $\mathbf{z}'_U = (z_1, \ldots, z_{N_o})$ of a single design variable, z, but choose to ignore this in the design and employ SRS. Consider a model in which the pairs of values (y_t, z_t), $t = 1, \ldots, N_o$, are IID outcomes of the random vector (Y, Z). We assume that the conditional distribution of the survey variable Y given the design variable $Z = z$ can be represented by the simple linear regression model

$$y_t = \alpha + \beta z_t + e_t, \qquad t = 1, \ldots, N_o, \tag{6.5}$$

where $E_{\text{true}}(e_t|Z = z_t) = 0$, $\text{var}_{\text{true}}(e_t|Z = z_t) = \sigma_e^2$.

If the joint distribution of (Y, Z) has mean vector $\boldsymbol{\mu}' = (\mu_y, \mu_z)$ then from equation (6.5) we have

$$E_{\text{true}}(Y) = \mu_y = \alpha + \beta\mu_z, \tag{6.6}$$

where $\mu_z = E_{\text{true}}(Z)$. If μ_y is the target parameter of interest and it is estimated by the unweighted mean $\bar{y}_s$, then under the model (6.5) the conditional expectation of $\bar{y}_s$ is

$$\begin{aligned} E_{\text{true}}(\bar{y}_s|\mathbf{z}_U, s) &= \alpha + \beta\bar{z}_s, \\ &\neq \mu_y, \end{aligned}$$

where $\bar{z}_s = \sum_s z_t/n_o$. So despite the fact that the sampling scheme is SRS the estimator $\bar{y}_s$ is conditionally biased under the model (6.5). The bias is

$$\begin{aligned} B(\bar{y}_s, \mu_y) &= E_{\text{true}}(\bar{y}_s|\mathbf{z}_U, s) - \mu_y \\ &= \beta(\bar{z}_s - \mu_z). \end{aligned} \tag{6.7}$$

Now μ_z will not be known, but $\bar{z}_U = \sum_U z_t/N_o$ is assumed known and will be a very good estimator of μ_z with error of $O_p(N_o^{-1/2})$. Thus a bias-corrected estimator of μ_y is

$$\begin{aligned} \hat{\mu}_y &= \bar{y}_s - \hat{\beta}(\bar{z}_s - \hat{\mu}_z) \\ &= \bar{y}_s + b(\bar{z}_U - \bar{z}_s), \end{aligned} \tag{6.8}$$

where b is the OLS regression estimator of β. We recognize equation (6.8) as the well-known regression estimator of the finite population mean $\bar{y}_U$ (Cochran 1977, Chapter 7), but note that it has been derived from a regression model conditional on the sample units, s, rather than under a randomization model averaging over all possible samples.

The meff of the SRS estimator $\bar{y}_s$ and its SRS variance estimator

$$v_0 = \sum (y_t - \bar{y}_s)^2/[n_o(n_o - 1)]$$

is

$$\begin{aligned} \text{meff}(\bar{y}_s, v_0) &= \text{var}_{\text{true}}(\bar{y}_s)/E_{\text{true}}(v_0) \\ &= \frac{\sigma_e^2/n_o}{(\sigma_e^2 + \beta^2 v_z)/n_o} \\ &= \frac{\sigma_e^2}{\sigma_e^2 + \beta^2 v_z}, \end{aligned} \tag{6.9}$$

where $v_z = \sum(z_t - \bar{z}_s)^2/(n_o - 1)$. As before, we note that the meff does not include a term in $(\bar{z}_s - \mu_z)^2$ and thus does not take into account the bias.

If for the joint distribution of (Y, Z) the covariance matrix is

$$\Sigma = \begin{bmatrix} \sigma_y^2 & \sigma_{yz} \\ & \sigma_z^2 \end{bmatrix},$$

then the unconditional variance from equation (6.5) is

$$\text{var}_{\text{true}}(Y) = \sigma_e^2 + \beta^2\sigma_z^2.$$

We can construct a bias-corrected estimator of $\text{var}_{\text{true}}(Y)$ using

$$\hat{\sigma}_e^2 = v_{yys} - b^2 v_{zzs},$$

where

$$v_{yys} = \sum_s (y_t - \bar{y}_s)^2/(n_o - 1),$$

similarly for v_{zzs}, and estimating β by b and σ_z^2 by

$$v_{zzU} = \sum_U (z_t - \bar{z}_U)^2/(N_o - 1),$$

we find

$$\widehat{\text{var}}_{\text{true}}(Y) = v_{yys} + b^2(v_{zzU} - v_{zzs}), \tag{6.10}$$

which is a regression-like estimator of σ_y^2 which corrects for the bias effects of selection even under SRS. Only if $v_{zzs} = v_{zzU}$ is there no bias due to selection.

We have demonstrated that for simple models based on the known values of design variables the effect of ignoring selection on $\mathbf{z}_U$ on model-based inferences is to introduce biases in addition to the effects on variances captured by meffs. In simple cases we have found bias-corrected estimators. We now generalize these results.

6.4. THE PEARSON ADJUSTMENT FOR MULTIVARIATE NORMAL MODELS

Suppose again that the values $(\mathbf{y}_t', \mathbf{z}_t')$, $t = 1, \ldots, N_0$, are IID outcomes of the random vector $(\mathbf{Y}', \mathbf{Z}')$. Consider the important case where $\mathbf{Z}$ is a design variable related to all the survey variables, $\mathbf{Y}$, but where the target parameter is defined in the marginal distribution of $\mathbf{Y}$, such as the mean $\boldsymbol{\mu}_y$ above or the correlation coefficient between two of the $\mathbf{Y}$ variables. This problem was first considered by Pearson (1903), who examined the effect of natural selection on biological populations. The current population values were viewed as the outcome of previous processes of natural selection, such as survival of the fittest. Pearson asked the question: What is the relationship between parameters, such as correlation coefficients, in the current population and those in the original population prior to selection?

Let the probability density function (p.d.f.) of $\mathbf{Y}$ and $\mathbf{Z}$ be

$$f(\mathbf{y}, \mathbf{z}; \boldsymbol{\phi}) = g(\mathbf{y} | \mathbf{z}; \boldsymbol{\theta}) h(\mathbf{z}; \boldsymbol{\eta}). \tag{6.11}$$

Consider representing the effect of selection on $\mathbf{Z}$ by changing $h(\mathbf{z}; \boldsymbol{\eta})$ to some other density $h^*(\mathbf{z}; \boldsymbol{\eta})$. For example, if the selection involves observing only those units for which $z > z_0$, where z is scalar, which is a truncation scheme, then

$$\begin{aligned} h^*(z; \boldsymbol{\eta}) &= h(z; \boldsymbol{\eta}) / \{1 - H(z_0)\}, & z > z_0, \\ &= 0, & z \leqslant z_0, \end{aligned} \tag{6.12}$$

where $H(z_0) = \int_{-\infty}^{z_0} h(z; \boldsymbol{\eta})\, dz$. Assuming that the conditional distribution of $\mathbf{Y}$ given $\mathbf{Z} = \mathbf{z}$ is unaffected by selection on $\mathbf{Z}$ the joint p.d.f. after selection is

$$f^*(\mathbf{y}, \mathbf{z}; \boldsymbol{\phi}) = g(\mathbf{y} | \mathbf{z}; \boldsymbol{\theta}) h^*(\mathbf{z}; \boldsymbol{\eta}), \tag{6.13}$$

see Birnbaum *et al.* (1950).

Pearson assumes that the joint distribution of $\mathbf{Y}$ and $\mathbf{Z}$ is multivariate normal with parameters

$$\boldsymbol{\mu} = \begin{bmatrix} \boldsymbol{\mu}_y \\ \boldsymbol{\mu}_z \end{bmatrix} \quad \text{and} \quad \boldsymbol{\Sigma} = \begin{bmatrix} \boldsymbol{\Sigma}_{yy} & \boldsymbol{\Sigma}_{yz} \\ & \boldsymbol{\Sigma}_{zz} \end{bmatrix}. \tag{6.14}$$

If the mean and covariance matrix of $(\mathbf{Y}', \mathbf{Z}')'$ after selection are

$$\boldsymbol{\mu}^* = \begin{bmatrix} \boldsymbol{\mu}_y^* \\ \boldsymbol{\mu}_z^* \end{bmatrix} \quad \text{and} \quad \boldsymbol{\Sigma}^* = \begin{bmatrix} \boldsymbol{\Sigma}_{yy}^* & \boldsymbol{\Sigma}_{yz}^* \\ & \boldsymbol{\Sigma}_{zz}^* \end{bmatrix}, \tag{6.15}$$

then he showed that the parameters before and after selection are related by

$$\boldsymbol{\mu} = \begin{bmatrix} \boldsymbol{\mu}_y \\ \boldsymbol{\mu}_z \end{bmatrix} = \begin{bmatrix} \boldsymbol{\mu}_y^* + \boldsymbol{\Sigma}_{yz}^* \boldsymbol{\Sigma}_{zz}^{*-1} (\boldsymbol{\mu}_z - \boldsymbol{\mu}_z^*) \\ \boldsymbol{\mu}_z \end{bmatrix} \tag{6.16}$$

and

$$\boldsymbol{\Sigma} = \begin{bmatrix} \boldsymbol{\Sigma}_{yy} & \boldsymbol{\Sigma}_{yz} \\ & \boldsymbol{\Sigma}_{zz} \end{bmatrix} = \begin{bmatrix} \boldsymbol{\Sigma}_{yy}^* + \boldsymbol{\Sigma}_{yz}^* \boldsymbol{\Sigma}_{zz}^{*-1} (\boldsymbol{\Sigma}_{zz} - \boldsymbol{\Sigma}_{zz}^*) \boldsymbol{\Sigma}_{zz}^{*-1} \boldsymbol{\Sigma}_{zy}^* & \boldsymbol{\Sigma}_{yz}^* \boldsymbol{\Sigma}_{zz}^{*-1} \boldsymbol{\Sigma}_{zz} \\ & \boldsymbol{\Sigma}_{zz} \end{bmatrix}. \tag{6.17}$$

Now $\boldsymbol{\Sigma}_{yz}^* \boldsymbol{\Sigma}_{zz}^{*-1} = \boldsymbol{\beta}_{yz}$, the regression coefficient matrix of the regression of $\mathbf{Y}$ on $\mathbf{Z}$, and so we can correct for selection bias using

$$\boldsymbol{\mu}_y = \boldsymbol{\mu}_y^* + \boldsymbol{\beta}_{yz} (\boldsymbol{\mu}_z - \boldsymbol{\mu}_z^*), \tag{6.18}$$

$$\boldsymbol{\Sigma}_{yy} = \boldsymbol{\Sigma}_{yy}^* + \boldsymbol{\beta}_{yz} (\boldsymbol{\Sigma}_{zz} - \boldsymbol{\Sigma}_{zz}^*) \boldsymbol{\beta}_{yz}'. \tag{6.19}$$

We call this the *Pearson adjustment*.

Lawley (1943) showed that the same result holds under the slightly weaker assumptions of linearity and homoscedasticity, namely when

$$E(\mathbf{Y} | \mathbf{Z} = \mathbf{z}) = \mathbf{C} + \mathbf{D}\mathbf{z} \quad \text{and} \quad \operatorname{Var}(\mathbf{Y} | \mathbf{Z} = \mathbf{z}) = \mathbf{K}. \tag{6.20}$$

By viewing the selected sample as a random sample from $f^*(\mathbf{y}, \mathbf{z}; \boldsymbol{\phi})$, the population after selection, the moments $\boldsymbol{\mu}_y^*$, $\boldsymbol{\Sigma}_{yy}^*$ in the selected population may be estimated by the simple unweighted sample moments $\bar{\mathbf{y}}_s$, $\mathbf{V}_{yys}$, etc. In addition, finite population estimates $\bar{\mathbf{z}}_U$ of $\boldsymbol{\mu}_z$ and $\mathbf{V}_{zzU}$ of $\boldsymbol{\Sigma}_{zz}$ are available when $\mathbf{z}$ is a design variable. Thus we can estimate all the terms on the right-hand sides of equations (6.16) and (6.17) or of equations (6.18) and (6.19), and we obtain the *Pearson adjusted estimators* of $\boldsymbol{\mu}_y, \boldsymbol{\Sigma}_{yy}$,

$$\hat{\boldsymbol{\mu}}_y = \bar{\mathbf{y}}_s + \mathbf{b}_{yz}(\bar{\mathbf{z}}_U - \bar{\mathbf{z}}_s), \tag{6.21}$$

and

$$\hat{\boldsymbol{\Sigma}}_{yy} = \mathbf{V}_{yys} + \mathbf{b}_{yz}(\mathbf{V}_{zzU} - \mathbf{V}_{zzs})\mathbf{b}_{yz}'. \tag{6.22}$$

These estimators turn out to be identical to the maximum likelihood estimators of $\boldsymbol{\mu}_y$ and $\boldsymbol{\Sigma}_{yy}$ but with $(n_o - 1)$ as the divisor of $\mathbf{V}_{yys}$ and $\mathbf{V}_{zzs}$ rather than n_0, see Anderson (1957), Smith (1981) and Skinner (1984).

For univariate $\mathbf{Y}$ and $\mathbf{Z}$ we recognize the expressions in equations (6.21) and (6.22) as the regression estimator of a mean and the regression-type estimator of a variance derived in equations (6.8) and (6.10). Thus the Pearson-adjusted estimators are essentially correcting for the bias due to selection on $\mathbf{Z}$ under assumptions of linearity and homoscedasticity.

If we return to the student selection problem referred to in Section 6.1 it can be shown under Lawley's assumption (see Lord and Novick, 1968, p. 143) that the correlations ρ^*, after selection, and ρ, before selection, are related by

$$\rho^2 = \left[1 + \frac{\sigma_z^{*2}}{\sigma_z^2}\left(\frac{1}{\rho^{*2}} - 1\right)\right]^{-1}.$$

For truncation selection schemes $\sigma_z^{*2} < \sigma_z^2$, and assuming that $\sigma_z^{*2} = 0.2\sigma_z^2$, $\rho^{*2} = 0.1$ (a poor predictor) we find that $\rho^2 = 0.36$. Thus selection on z explains 36% of the variation in performance rather than the perceived 10%. Perhaps SAT scores and A-level scores are not such bad selection variables after all!

The properties of the Pearson adjusted estimators are examined in Chapter 8 both theoretically and empirically for various survey applications. The assumptions of linearity and homoscedasticity turn out to be vital to the validity of the estimators and so some alternatives are also considered.

6.5 ALTERNATIVES TO THE PEARSON ADJUSTMENTS

The Pearson adjustments, equations (6.18) and (6.19), and the corresponding estimators, equations (6.21) and (6.22), are based on strong assumptions of linearity and homoscedasticity. Can such models capture adequately the complex population structure and the relationships between the survey variables, $\mathbf{y}$, and the design variables, $\mathbf{z}$? Both theoretical and empirical evidence, some of which is presented in Chapter 8, suggest that the Pearson adjusted estimators may not be robust. Thus we should look for more robust alternatives.

Randomization inferences do not depend on models and so in some sense are robust to model specification. Like model-based procedures random sampling designs, such as stratified multi-stage sampling, make explicit use of known population structure. Thus they offer a seemingly robust alternative to model-based procedures. We say 'seemingly robust' because the concept of robustness is not well defined in randomization inference. For example if a population has a natural stratification but this is ignored by the surveyor and a SRS is chosen, then the only apparent effect of using the SRS design and the corresponding SRS estimator $\bar{\mathbf{y}}_s$ is a loss of efficiency relative to a stratified design. The bias due to misspecification under the model-based approach which was identified in Section 6.2, has been subsumed into an increase in variance under the randomization approach. As long as the estimator is unbiased relative to the chosen design, randomization inference is independent of the population structure. Even the most inefficient sample design can claim to be robust! We find this concept of robustness unattractive.

Another problem with randomization inferences is that strictly speaking they are only applicable for descriptive problems, for the estimation of finite population parameters such as $\sum y_t/N_o$ or $\sum y_t^2/N_o$. In Section 1.6.1 we discussed this issue and pointed out that if the finite population values $y_1, \ldots, y_{N_o}$ could be viewed as the outcomes of random variables, $Y_1, \ldots, Y_{N_o}$, with distribution ξ, then the finite population parameter θ_U, based on the N_o values, differs from the model parameter θ by a term of $O_p(N_o^{-1/2})$. Thus for large N_o there will be little difference between estimating the analytic parameter θ and the equivalent descriptive parameter θ_U.

If θ_U can be expressed as a function of the finite population means of suitably defined y variables (see equation (2.34)), that is

$$\theta_U = g(\bar{\mathbf{y}}_U), \tag{6.23}$$

then, provided $g(\ )$ is continuous, a design-consistent estimator of θ_U (Section 1.6.4) is given by

$$\hat{\theta}_U = g(\bar{\mathbf{y}}_{pw}) \tag{6.24}$$

where $\bar{\mathbf{y}}_{pw}$ is a design consistent estimator of $\bar{\mathbf{y}}_U$. Now for a general design $p(S)$ with inclusion probabilities π_t, $t = 1, \ldots, N_o$, the *probability weighted estimator*

$$\bar{\mathbf{y}}_{pw} = N_o^{-1} \sum_{t \in s} \mathbf{y}_t/\pi_t, \tag{6.25}$$

is design-unbiased and, under regularity conditions design-consistent for $\bar{\mathbf{y}}_U$. Thus we have

$$g(\bar{\mathbf{y}}_{pw}) = g(\bar{\mathbf{y}}_U) + O_p(n_o^{-1/2}) = \theta + O_p(n_o^{-1/2}), \tag{6.26}$$

since $N_o^{-1/2} < n_o^{-1/2}$, so that $\hat{\theta}_U$ is a design-consistent estimator of the target parameter θ. This is one alternative to the Pearson adjusted estimator, but note that the inferential framework is quite different.

The Pearson adjusted parameters (6.18) and (6.19) could themselves be used to define descriptive target parameters $\boldsymbol{\theta}_U$, which could be expressed as functions of means, $\bar{\mathbf{y}}_U$. Estimating $\bar{\mathbf{y}}_U$ by $\bar{\mathbf{y}}_{pw}$ leads to a p-weighted version of the Pearson adjustments. Nathan and Holt (1980) suggest an estimator of this form which provides yet another alternative. The properties of all these estimators are compared in Chapters 7 and 8.

The main problem with comparing these alternative estimators is that they have been derived from different statistical frameworks. The Pearson adjustments are model-based while the p-weighted estimators are randomization-based. Within which framework should the estimators be compared? We have seen that it is sensible to talk about the robustness of an estimator to model misspecification within a model-based framework but not within a randomization framework. The post-stratified estimator (6.4) removes the model bias but has no such justification within randomization inference. The reason for this difference is that model-based inferences condition on the sample, $S = s$, which has actually been drawn, while randomization inferences average over S using the design $p(S)$. Within a model-based framework the sample units, s, can be shown to be an ancillary statistic (Cox and Hinkley, 1974), and so it is natural within this framework to make inferences conditional on $S = s$. A bias within this conditional framework becomes a variance when it is averaged over all S under the randomization framework. The two frameworks for inference are quite different and are incompatible with one another. You can use one or the other, but not both. We favour the model-based framework for comparing alternative estimators.

To achieve design-unbiasedness, randomization estimators use the inverse probability weights, π_t^{-1}, as in equation (6.25), whereas these have no obvious role in model-based inference. This causes yet another conflict between the two approaches to inference which can be summarized in the question: 'To weight or not to weight?' There is no simple answer but Du Mouchel and Duncan (1983), and Fuller (1984) propose that the starting-point should be to calculate both the model-based estimator $\hat{\theta}_{\mathrm{m}}$, say, and the randomization estimator $\hat{\theta}_p$, say, and to test whether the corresponding parameters differ significantly from one another. If there is no significant difference then either approach may be used, and basically the weighting does not matter. If they do differ then the position is open. Du Mouchel and Duncan suggest using a model-based approach incorporating the weight π_t say, or the design variable $\mathbf{z}_t$, into the model. They show in an example that this leads to reasonable inferences which also make substantive sense. Our problem with this solution is that the target parameter, $\boldsymbol{\theta}$, is implicitly changed by the change in model. This approach would be useful in any situation where the model and target parameters are not tightly specified and the analysis is exploratory.

Rubin (1985) suggests replacing $\mathbf{z}$ in the model (6.11) by the probability weight π_t. He shows that under certain conditions π_t is an adequate summary of the information in $\mathbf{z}$, and he suggests that it may be easier to construct the conditional distribution of $\mathbf{y}$ given π_t, than of $\mathbf{y}$ given $\mathbf{z}$. Smith (1988) shows that if the probability weights are measures of size then Rubin's approach can be interpreted as size-biased sampling (Cox, 1969; Patil and Rao, 1978) and further that if the form of the model distribution in equation (6.11) is not known exactly methods of moments estimators lead to p-weighted estimators similar to those used in randomization inference. Although the estimators are the same the framework for inference is still model-based and conditions on $S = s$. We show in Chapter 8 that the probability weighted estimators often have poor properties in the conditional distribution given s.

6.6 IGNORABLE AND UNINFORMATIVE SAMPLING SCHEMES

Since model-based analytic inferences condition on the sample units, $S = s$, actually drawn, they do not depend on the randomization distribution $p(S)$ which averages over repeated random samples. Such inferences apparently ignore the sample design and in a model-based framework, provided certain conditions are satisfied, see Sugden and Smith (1984), the sample design $p(S)$ is called *ignorable*. A condition for a design to be ignorable is that it be *uninformative*. A design is uninformative if it is of known form and depends solely on the known values, $\mathbf{z}_U$, of the design variables.

The choice of the emotive words, ignorable and uninformative, is unfortunate. The assertion is not that the design carries no useful information, it is that if both the surveyor and the analyst have access to all the information about stratification factors, clusters, measures of size and so on, then the actual *selection probabilities* $p(S)$ carry no extra information. Both models and sampling schemes should reflect the underlying population structure.

Random sampling schemes can be written $p(s|\mathbf{z}_U)$ to show their dependence on $\mathbf{z}_U$ and if $\mathbf{z}_U$ is known they are ignorable for model-based inference. This statement is anathema to those who believe in randomization inference since for them $p(s|\mathbf{z}_U)$ contains all the useful information. A few words of explanation may help. We are assuming that the values $\mathbf{z}_U$ for all the units in the population are known prior to sampling. We are also assuming that the form of the scheme $p(s|\mathbf{z}_U)$ is known. Thus given $\mathbf{z}_U$ we can construct the values of the sampling probabilities, $p(s|\mathbf{z}_U)$ and hence of the unit inclusion probabilities π_t, $\pi_{tt'}$, etc., from our knowledge of $\mathbf{z}_U$. The converse is not necessarily true. Given the numerical values of the probabilities $p(s|\mathbf{z}_U)$ we may not be able to determine the values of $\mathbf{z}_U$. For example, if $\mathbf{z}$ is a quantitative variable and is used to construct two strata; stratum 1 containing all units with $z \leqslant z_0$, stratum 2 containing units with $z > z_0$, and if we select n_1 units by SRS from stratum 1 and n_2 by SRS from stratum 2, then knowing only the values of n_1, n_2 and the stratum sizes N_1 and N_2 we cannot reconstruct $\mathbf{z}_U$, or even the value of z_0. Thus there is less information in the values of the probabilities $p(s|\mathbf{z}_U)$ than in the known values of $\mathbf{z}_U$ and knowledge of the form of $p(s|\mathbf{z}_U)$. It is in this sense that the sampling scheme is uninformative. Note, however, that if $\mathbf{z}_U$ is unknown, then knowledge of the values $p(s|\mathbf{z}_U)$ and how they were formed may allow us to make useful inferences about the unknown values of $\mathbf{z}_U$. So ignorability depends crucially on knowing the values of the design variable (Sugden and Smith, 1984).

The fact that random sampling schemes can be ignored for model-based inference when the values of $\mathbf{z}_U$ are known, does not mean that randomization should not be employed. Rather it provides a strong case for randomization since it is a scheme which guarantees ignorability and is scientifically and publicly acceptable (Little, 1982a; Smith, 1983). Non-random schemes, such as balanced sampling (Royall and Herson, 1973), also satisfy the condition for ignorability and also by being balanced on z have a claim to give 'representative' samples. However, these schemes have yet to gain wide public acceptance. We conclude that the ignorability argument lends support to randomization but is not conclusive. The issue for analysis is not the use of random sampling it is the relevance of the randomization distribution, of averaging over s rather than conditioning on s, for analytic inference.

In general a scheme for selecting a sample could depend not only on a known design variable, $\mathbf{z}$, but also on the survey variables $\mathbf{y}$. Thus in quota sampling the interviewers have some choice about whom to interview and may base their decision, either consciously or unconsciously, on perceived values of $\mathbf{y}$ (Smith, 1983). Non-response is a selection mechanism in which the respondent chooses whether to reply or not, and this could clearly be related to the values of important survey variables such as income or social class (Little, 1982a). In retrospective case-control studies selection is based on the dependent variable, y, as well as on known design variables $\mathbf{z}$. Designs which depend on $\mathbf{y}$ are informative and cannot necessarily be ignored for model-based inference, as we shall see in Chapters 7 and 9.

CHAPTER 7

The Effect of Selection on Regression Analysis

G. Nathan and T. M. F. Smith

7.1 INTRODUCTION

The standard estimation procedure of ordinary least squares (OLS) implicitly assumes IID errors. So OLS does not incorporate any probability weighting or population structure, such as clustering, stratification or measures of size, into the analysis. If we assume, as in Part A, that selection and population structure affect only the covariances of the error terms then the effect of misspecifying the regression model will be reflected only in the covariance matrix of the estimators. If the true regression model for the sample data $(\mathbf{y}_s, \mathbf{X}_s)$ is

$$\mathbf{y}_s = \mathbf{X}_s\boldsymbol{\beta} + \mathbf{u}_s, \tag{7.1}$$

where $E_{\text{true}}(\mathbf{u}_s|\mathbf{X}_s) = \mathbf{0}$, $\text{Var}_{\text{true}}(\mathbf{u}_s|\mathbf{X}_s) = \sigma^2\mathbf{V}_s$, then the OLS estimator of $\boldsymbol{\beta}$ is

$$\mathbf{b} = (\mathbf{X}_s'\mathbf{X}_s)^{-1}\mathbf{X}_s'\mathbf{y}_s, \tag{7.2}$$

and is unbiased, with

$$\text{Var}_{\text{true}}(\mathbf{b}|\mathbf{X}_s) = (\mathbf{X}_s'\mathbf{X}_s)^{-1}\mathbf{X}_s'\mathbf{V}_s\mathbf{X}_s(\mathbf{X}_s'\mathbf{X}_s)^{-1}\sigma^2. \tag{7.3}$$

Statistical packages give estimates of the OLS variance,

$$\text{Var}_{\text{OLS}}(\mathbf{b}|\mathbf{X}_s) = (\mathbf{X}_s'\mathbf{X}_s)^{-1}\sigma^2, \tag{7.4}$$

not of equation (7.3), and this leads to a misspecification error. Thus use of the OLS option under model (7.1) leads to an unbiased point estimator of $\boldsymbol{\beta}$ but to incorrect inferences about the precision of $\mathbf{b}$. See also Section 3.3.4.

We have seen in Chapter 6 that failing to take account of selection effects and their relationship to the population structure can lead to biases as well as to misspecified variances. We now examine this further.

7.2 SOME EXAMPLES OF SELECTION SCHEMES

To illustrate how a selection scheme may bias regression analysis consider the bivariate finite population with distribution shown in Figure 7.1. The corresponding population distributions under a variety of selection schemes are shown in Figures 7.2–7.6. The lines shown are straight lines fitted to the

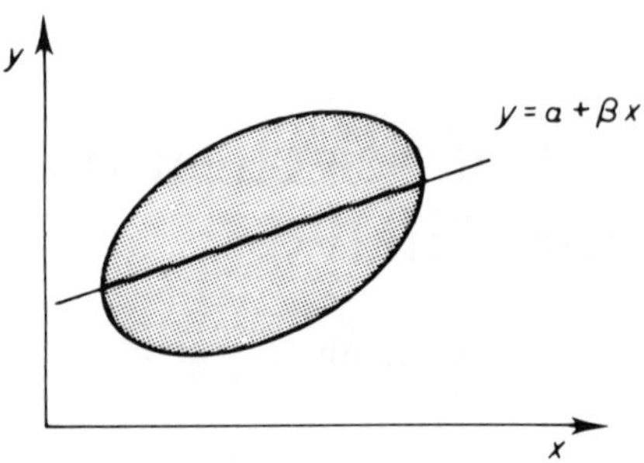

Figure 7.1 Bivariate population distribution before selection.

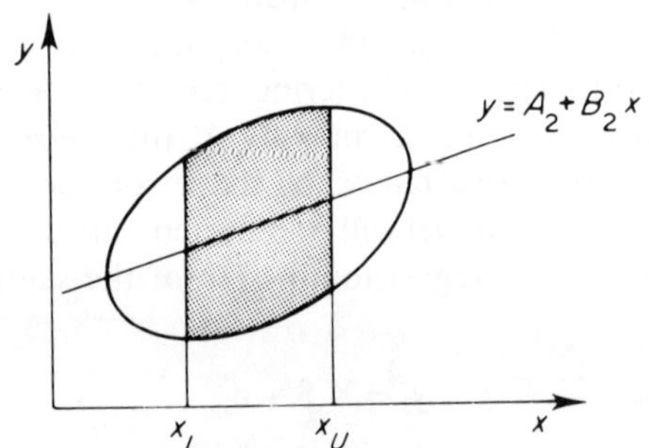

Figure 7.2 Selection on x:$x_L < x < x_U$,

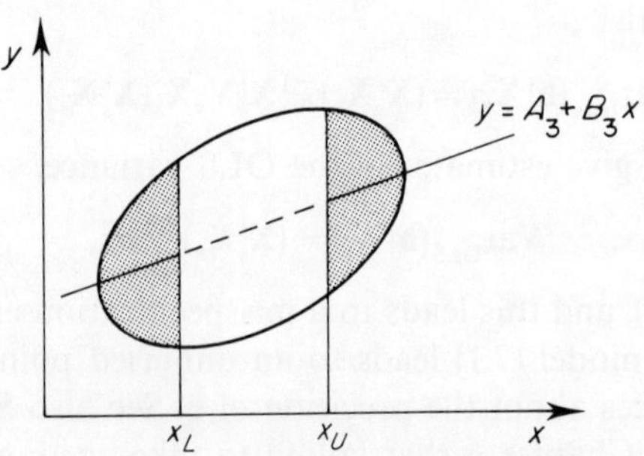

Figure 7.3 Selection on x:$x < x_L$, $x > x_U$

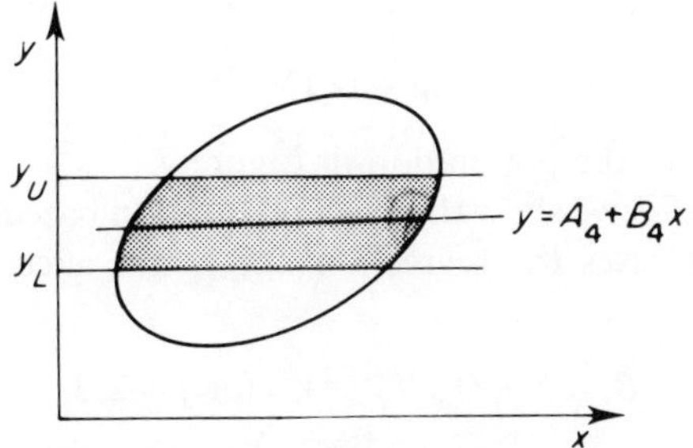

Figure 7.4 Selection on y:$y_L < y < y_U$.

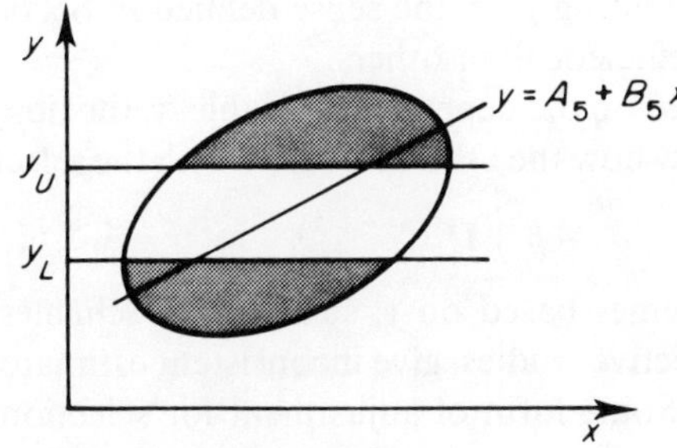

Figure 7.5 Selection on y:$y < y_L$, $y > y_U$.

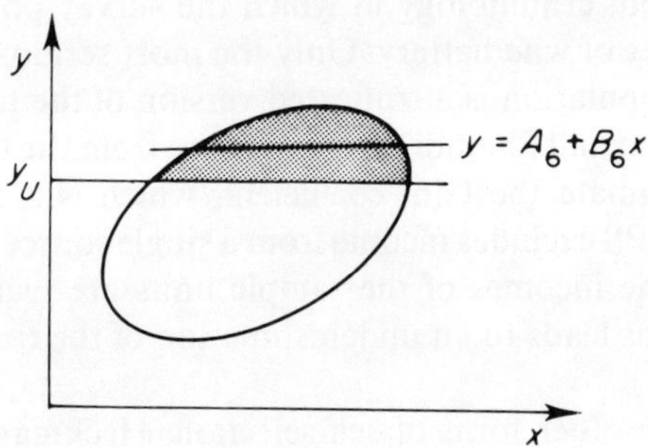

Figure 7.6 Selection on y:$y > y_U$.

distributions; they are not necessarily the regression functions $E(y|x$ and selection). These finite population OLS regression lines correspond to the minimum mean squared error (MSE) straight lines with slope parameter, B_U, as defined in Section 3.4.2. The target parameter is β, the infinite population analogue of B_U defined for the complete population in Figure 7.1. We assume that

$$B_U = \beta + O_p(N_o^{-1/2}), \tag{7.5}$$

where N_o is the size of the population in Figure 7.1.

In Figures 7.2 and 7.3 we show the effect of selection based on the independent variable, x. In both cases the regression line is not affected by the selection scheme and thus

$$B_j = \beta + O_p(N_j^{-1/2}), \quad \text{for } j = 2, 3,$$

where N_j is the size of the population after selection. We note that a selection scheme like that in Figure 7.3 would be much more efficient for estimating β than one like that in Figure 7.2. This demonstrates that selection based on x is ignorable for estimating β, in the sense defined in Section 6.6, but that some schemes are more efficient than others.

If selection is based on the dependent variable, y, the position is quite different. Figures 7.4–7.6 show how the OLS line varies with the selection scheme and that

$$B_j \neq \beta + O_p(N_j^{-1/2}), \quad \text{for } j = 4, 5, 6.$$

Thus sampling schemes based on y, such as the schemes used in educational testing and retrospective studies, give inconsistent estimators of β if the selection scheme is ignored. Some form of adjustment for selection is essential for valid inference.

An insidious form of selection bias is that of self-selection. Survey populations based on administrative records are often affected in this way and Berk (1983) gives an example from criminology in which the survey population is a set of arrest reports of cases of wife battery. Only the most serious cases are reported and so the survey population is a truncated version of the target population of all cases. Fichtenbaum and Shahidi (1988) use data from the Current Population Survey (CPS) to estimate the Gini coefficient, which is a measure of income inequality. But the CPS excludes income from a single source exceeding a certain amount, and thus the incomes of the sample units are truncated. They show that this selection bias leads to an underestimation of the true value of the Gini coefficient.

Non-response is another form of self-selection. Heckman (1979) and Little (1982a) model non-response by the value of an unobserved variable, w say, such that if $w > w_0$ a response is made and if $w \leqslant w_0$ there is no response. This truncation scheme is represented by a second equation related to a target regression equation such as (7.1). After selection on the unobserved variable w

$$E(\mathbf{y}_s|\mathbf{X}_s \text{ and selection}) = \mathbf{X}_s\boldsymbol{\beta} + E(\mathbf{u}_s|\mathbf{X}_s \text{ and selection}), \tag{7.6}$$

and $E(\mathbf{u}_s|\mathbf{X}_s \text{ and selection}) \neq \mathbf{0}$ due to the selection scheme. A truncation selection scheme is sometimes represented by a probit or a logit model, but Little (1982a) shows that these models are not robust.

Much recent work on selection effects has been in the area of the evaluation of economic and social experiments in the USA. These studies are quasi-experiments since the allocation of treatments to experimental units (people or families) has not been strictly controlled and randomized. In retrospective studies the selection of units for measurement may be based on the dependent variable leading to the potential biases shown in Figures 7.4–7.6, if the selection effect is ignored. In prospective studies selection is not based on the dependent variable, y, but on a known design variable, z, which may be highly correlated with either $\mathbf{x}$ or y or both, see Section 6.2. In the Framingham heart study (DeMets and Halperin, 1977), the effect of dietary cholesterol on serum cholesterol was of interest. This was studied using a truncated sample based on those with high and low levels of initial serum cholesterol. Kotchen *et al.* (1980) describe a similar study in which only those with high and low initial blood pressures were selected. In the New Jersey negative income tax experiment (Hausman and Wise, 1977), no families with initial incomes greater than one and one-half times the poverty level were selected.

There is no need for selection on a design variable, z, to be a truncation scheme. In the Gary income maintenance experiment (Hausman and Wise, 1981), selection is based on a non-proportional stratified sample, where the strata are initial income groups. We showed in Section 6.2, that if the estimation weights, w_h, differ from the weights, w_h^*, which define the target parameter, then this leads to an estimation bias. McKennell (1969) describes the aircraft noise survey around Heathrow, which was used as an illustrative example in Chapter 6. A non-proportional stratified sample based on prior measurements of aircraft numbers and loudness was used for the design. In a study of the profitability of wine-grape growing in Israel, vineyards were selected with probabilities proportional to a measure relating to the previous year's production (IFIR and CBS, 1979).

These examples illustrate some of the ways in which selection schemes can be based on the dependent variable, y, or on an auxiliary design variable, z. These schemes usually lead to bias if they are ignored and OLS methods are used. We consider the bias of using OLS for regression in Section 7.3 and some alternatives to OLS in Section 7.4.

7.3 THE BIAS OF OLS

Selection on the dependent variable, y, or on a correlated design variable, z, leads to bias in the estimation of β if the OLS estimator, b (equation (7.2)) is used. DeMets and Halperin (1977) show that if X, Y, Z are scalar random variables with a trivariate normal distribution then the conditional expectation of $b = v_{xys}/v_{xxs}$, given $\mathbf{Z}_U = \mathbf{z}_U$ and the sample s, is

$$E_{\text{true}}(b \mid \mathbf{z}_U, s) = \frac{\beta_{yx} + \beta_{yz}\beta_{zx}(v_{zzs}/\sigma_z^2 - 1)}{1 + \rho_{xz}^2(v_{zzs}/\sigma_z^2 - 1)}, \tag{7.7}$$

where $\beta_{yx} = \beta$, and β_{yz}, β_{zx}, σ_z^2, ρ_{xz} are parameters in the trivariate normal distribution defined in the usual way. This result is closely related to the Pearson adjustment, Sections 6.3 and 6.4. Nathan and Holt (1980) show that equation (7.7) holds to $O(n_o^{-1})$ under less restrictive conditions than those of multivariate normality, or those of Lawley (1943), equation (6.20).

From equation (7.7) it can be seen that the OLS estimator is conditionally unbiased (and consistent) for β if the sample variance of z, v_{zzs}, equals the population variance, σ_z^2. A sample with this property is *balanced* on the second moment. An alternative condition for unbiasedness is that $\rho_{xz} = 0$.

Nathan and Holt (1980) show that under their model assumptions the unconditional expectation of b to $O(n_o^{-1})$ is

$$E(b|\mathbf{z}_U) = \beta + \frac{\sigma_y}{\sigma_x}\left\{\frac{\rho_{yz\cdot x}\rho_{xz}(1-\rho_{xy}^2)(1-\rho_{xz}^2)(Q-1)}{1+\rho_{xz}^2(Q-1)}\right\}, \tag{7.8}$$

where $Q = E(v_{zzs}|\mathbf{z}_U)/\sigma_z^2$. Thus b will be conditionally consistent if one of the following conditions holds:

$$Q = 1; \quad \rho_{yz\cdot x} = 0; \quad \rho_{xz} = 0; \quad |\rho_{xz}| = 1; \text{ or } |\rho_{xy}| = 1.$$

The balance condition, $Q = 1$, is now a balance in design expectation averaged over all samples rather than balance for the particular sample drawn.

For non-proportional stratified sampling expressions for the bias of the OLS estimator under different conditions are given by Jewell (1985) and by Hausman and Wise (1981). These theoretical results show that for many selection schemes the OLS estimator of β will be biased. The empirical studies in Sections 7.5 and 8.5, confirm these theoretical results.

7.4 ALTERNATIVES TO OLS

In Section 6.4, we introduced the Pearson adjustments, equations (6.18) and (6.19), and their associated estimators, equations (6.21) and (6.22). Under multivariate normality these are MLEs and they constitute our first alternative to OLS. If we let the vector $\mathbf{y}$ in Section 6.4 be $(yx)'$ then the MLE of the covariance matrix $\mathbf{\Sigma}$ of $(yx)'$ is obtained from equation (6.22) as

$$\hat{\mathbf{\Sigma}} = \begin{pmatrix} \hat{\sigma}_y^2 & \hat{\sigma}_{yx} \\ & \hat{\sigma}_x^2 \end{pmatrix}, \tag{7.9}$$

where $\hat{\sigma}_{yx}$ is the MLE of $\sigma_{yx} = \text{cov}(y, x)$, etc. If the object of interest is the regression of y on x then the target regression parameter is

$$\beta = \sigma_{yx}/\sigma_x^2. \tag{7.10}$$

The MLE of β from equation (7.9) is

$$\hat{\beta} = \hat{\sigma}_{yx}/\hat{\sigma}_x^2. \tag{7.11}$$

DeMets and Halperin (1977) derived this estimator, which is a particular case of the more general MLE derived by Anderson (1957). Nathan and Holt (1980) show that $\hat{\beta}$ is consistent for β under less stringent conditions. Pfeffermann (1982) suggests an alternative to the MLE.

In Section 6.5, we considered some other alternatives to $\hat{\beta}$, which are clearly also alternatives to OLS. If we follow Kish and Frankel (1974) and let B_U in equation (7.5) be the target parameter of interest then

$$B_U = \frac{\sum_U (y_t - \bar{y}_U)(x_t - \bar{x}_U)}{\sum_U (x_t - \bar{x}_U)^2}$$

$$= \frac{N_o \sum_U y_t x_t - \sum_U y_t \sum_U x_t}{N_o \sum_U x_t^2 - \left(\sum_U x_t\right)^2} \tag{7.12}$$

We see that this is now written in the form $g(\bar{\mathbf{y}}_U)$ of equation (6.23), namely as a function of means. Estimating $\sum_U y_t x_t$ by the weighted sum $\sum_s y_t x_t / \pi_t$, with similar expressions for the other totals, leads to the design-consistent p-weighted estimator,

$$\hat{B}_w = g(\bar{\mathbf{y}}_{pw}) \tag{7.13}$$

as in equation (6.24). If we also estimate N_o in equation (7.12), even though it is known, by $\hat{N}_o = \sum_s \pi_t^{-1}$, then this gives another p-weighted estimator which has a particularly convenient form for its Taylor series variance (see Rao, 1975b; Smith, 1981).

Weighting by the sample inclusion probabilities $\boldsymbol{\pi}_s' = (\pi_1, \ldots, \pi_{n_o})$ leads to a weighted least-squares (WLS) analysis of the model in equation (7.1). If

$$\mathbf{\Pi} = \operatorname{diag}(\boldsymbol{\pi}_s) = \begin{pmatrix} \pi_1 & & & & 0 \\ & \cdot & & & \\ & & \cdot & & \\ & & & \cdot & \\ 0 & & & & \pi_{no} \end{pmatrix}, \tag{7.14}$$

then the WLS estimator of $\boldsymbol{\beta}$ is

$$\hat{\boldsymbol{\beta}}_{\text{WLS}} = (\mathbf{X}_s' \mathbf{\Pi}^{-1} \mathbf{X}_s)^{-1} \mathbf{X}_s' \mathbf{\Pi}^{-1} \mathbf{y}_s, \tag{7.15}$$

which under equation (7.1) is also a model-unbiased estimator of $\boldsymbol{\beta}$. The WLS option in a statistical program package estimates the WLS variance

$$\text{Var}_{\text{WLS}}(\hat{\boldsymbol{\beta}}_{\text{WLS}} | \mathbf{X}_s) = (\mathbf{X}_s' \mathbf{\Pi}^{-1} \mathbf{X}_s)^{-1} \sigma^2, \tag{7.16}$$

which is different from the true variance under model (7.1), namely

$$\mathrm{Var}_{\mathrm{true}}(\hat{\boldsymbol{\beta}}_{\mathrm{WLS}}|\mathbf{X}_s) = (\mathbf{X}_s'\boldsymbol{\Pi}^{-1}\mathbf{X}_s)^{-1}\mathbf{X}_s'\boldsymbol{\Pi}^{-1}\mathbf{V}_s\boldsymbol{\Pi}^{-1}\mathbf{X}_s(\mathbf{X}_s'\boldsymbol{\Pi}^{-1}\mathbf{X}_s)^{-1}\sigma^2. \quad (7.17)$$

So inferences about the precision of $\hat{\boldsymbol{\beta}}_{\mathrm{WLS}}$ will be incorrect except when $\mathbf{V}_s = \boldsymbol{\Pi}$, which is unlikely in practice. Thus the issue of weighting is more complex than that of whether or not to use weights within a standard statistical program package, the effect on variances must also be considered, as must the possibility of biases if equation (7.1) does not hold.

Nathan and Holt (1980) propose a *p*-weighted version, $\hat{\beta}^*$, of the MLE in equation (7.11) as a compromise between the model-based approach and the randomization approach. They also show that under their assumptions

$$\mathrm{var}(\hat{\beta}) < \mathrm{var}(\hat{\beta}^*) < \mathrm{var}(\hat{B}_w), \quad (7.18)$$

where var( ) denotes the unconditional variance under both their model and repeated random sampling. Some additional comparisons are made in Chapter 8.

In the alternatives above the target parameter has been the regression slope, β, defined in the marginal distribution of y and x. If the analyst is prepared to weaken this assumption and to consider wider classes of regression models including z then the possibilities become very large. In Part C we consider alternatives based on disaggregated regression models. The argument is that if the structure of the population, as reflected in the selection scheme used or the values of the design variable, $\mathbf{z}_U$, affects the regression relationship then this should be incorporated into the model. A disaggregated model is one way of conditioning on the values, $\mathbf{z}_U$. We think that frequently this is the best way to proceed.

If the analyst wishes to remain with models at the aggregate level then a model-based alternative is to modify the regression equation (7.1) by including the design variable z in the equation. The new model has mean

$$E(y_t|x_t, z_t) = \alpha + \beta_{yx\cdot z}x_t + \beta_{yz\cdot x}z_t. \quad (7.19)$$

The target parameters are now $\beta_{yx\cdot z}$ and $\beta_{yz\cdot x}$ rather than β. Whether these parameters are of interest depends on the problem being studied. Du Mouchel and Duncan (1983) propose using the model (7.19) with $z_t = x_t/\pi_t$ as an alternative to the OLS model (7.1) (with $\mathbf{V}_s = \mathbf{I}_s$) in a test of the equality of $E(b)$ and $E(\hat{B}_w)$, where b is the unweighted OLS estimator of β and $\hat{B}_w$ is the *p*-weighted estimator. Under the OLS model they derive a test based on the distribution of $\hat{\Delta} = b - \hat{B}_w$. If this test is not significant and the usual regression diagnostics support the OLS model then there is a strong case for using the unweighted estimator, b. If the hypothesis is rejected then the weights affect the regression relationship and the model should be changed. In an example they show how an understanding of the design groups and certain interaction effects leads to a new OLS model, adjusted for the selection effect, which makes substantive sense.

An alternative modification of the regression equation (7.1) to take account of selection effects is that proposed in economics (see Maddala, 1983, for full details). The regression equation of interest is

$$y_t = \mathbf{X}_{1t}\boldsymbol{\beta} + U_{1t}, \tag{7.20}$$

and this is supplemented by the selection equation

$$w_t = \mathbf{X}_{2t}\boldsymbol{\gamma} + U_{2t}, \tag{7.21}$$

where $E(U_{jt}) = 0$, $E(U_{jt}U_{j't'}) = \sigma_{jj'}$ for $t = t'$, and is zero otherwise, for $j = 1, 2$. Then if y_t is subject to self-selection, such as non-response, or any form of selection on w_t then the covariance between U_{1t} and U_{2t} means that

$$E(y_t|\mathbf{X}_{1t} \text{ and selection}) = \mathbf{X}_{1t}\boldsymbol{\beta} + E(U_{1t}|\text{selection}). \tag{7.22}$$

It is clear from equation (7.22) that if $E(U_{1t}|\text{selection}) \neq 0$ then ignoring the selection rule will lead to bias in the estimation of $\boldsymbol{\beta}$. For truncation schemes Heckman (1979) has proposed selection adjustments based on probit models.

All of the alternative estimators which have been proposed are consistent within their own specific inferential framework. Because they are derived under different assumptions there are few analytical results available. We have already referred to the results of Nathan and Holt (1980) within an unconditional inferential framework. In the absence of a common theoretical framework which embraces all procedures one has to resort to Monte Carlo sampling experiments. We report the results of several studies in Section 7.5 and some new results in Chapter 8.

7.5 SOME EMPIRICAL STUDIES OF THE EFFECTS OF SELECTION

Since theoretical comparisons between the estimators are difficult to obtain, several comparisons based on simulation studies and on empirical data have been carried out. We now review the main results.

Demets and Halperin (1977) consider a restricted sampling scheme in which a subsample of proportion λ is obtained by taking only extreme values of the design variable, z. For simulations based on 500 samples each of sizes ranging from 50 to 400 and subsampling proportions ranging from 0.1 to 0.5, the ratio of the average value of the OLS estimator, b, and of the MLE, $\hat{\beta}$, to the true value of β are given in Table 7.1. The simulations were performed under a trivariate normal distribution with $\rho_{yz} = 0.9$ and $\rho_{xy} = \rho_{xz}$ ranging from 0.1 to 0.9.

The results clearly show the severe effect of restricted selection on the OLS estimator and the lessening of this effect by use of the MLE. Results reported, but not presented, indicate also considerable reduction in the mean square error.

***Table 7.1** Ratios of average b and average $\hat{\beta}$ to β under restricted sampling for various subsampling proportions λ, total sample size N (500 samples of size N being drawn) and correlations ρ_{xy} where $\rho_{xy} = \rho_{xy}$ and $\rho_{yz} = 0.9$*

		$\lambda = 0.1$			$\lambda = 0.3$			$\lambda = 0.5$		
N	$\rho_{xy} =$	0.1	0.5	0.9	0.1	0.5	0.9	0.1	0.5	0.9
50	b	4.27	2.48	1.09	2.41	1.88	1.07	1.68	1.46	1.04
	$\hat{\beta}$	1.28	1.21	1.00	0.91	1.04	0.99	0.95	1.01	0.99
100	b	3.89	2.36	1.10	2.41	1.81	1.07	1.72	1.49	1.05
	$\hat{\beta}$	1.03	1.08	1.01	0.99	1.03	1.00	0.96	1.02	1.00
200	b	4.21	2.29	1.09	2.48	1.78	1.07	1.81	1.48	1.05
	$\hat{\beta}$	1.16	1.04	1.00	1.01	1.02	1.00	1.03	1.01	1.00
400	b	4.19	2.24	1.09	2.46	1.75	1.07	1.79	1.46	1.05
	$\hat{\beta}$	1.06	1.02	1.00	1.01	1.00	1.00	1.02	1.00	1.00

Source: Demets and Halperin (1977).

Nathan and Holt (1980) used data from 3850 farms on cropland acreage (y), total acreage (x) and total value of products sold in the previous year (z), to estimate the biases and variances of the four estimators they considered, $b, \hat{\beta}, \hat{B}_w$ and $\hat{\beta}^*$, under five different sample designs with sample size 400. The designs considered were: (A) simple random sampling; (B) proportional stratified sampling; (C) equal allocation stratified sampling; (D) non-proportional stratified sampling with over-sampling for high values of the design variable; (E) non-proportional stratified sampling with over-sampling for extreme values of the design variables. The results are shown in Table 7.2 and clearly show the relationship (7.18) and the sensitivity of the unweighted MLE, $\hat{\beta}$, relative to that of the weighted estimators, $\hat{B}_w$ and $\hat{\beta}^*$.

Holt *et al.* (1980) simulated several finite populations, of size 10 000 each, for

***Table 7.2** Bias and mean square error of ordinary least squares estimator and variances of unbiased estimators for population of 3850 farms using various survey designs*

Survey design	$E(b) - \beta$	MSE(b)	$V(\hat{\beta})$	$V(\hat{B}_w)$	$V(\hat{\beta}^*)$
A	0.000	0.000 214	0.000 197	0.000 226	0.000 197
B	0.000	0.000 200	0.000 198	0.000 222	0.000 196
C	0.031	0.001 102	0.000 160	0.000 222	0.000 196
D	0.027	0.000 879	0.000 163	0.000 220	0.000 195
E	0.042	0.001 887	0.000 152	0.000 225	0.000 195

Source: Nathan and Holt (1980).

the values of z and then selected samples of size 1000 (by sample designs based on the z-values) for values of x and y, on the basis of conditional normality. The parameters used for each population were derived from sets of real data—the data on farms described above and three sets of data from a study of social deprivation (Runciman, 1972).

Six sample designs were considered: D_1—simple random sampling; D_2—proportional stratified sampling; D_3 and D_4—two variants of increasing allocation to strata; D_5 and D_6—two variants of U-shaped allocation to strata.

The results from the simulations based on the farm data are given in Tables 7.3 and 7.4, comparing the average values, standard errors and coverages of the OLS estimator, b, the MLE, $\hat{\beta}$, and the weighted estimator $\hat{B}_w$.

The main points to note are that the bias of b distorts coverage considerably for unequal probability designs, whereas the other two estimators perform equally well, except for the extreme U-shaped allocation design, D_6, for which the weighted estimator, $\hat{B}_w$, performs poorly. Similar patterns were obtained from the simulations based on the Runciman data.

Hausman and Wise (1981) present data from the Gary income maintenance experiment based on a subsample of 585 households, selected as a non-proportional

Table 7.3 Average values and standard errors for the three procedures for the designs considered

Design	Ordinary least squares, b	Maximum likelihood, $\hat{\beta}$	Probability weighted estimation, $\hat{B}_w$
D1	0.308	0.308	0.308
	0.825(−2)	0.826(−2)	0.825(−2)
D2	0.309	0.307	0.309
	0.795(−2)	0.794(−2)	0.795(−2)
D3	0.298	0.308	0.306
	0.839(−2)	0.849(−2)	0.108(−1)
D4	0.296	0.307	0.306
	0.821(−2)	0.863(−2)	0.123(−1)
D5	0.321	0.307	0.308
	0.760(−2)	0.770(−2)	0.827(−2)
D6	0.332	0.307	0.309
	0.714(−2)	0.752(−2)	0.235(−1)

Notes:

1. Population value estimated = 0.307 2.
2. First entry is average value of estimator and second is average conditional standard error.
3. Standard error given with number of extra decimal places in brackets. Thus 0.825(−2) = 0.008 25.

Source: Holt *et al.* (1980).

Table 7.4 Confidence interval coverage for the three procedures for the designs considered

Design	Estimation procedure	Expected frequencies									
		5	5	15	25	450	450	25	15	5	5
D1	b	7	3	19	32	500	394	28	9	4	4
	$\hat{\beta}$	8	4	18	31	465	419	30	13	4	8
	$\hat{B}_w$	8	5	25	31	492	387	26	16	4	6
D2	b	11	5	17	30	517	394	14	11	1	0
	$\hat{\beta}$	9	6	9	23	433	476	19	15	8	2
	$\hat{B}_w$	14	5	24	36	501	384	20	11	4	1
D3	b	0	0	2	1	142	618	79	68	42	48
	$\hat{\beta}$	5	8	16	26	464	431	27	16	3	4
	$\hat{B}_w$	4	4	11	18	391	494	33	24	6	15
D4	b	0	0	0	2	98	577	97	90	49	87
	$\hat{\beta}$	11	2	16	22	455	443	25	21	2	3
	$\hat{B}_w$	6	6	17	21	401	462	44	23	10	10
D5	b	185	69	117	130	458	41	0	0	0	0
	$\hat{\beta}$	11	7	11	32	432	454	26	18	3	6
	$\hat{B}_w$	12	12	21	40	460	409	25	13	2	6
D6	b	777	83	74	31	34	1	0	0	0	0
	$\hat{\beta}$	9	5	18	24	422	461	32	18	5	6
	$\hat{B}_w$	48	23	35	50	410	340	33	14	9	38

Source: Holt *et al.* (1980).

stratified sample. The design variable, z, was income (as a multiple of the poverty line). The dependent variable was the logarithm of labour income and the independent variables were years of education, years of work experience, log of non-labour family income and union membership (a dummy variable). The OLS estimator, the weighted estimator and two versions of the MLE (for known and unknown sampling ratios) were compared. The results are given in Tables 7.5 and 7.6.

The results again indicate serious biases in the OLS estimators. Although the three estimators which take sample design into account differ little from each other for most independent variables, the weighted estimator for the non-labour income coefficient is considerably smaller than the MLE.

Jewell (1985) considers a two-strata variable probability sampling scheme where stratification is based on the dependent variable and subsampling at the rate of one-half in one of the strata is used. Samples of size 50 from normal

Table 7.5 Parameter estimates (and standard errors) by method of estimation

Variable	Least squares	Weighted least squares[a]	Maximum likelihood (ratios known)	Maximum likelihood (ratios unknown)
Constant	5.916	5.8424	5.9300	5.7355
	(0.0879)	(0.0899)	(0.1047)	(0.1196)
Education	0.0190	0.0270	0.0252	0.0281
	(0.0068)	(0.0068)	(0.0079)	(0.0083)
Experience	0.0042	0.0048	0.0050	0.0053
	(0.0018)	(0.0018)	(0.0020)	(0.0023)
Income	−0.0162	−0.0055	−0.0189	−0.0231
	(0.0068)	(0.0069)	(0.0092)	(0.0101)
Union	0.2596	0.2314	0.2021	0.2881
	(0.0519)	(0.0407)	(0.0386)	(0.0647)

[a]The standard errors shown in this column are those reported from a regression program. They underestimate the true standard errors.
Source: Hausman and Wise (1981).

Table 7.6 Ratios of other estimates to least-squares estimates by method of estimation

Variable	Weighted least squares	Maximum likelihood (ratios known)	Maximum likelihood (ratios unknown)
Constant	0.99	1.00	0.97
Education	1.42	1.33	1.48
Experience	1.14	1.19	1.23
Income	0.35	1.17	1.43
Union	0.89	0.78	1.11

Source: Hausman and Wise (1981).

distributions were simulated 100 times to provide estimates of biases and variances of the probability weighted estimator, $\hat{B}_w$, of three variants of an iterative least-squares estimator (sub-indexed by w, c and c^*) and of the MLEs. The results are given in Table 7.7.

The results indicate that in this situation one of the iterative least-squares estimators, $\hat{\beta}_w$, performs slightly better than the others and another, $\hat{\beta}_c^*$, performs considerably worse. The bias of OLS is again apparent.

Finally, Nathan (1983), presents results of a simulation study for the case where selection (response) is determined by the value of an unobservable design variable, z_2, which is correlated with the observed design variable, z_1. Ten samples of 2000 values from a multivariate normal distribution were simulated

Table 7.7 Example with $\alpha = 0$ and $\alpha = 1$. Simulation estimates of bias and variance of various estimators of α and β

	Example estimate of		Simulation results			
			Estimation of α		Estimation of β	
Estimator	α	β	Bias	Variance	Bias	Variance
OLS	+4.1702	0.7559	4.1779	1.9191	−0.1768	0.0298
Prob. weighted	+0.0221	1.2577	0.4734	3.1955	−0.0305	0.0413
$\hat{a}_w, \hat{\beta}_w$	+0.1575	1.0150	0.4569	3.0788	−0.0425	0.0293
$\hat{a}_c, \hat{\beta}_c$	−0.0373	1.0857	0.6072	3.0396	−0.1184	0.0393
$\hat{a}_{ML}, \hat{\beta}_{ML}$	+0.2424	0.9260	0.3777	2.5699	−0.0445	0.0290

Note: Simulation estimates from 100 simulations, OLS, ordinary least squares.
Source: Jewell (1985).

Table 7.8 Simulation study results under the overall normal model:
$$(X_i, Y_i, Z_{1i}, Z_{2i}) \sim N(\mu, \Sigma)$$
$$N = 2000;\ E(n) = NXP(Z_{2i} \geqslant 0) = 1200$$

		Parameter set				
		1	2	3	4	5
	Estimator	General	Exact sample theory conditions	Exact econometeric conditions	Approximate econometric conditions	Sample theory and econometric conditions
Relative bias	$\hat{\beta}$	0.038	(−0.002)	0.046	0.040	0
	$\hat{\beta}(P)$	0.036	(−0.001)	0.046	0.040	0
	$\hat{\beta}(D)$	0.230	0.210	(0.000)	0.180	0
	$\hat{\beta}(H)$	0.226	0.206	(0.000)	0.176	0
	b	0.045	0.006	0.045	0.044	0
Relative standard error	$\hat{\beta}$	0.024	0.024	0.024	0.025	0.025
	$\hat{\beta}(P)$	0.024	0.024	0.025	0.025	0.025
	$\hat{\beta}(D)$	0.028	0.029	0.041	0.033	0.043
	$\hat{\beta}(H)$	0.028	0.029	0.041	0.033	0.025
	b	0.024	0.024	0.024	0.025	0.025
Relative root mean square error	$\hat{\beta}$	0.045	0.024	0.052	0.047	0.025
	$\hat{\beta}(P)$	0.043	0.024	0.053	0.048	0.025
	$\hat{\beta}(D)$	0.231	0.212	0.041	0.183	0.043
	$\hat{\beta}(H)$	0.228	0.208	0.041	0.179	0.042
	b	0.051	0.025	0.051	0.051	0.025

Note: ( ) Not significantly different from zero ($\alpha = 0.01$). Simulation sample errors of above do not exceed 0.002.
Source: Nathan (1983).

for each of five parameters sets, corresponding to various assumptions on the underlying structure, which served as the basis for the derivations of the estimators. The estimators compared were: (i) $\hat{\beta}$, the MLE; (ii) $\hat{\beta}(P)$, a modification of $\hat{\beta}$ due to Pfeffermann (1982); (iii) $\hat{\beta}(D)$, a modification of $\hat{\beta}(P)$; (iv) $\hat{\beta}(H)$, Heckman's (1979) two-stage estimator; and (v) b, the OLS estimator. The results given in Table 7.8 show that the MLE-based estimators $\hat{\beta}(D)$ and $\hat{\beta}(H)$ provide some improvement for the specific conditions under which they are derived, but perform considerably worse even than OLS under other conditions.

These empirical and simulation results show conclusively that the OLS estimator $\mathbf{b}$ in equation (7.2) of the regression parameter $\boldsymbol{\beta}$ in equation (7.1) is badly biased whenever the selection scheme is not self-weighting. Thus the default option of using OLS methods for the analysis of survey data is not appropriate for regression analysis. In Chapter 8 we show that this conclusion holds for other methods of multivariate analysis. Beyond this it is difficult to draw any firm conclusions. Each method does well under its own specific set of assumptions but can break down under alternative assumptions. Thus randomization inference, which is always consistent within the full randomization distribution, exhibits serious biases under the conditional framework which would usually be considered appropriate for regression analysis (Holt *et al.*, 1980). The robustness of some model-based procedures is examined in Chapter 8.

CHAPTER 8

Multivariate Analysis

T. M. F. Smith and D. J. Holmes

8.1 INTRODUCTION

Standard errors and test procedures for some multivariate methods for continuous variables were considered in Chapter 3. In this chapter we extend the work on regression analysis, reviewed in Chapter 7, by examining the robustness of alternative multivariate procedures to departures from model assumptions. We assume that the design variables, **z**, are continuous and that if they have been used to form population design groups, such as strata (see Section 6.2), then these groups are not intrinsic to the analysis. Under these conditions there is usually no case for a disaggregated analysis within and between groups and so we consider only aggregate analyses. If in addition to the continuous design variables, **z**, there is a further categorical design variable leading to natural population groups, such as types of business, farm or school, then the methods that we propose can be employed for analysis within these natural groups provided that the sample sizes within groups are sufficiently large, see Krzanowski (1979, 1984).

Most classical methods of multivariate analysis of continuous data are based on an examination of the structure of population mean vectors and covariance matrices. We consider the problem of estimating $\boldsymbol{\mu}_y$ and $\boldsymbol{\Sigma}_{yy}$, the mean and covariance matrix of a vector of survey variables **y**, from a complex sample and the related problems of estimating known functions of $\boldsymbol{\Sigma}_{yy}$ such as regression and correlation coefficients and principal components. The original theoretical basis for these methods was the multivariate normal distribution and this assumption influences the choice of some of our estimators. Clearly this is a strong assumption which is unlikely to be satisfied by most survey variables and so we also examine alternative assumptions and estimators. We do not address the question of whether multivariate methods based on the covariance matrix $\boldsymbol{\Sigma}_{yy}$ are relevant for non-normal data, rather we take the pragmatic view that these are the methods that many survey analysts employ.

As in Chapters 6 and 7 we are concerned with the bias effects due to selection based on the design variables $\mathbf{z}$. We examine properties of alternative procedures for taking into account the effect of selection on $\mathbf{z}$, such as the Pearson adjustments and probability weighting (see Section 7.4), within both a conditional and an unconditional inferential framework.

8.2 ALTERNATIVE ESTIMATORS

In order to compare randomization methods with other methods we assume that the sampling scheme, $p(s|\mathbf{z}_U)$, is a random sampling scheme. A random sampling scheme is just one example of the class of uninformative selection methods (see Section 6.6) which depend only on the values of the known design variables, $\mathbf{z}_U$. On each unit, $t \in s$, we measure the survey variables $\mathbf{y}_t$. The data on which we will base our estimators comprise the labels, s, the values $\mathbf{y}_t$, $t \in s$, the known values of the design variables $\mathbf{z}_t$, $t \in U$, where U denotes the population of all units, and our knowledge of the sampling scheme $p(s|\mathbf{z}_U)$.

The target parameter of interest is some function of the superpopulation mean vector $\boldsymbol{\mu}_y$ or the covariance matrix, $\boldsymbol{\Sigma}_{yy}$, for example the eigenvalues and eigenvectors (principal components) of $\boldsymbol{\Sigma}_{yy}$. We assume that the finite population is generated from the superpopulation by the equivalent of simple random sampling (SRS) so that the finite population mean $\bar{\mathbf{y}}_U$ and covariance matrix $\mathbf{V}_{yyU}$ differ from $\boldsymbol{\mu}_y$ and $\boldsymbol{\Sigma}_{yy}$ only by terms of $O_p(N_o^{-1/2})$. Since the finite population size N_o will usually be large this means that inferences about $\boldsymbol{\Sigma}_{yy}$ and $\mathbf{V}_{yyU}$ will be similar. Thus randomization inference about $\mathbf{V}_{yyU}$ can be employed for inference about the superpopulation parameter $\boldsymbol{\Sigma}_{yy}$ under the above assumptions.

Before defining the alternative estimators we introduce some preliminary notation. The sample inclusion probability for unit t is

$$\pi_t = \sum_{s \supset t} p(s|\mathbf{z}_U).$$

Let

$$w_t = (N_o \pi_t)^{-1}, \qquad w(s) = \sum_s w_t.$$

We define sample values and statistics,

$$\mathbf{x}_t = (\mathbf{y}_t', \mathbf{z}_t')', \qquad \bar{\mathbf{x}}_s = n_o^{-1} \sum_s \mathbf{x}_t = (\bar{\mathbf{y}}_s', \bar{\mathbf{z}}_s')',$$

$$\bar{\mathbf{x}}_s^* = \sum_s w_t \mathbf{x}_t = (\bar{\mathbf{y}}_s^{*\prime}, \bar{\mathbf{z}}_s^{*\prime})',$$

$$\mathbf{V}_{xxs} = (n_o - 1)^{-1} \sum_s (\mathbf{x}_t - \bar{\mathbf{x}}_s)(\mathbf{x}_t - \bar{\mathbf{x}}_s)' = \begin{pmatrix} \mathbf{V}_{yys} & \mathbf{V}_{yzs} \\ \mathbf{V}_{zys} & \mathbf{V}_{zzs} \end{pmatrix},$$

$$\mathbf{V}_{xxs}^* = \sum_s w_t \mathbf{x}_t \mathbf{x}_t' - w(s)^{-1} \bar{\mathbf{x}}_s^* \bar{\mathbf{x}}_s^{*\prime} = \begin{pmatrix} \mathbf{V}_{yys}^* & \mathbf{V}_{yzs}^* \\ \mathbf{V}_{zys}^* & \mathbf{V}_{zzs}^* \end{pmatrix}.$$

The finite population mean and covariance matrix of $\mathbf{z}$ are

$$\bar{\mathbf{z}}_U = N_o^{-1} \sum_U \mathbf{z}_t \quad \text{and} \quad \mathbf{V}_{zzU} = N_o^{-1} \sum_U \mathbf{z}_t \mathbf{z}_t' - \bar{\mathbf{z}}_U \bar{\mathbf{z}}_U'.$$

The four estimators of $\boldsymbol{\mu}_y$ and $\boldsymbol{\Sigma}_{yy}$ which we consider are:

1. The unweighted (SRS) estimators:

$$\hat{\boldsymbol{\mu}}_{y0} = \bar{\mathbf{y}}_s, \qquad \hat{\boldsymbol{\Sigma}}_{yy0} = \mathbf{V}_{yys}.$$

2. The probability weighted estimators:

$$\hat{\boldsymbol{\mu}}_{yw} = \bar{\mathbf{y}}_s^*, \qquad \hat{\boldsymbol{\Sigma}}_{yyw} = \mathbf{V}_{yys}^*.$$

3. The maximum likelihood estimators (MLEs) based on the Pearson adjustments (equations (6.21) and (6.22)):

$$\hat{\boldsymbol{\mu}}_{ym} = \bar{\mathbf{y}}_s + \mathbf{b}_{yz}(\bar{\mathbf{z}}_U - \bar{\mathbf{z}}_s),$$
$$\hat{\boldsymbol{\Sigma}}_{yym} = \mathbf{V}_{yys} + \mathbf{b}_{yz}(\mathbf{V}_{zzU} - \mathbf{V}_{zzs})\mathbf{b}_{yz}',$$

where $\mathbf{b}_{yz} = \mathbf{V}_{yzs} \mathbf{V}_{zzs}^{-1}$.

4. The probability weighted MLEs:

$$\hat{\boldsymbol{\mu}}_{ywm} = \bar{\mathbf{y}}_s^* + \mathbf{b}_{yz}^*(\bar{\mathbf{z}}_U - \bar{\mathbf{z}}_s^*),$$
$$\hat{\boldsymbol{\Sigma}}_{yywm} = \mathbf{V}_{yys}^* + \mathbf{b}_{yz}^*(\mathbf{V}_{zzU} - \mathbf{V}_{zzs}^*)\mathbf{b}_{yz}^{*\prime},$$

where $\mathbf{b}_{yz}^* = \mathbf{V}_{yzs}^* \mathbf{V}_{zzs}^{*-1}$.

The unweighted estimators $\hat{\boldsymbol{\mu}}_{y0}$ and $\hat{\boldsymbol{\Sigma}}_{yy0}$ are chosen because they are the estimators most widely used in practice. They correspond to the default option in many statistical packages, and in the absence of clear advice to the contrary these estimators will be chosen by most data analysts. The weighted estimators $\hat{\boldsymbol{\mu}}_{yw}$ and $\hat{\boldsymbol{\Sigma}}_{yyw}$ are the classical randomization estimators and can be computed using packages that offer a weighting option. Unfortunately these packages usually employ a generalized least-squares program which gives least-squares standard errors which may bear no relation to the true randomization standard errors (see Section 7.4). The MLEs $\hat{\boldsymbol{\mu}}_{ym}$ and $\hat{\boldsymbol{\Sigma}}_{yym}$ are model-based estimators derived from multivariate normal assumptions. These estimators use the assumed linear homoscedastic relationship between $\mathbf{y}$ and $\mathbf{z}$ to perform a regression-type adjustment, and as we shall see they are not robust to departures from linearity and homoscedasticity. The weighted adjusted estimators $\hat{\boldsymbol{\mu}}_{ywm}$ and $\hat{\boldsymbol{\Sigma}}_{yywm}$ are compromise estimators proposed by Nathan and Holt (1980) to try to capture the benefits of both the randomization approach and the model-based adjustment procedure by using the auxiliary information in $\mathbf{z}$ within a probability weighted procedure.

8.3 PROPERTIES OF THE ESTIMATORS UNDER LINEARITY AND HOMOSCEDASTICITY CONDITIONS

The important properties of the multivariate normal distribution that affect the bias of the MLEs of $\boldsymbol{\mu}_y$ and $\boldsymbol{\Sigma}_{yy}$ are those of the linearity of the regression of $\mathbf{y}$ on $\mathbf{z}$ and of the constant variance of the regression residuals. If we assume a superpopulation model in which the finite population values $(\mathbf{y}_t', \mathbf{z}_t')'$, $t \in U$ are IID realizations of random variables $(\mathbf{Y}', \mathbf{Z}')'$ which have the following properties:

$$E_\xi(\mathbf{Y} \mid \mathbf{Z} = \mathbf{z}) = \mathbf{C} + \mathbf{D}\mathbf{z},$$

and

$$V_\xi(\mathbf{Y} \mid \mathbf{Z} = \mathbf{z}) = \mathbf{K}; \tag{8.1}$$

then it follows that $\mathbf{D} = \boldsymbol{\beta}_{yz} = \boldsymbol{\Sigma}_{yz}\boldsymbol{\Sigma}_{zz}^{-1}$ and $\mathbf{C} = \boldsymbol{\mu}_y - \boldsymbol{\beta}_{yz}\boldsymbol{\mu}_z$, where $E_\xi(\cdot)$ and $V_\xi(\cdot)$ denote the expectation and covariance matrix, respectively, under the model. With these assumptions we can examine the bias of the various estimators either conditionally on both s and $\mathbf{z}$ or conditionally on $\mathbf{z}$ alone. We will refer to the latter as the unconditional bias since it averages over all samples. The theoretical details are given in Skinner (1982) and Holmes (1987).

8.3.1 Model-based Biases Conditional on s and $\mathbf{z}$

In Table 8.1 we show the expectations of the four estimators of $\boldsymbol{\mu}_y$ conditional on s and $\mathbf{z}$ under the model (8.1).

We see that all of the estimators of $\boldsymbol{\mu}_y$ are conditionally biased, but that in some cases the bias should be small. For designs which are not equally weighted the bias in $\hat{\boldsymbol{\mu}}_{y0}$ could be very large and even for an equally weighted design (i.e. π_t = constant) it will be of $O_p(n_o^{-1/2})$. Since $w(s) = \sum_s (N_0 \pi_t)^{-1}$ is usually close to unity the conditional bias of $\hat{\boldsymbol{\mu}}_{yw}$ is also $O_p(n_o^{-1/2})$, but in any particular sample it could be very large. For the adjusted estimators $\hat{\boldsymbol{\mu}}_{ym}$ and $\hat{\boldsymbol{\mu}}_{ywm}$ the bias depends on the deviation of $\bar{\mathbf{z}}_U$ from $\boldsymbol{\mu}_z$ and since this is $O_p(N_o^{-1/2})$ it should be small. Thus both the adjusted estimators are approximately conditionally unbiased and model-consistent.

Table 8.2 gives the corresponding conditional expectations for the four estimators of $\boldsymbol{\Sigma}_{yy}$.

***Table 8.1** The conditional expectations of estimators of the mean, $\boldsymbol{\mu}_y$*

$$E_\xi(\hat{\boldsymbol{\mu}}_{y0} \mid s, \mathbf{z}) = \boldsymbol{\mu}_y + \mathbf{D}(\bar{\mathbf{z}}_s - \boldsymbol{\mu}_z)$$
$$E_\xi(\hat{\boldsymbol{\mu}}_{yw} \mid s, \mathbf{z}) = w(s)\boldsymbol{\mu}_y + \mathbf{D}(\bar{\mathbf{z}}_s^* - w(s)\boldsymbol{\mu}_z)$$
$$E_\xi(\hat{\boldsymbol{\mu}}_{ym} \mid s, \mathbf{z}) = \boldsymbol{\mu}_y + \mathbf{D}(\bar{\mathbf{z}}_U - \boldsymbol{\mu}_z)$$
$$E_\xi(\hat{\boldsymbol{\mu}}_{ywm} \mid s, \mathbf{z}) = w(s)\boldsymbol{\mu}_y + \mathbf{D}(\bar{\mathbf{z}}_U - w(s)\boldsymbol{\mu}_z)$$

Table 8.2 The conditional expectations of estimators of the covariance matrix, Σ_{yy}

$$E_\xi(\hat{\mathbf{\Sigma}}_{yy0}|s,\mathbf{z}) = \mathbf{\Sigma}_{yy} + \mathbf{D}(\mathbf{V}_{zzs} - \mathbf{\Sigma}_{zz})\mathbf{D}'.$$
$$E_\xi(\hat{\mathbf{\Sigma}}_{yyw}|s,\mathbf{z}) = \alpha_w\mathbf{\Sigma}_{yy} + \mathbf{D}(\mathbf{V}^*_{zzs} - \alpha_w\mathbf{\Sigma}_{zz})\mathbf{D}',$$

where $\alpha_w = w(s) - \sum_s w_t^2/w(s)$.

$$E_\xi(\hat{\mathbf{\Sigma}}_{yym}|s,\mathbf{z}) = \alpha\mathbf{\Sigma}_{yy} + \mathbf{D}(\mathbf{V}_{zzU} - \alpha\mathbf{\Sigma}_{zz})\mathbf{D}',$$

where $\alpha = n_o^{-1}[n_o - q - 1 + \text{tr}(\mathbf{V}_{zzU}\mathbf{V}_{zzs}^{-1})]$, and q = number of design variables.

$$E_\xi(\hat{\mathbf{\Sigma}}_{yywm}|s,\mathbf{z}) = \alpha^*\mathbf{\Sigma}_{yy} + \mathbf{D}(\mathbf{V}_{zzU} - \alpha^*\mathbf{\Sigma}_{zz})\mathbf{D}',$$

where α^* is a weighted version of α.

For most designs α_w, α, and α^* (see Holmes, 1987) should all be equal to $1 + O_p(n_o^{-1})$ and so we see that the adjusted estimators, $\hat{\mathbf{\Sigma}}_{yym}$ and $\hat{\mathbf{\Sigma}}_{yywm}$, have conditional biases of $O_p(n_o^{-1})$, the weighted estimator $\hat{\mathbf{\Sigma}}_{yyw}$ should have a conditional bias of $O_p(n_o^{-1/2})$ and the conditional bias of $\hat{\mathbf{\Sigma}}_{yy0}$ for unequally weighted designs could be of $O_p(1)$.

These theoretical results are expressed in terms of orders of magnitude and thus depend on unknown constant factors. The simulation study in Section 8.5 provides evidence that the theoretical results give good guidance for the choice of estimator when the linearity and homoscedasticity assumptions are satisfied.

8.3.2 Model-based Biases Conditional on z only (Unconditional Biases)

If we are considering an inferential procedure before sampling then it is natural to examine its performance averaged over all possible samples, that is, the expectations in Tables 8.1 and 8.2 should be averaged over s. After the sample is drawn it is more natural to condition on s and to average only over samples like the one actually drawn, in which case the results in Tables 8.1 and 8.2 are appropriate. Unconditionally we find that the only estimators with significant biases are $\hat{\boldsymbol{\mu}}_{y0}$ and $\hat{\mathbf{\Sigma}}_{yy0}$, which are based on the SRS assumption. If an unequal probability design is employed then $E_p(\bar{\mathbf{z}}_s) \neq \bar{\mathbf{z}}_U$ and $E_p(\mathbf{V}_{zzs}) \neq \mathbf{V}_{zzU}$, where $E_p(.)$ denotes averaging with respect to $p(s|\mathbf{z}_U)$. However, for equal probability designs $E_p(\bar{\mathbf{z}}_s) = \bar{\mathbf{z}}_U$ and $E_p(\mathbf{V}_{zzs}) = \mathbf{V}_{zzU}$, and in this special case all the procedures are unconditionally unbiased in large samples. This lends support to the common practice in social surveys of employing self-weighting probability designs wherever possible. The empirical results in Section 8.5 confirm these conclusions.

8.4 APPLICATIONS IN MULTIVARIATE ANALYSIS

It is difficult to measure the accuracy of an estimator $\hat{\mathbf{\Sigma}}_{yy}$ of $\mathbf{\Sigma}_{yy}$ since general-purpose distance measures for matrices may be poor for a particular application of interest. Rather than follow the general route we have chosen to examine the accuracy of estimators of particular functions of $\boldsymbol{\Sigma}_{yy}$ of interest to applied statisticians. Two of the most common multivariate procedures used are regression analysis and principal component analysis.

For regression analysis the vector $\mathbf{y}$ is partitioned into $(\mathbf{y}'_1, \mathbf{y}'_2)'$, with

$$\boldsymbol{\mu}_y = \begin{pmatrix} \boldsymbol{\mu}_1 \\ \boldsymbol{\mu}_2 \end{pmatrix} \quad \text{and} \quad \boldsymbol{\Sigma}_{yy} = \begin{pmatrix} \boldsymbol{\Sigma}_{11} & \boldsymbol{\Sigma}_{12} \\ \boldsymbol{\Sigma}_{21} & \boldsymbol{\Sigma}_{22} \end{pmatrix}.$$

The linear regression of $\mathbf{y}_1$ on $\mathbf{y}_2$ leads to the target parameter

$$\boldsymbol{\beta} = \boldsymbol{\Sigma}_{12}\boldsymbol{\Sigma}_{22}^{-1}. \tag{8.2}$$

The principal components of $\mathbf{y}$ are based on the standardized eigenvectors and eigenvalues of $\boldsymbol{\Sigma}_{yy}$. These are obtained from solutions to the equations

$$\boldsymbol{\Sigma}_{yy}\boldsymbol{\gamma}_i = \lambda_i\boldsymbol{\gamma}_i, \qquad i = 1, \ldots, p,$$

subject to

$$\begin{aligned} \boldsymbol{\gamma}'_i\boldsymbol{\gamma}_j &= 1, \qquad i = j, \\ &= 0, \qquad i \neq j, \end{aligned}$$

where $\boldsymbol{\Sigma}_{yy}$ is of dimension p.

If we order the eigenvalues as $\lambda_1 > \lambda_2 > \cdots > \lambda_p$, and assume them to be distinct, then the *ith principal component* of $\mathbf{y}$ is

$$u_i = \boldsymbol{\gamma}'_i\mathbf{y}, \qquad i = 1, \ldots, p. \tag{8.3}$$

Since $\text{var}(u_i) = \lambda_i$, principal component analysis concentrates on the first two or three components which usually account for a large proportion of the total variance, $\sum_{i=1}^{p} \lambda_i$.

We estimate functions of $\boldsymbol{\Sigma}_{yy}$ by the corresponding functions of our estimate $\hat{\boldsymbol{\Sigma}}_{yy}$. Thus if

$$\hat{\boldsymbol{\Sigma}}_{yy} = \begin{pmatrix} \hat{\boldsymbol{\Sigma}}_{11} & \hat{\boldsymbol{\Sigma}}_{12} \\ \hat{\boldsymbol{\Sigma}}_{21} & \hat{\boldsymbol{\Sigma}}_{22} \end{pmatrix},$$

then

$$\hat{\boldsymbol{\beta}} = \hat{\boldsymbol{\Sigma}}_{12}(\hat{\boldsymbol{\Sigma}}_{22})^{-1}, \tag{8.4}$$

and similarly $\hat{\lambda}_i$ and $\hat{\boldsymbol{\gamma}}_i$ are obtained by solving

$$\hat{\boldsymbol{\Sigma}}_{yy}\hat{\boldsymbol{\gamma}}_i = \hat{\lambda}_i\hat{\boldsymbol{\gamma}}_i, \qquad i = 1, \ldots, p, \tag{8.5}$$

subject to

$$\begin{aligned} \hat{\boldsymbol{\gamma}}'_i\hat{\boldsymbol{\gamma}}_j &= 1, \qquad i = j, \\ &= 0, \qquad i \neq j. \end{aligned} \tag{8.6}$$

Thus

$$\hat{u}_i = \hat{\boldsymbol{\gamma}}'_i\mathbf{y}, \qquad i = 1, \ldots, p. \tag{8.7}$$

The biases of these estimators can be evaluated theoretically using Taylor series expansions for each of the four estimators of $\boldsymbol{\Sigma}_{yy}$ together with the results in Section 8.3. Details can be found in Skinner (1982), Skinner *et al.* (1986) and Holmes (1987). Simulation studies were carried out to determine the actual magnitude of the biases rather than just the order of magnitude, and some results are reported in the next section.

8.5 STUDY I: MULTIVARIATE NORMAL DATA

8.5.1 Introduction

This simulation study differs from the model-based studies considered in Section 7.5, by being based on repeated random samples from a fixed finite population. It is a model-based simulation only in the sense that the finite population is generated from a model. We follow the approach adopted by Royall and Cumberland (1981) in their simulation studies which allow both conditional and unconditional properties of statistical procedures to be examined, and, therefore, is 'fair' to both model-based and randomization-based procedures.

In order to assess the accuracy of the procedures when the linearity and homoscedasticity assumptions are satisfied, population data was generated from a multivariate normal distribution for $\mathbf{x} = (\mathbf{y}', \mathbf{z}')'$. The mean vector $\boldsymbol{\mu}_x$ and covariance matrix $\boldsymbol{\Sigma}_{xx}$ were chosen to be those estimated for a set of variables from the Family Expenditure Survey (FES) for 1975 (see Section 3.2). The vector of $\mathbf{y}$ variables comprised the logarithms of household expenditures on six items, and we use a single design variable z the logarithm of expenditure on all items. The details are given in Table 8.3 together with the correlation matrix and standard deviations of the variables; 10 000 values were generated independently

Table 8.3 Parameter values used in simulation

Label	Variable: Log(household expenditure on...)	Standard deviation	Correlation matrix						
y_1	Housing	0.63	1						
y_2	Fuel, light and power	0.62	0.16	1					
y_3	Food	0.52	0.24	0.33	1				
y_4	Clothing and footwear	1.41	0.16	0.13	0.35	1			
y_5	Transport and vehicles	1.47	0.28	0.16	0.39	0.26	1		
y_6	Services	1.02	0.31	0.18	0.33	0.26	0.34	1	
z	All items	0.54	0.48	0.36	0.71	0.52	0.64	0.60	1

Note: Units are log(£10^{-3} per week).

from the seven-dimensional multivariate normal distribution of **x**, and these values comprise a target finite population. This finite population was stratified into five equal-sized strata based on the ordered z-values; the first stratum containing the 2000 units with smallest z-values, and so on.

Samples of size 200 were selected from this stratified population using the designs indicated in Table 8.4. The set of designs D_3 to D_6 have allocations which increase with the size of z and hence are related to probability proportional to size sampling. The designs D_7 to D_9 are U-shaped designs which should be optimal for linear regressions.

The sampling procedure was repeated independently 1000 times for each design. Averaging over all the 1000 samples for a particular design is thus equivalent to finding properties of the procedures given only the values, $\mathbf{z}_U$, of the design variable.

We evaluated conditional properties in the manner of Royall and Cumberland (1981) by ordering the 1000 samples for a particular design by the value of some function of z of interest (this depended on the estimator) and dividing them into 20 groups, each of 50 samples. These results are presented graphically for each procedure.

8.5.2 Principal Component Analysis

We first consider a principal component analysis based on the covariance matrix. For each sample the estimators $(\hat{\lambda}_{i,0}, \hat{\gamma}_{i,0})$, $(\hat{\lambda}_{i,m}, \hat{\gamma}_{i,m})$, $(\hat{\lambda}_{i,w}, \hat{\gamma}_{i,w})$, $(\hat{\lambda}_{i,wm}, \hat{\gamma}_{i,wm})$ of the superpopulation eigenvalues and eigenvectors were obtained using equations (8.5) and (8.6) and the appropriate estimator of $\mathbf{\Sigma}_{yy}$ from Section 8.2. Table 8.5

Table 8.4 Sample designs employed

Design		Stratum sample sizes			
	n_1	n_2	n_3	n_4	n_5
D_1 SRS	←		200		→
D_2 Proportionate	40	40	40	40	40
D_3 Increasing allocations	25	30	40	50	55
D_4 Increasing allocations	10	30	40	50	70
D_5 Increasing allocations	10	10	20	60	100
D_6 Increasing allocations	2	5	10	33	150
D_7 U-shaped allocations	60	30	20	30	60
D_8 U-shaped allocations	90	9	2	9	90
D_9 U-shaped allocations	97	2	2	2	97

Table 8.5 Principal component analysis of superpopulation covariance matrix Σ_{yy} and finite population covariance matrix V_{yyU}

	Analysis of $\mathbf{\Sigma}_{yy}$		Analysis of $\mathbf{V}_{yyU}$	
	Eigenvalue	Proportion of variance	Eigenvalue	Proportion of variance
Component i	λ_i	$\lambda_i \Big/ \sum_i \lambda_i$	L_i	$L_i \Big/ \sum_i L_i$
1	2.965	0.477	2.980	0.478
2	1.526	0.246	1.555	0.250
3	0.840	0.135	0.820	0.132
4	0.392	0.063	0.387	0.062
5	0.320	0.051	0.315	0.051
6	0.172	0.028	0.173	0.028

displays the eigenvalues of the superpopulation covariance matrix $\mathbf{\Sigma}_{yy}$. Since samples are drawn from the finite population we also show the eigenvalues, L_i, of the finite population covariance matrix $\mathbf{V}_{yyU}$. The results are very similar.

Unconditional Results

In Table 8.6 we present the means and standard deviations of the sampling distribution of the first eigenvalue, $\hat{\lambda}_1$, for each of the four estimation procedures described in Section 8.2, based on 1000 independent replications for each

Table 8.6 Unconditional means and standard deviations of the estimators of λ_1 (= 2.97) over 1000 replications for each design

	Means				Standard deviations			
Design	$\hat{\lambda}_{1,0}$	$\hat{\lambda}_{1,m}$	$\hat{\lambda}_{1,w}$	$\hat{\lambda}_{1,wm}$	$\hat{\lambda}_{1,0}$	$\hat{\lambda}_{1,m}$	$\hat{\lambda}_{1,w}$	$\hat{\lambda}_{1,wm}$
D_1	3.00	2.99	2.99	2.99	0.29	0.21	0.29	0.21
D_2	3.01	3.00	3.00	3.00	0.26	0.22	0.26	0.22
D_3	2.87	3.00	3.00	3.00	0.26	0.22	0.29	0.23
D_4	2.57	3.00	3.02	3.01	0.24	0.25	0.39	0.31
D_5	2.46	3.01	3.02	3.02	0.23	0.25	0.39	0.32
D_6	2.00	3.02	3.16	3.17	0.18	0.30	0.83	0.69
D_7	3.84	3.00	3.00	3.01	0.31	0.19	0.24	0.20
D_8	5.05	2.98	3.04	3.04	0.35	0.16	0.35	0.33
D_9	5.28	2.96	3.09	3.09	0.35	0.16	0.56	0.55

sampling design. Thus each entry in Table 8.6 is an average or standard deviation of 1000 values.

For the self-weighting designs, D_1 and D_2, all four procedures give approximately unbiased estimators in accord with the theoretical results in Section 8.3.2, with the adjusted estimators showing some useful gains in efficiency. For the unequally weighted designs, D_3 to D_9, the 'SRS' estimator, $\hat{\lambda}_{1,0}$, is severely biased as predicted by the results in Table 8.2. This bias is not accompanied by an increase in the standard deviation which means that interval estimates would also be unreliable. The two weighted estimators show evidence of bias for the extreme designs D_6 and D_9. These are accompanied by increases in standard deviations, but these are not sufficient to account for all the bias. For all designs apart from D_6 and D_9, all the estimators except the 'SRS' estimator perform very well on average over all samples (unconditionally) with the adjusted estimator, $\hat{\lambda}_{1,m}$, giving significant gains in efficiency. A surprising result is that the weighted MLE, $\hat{\lambda}_{1,wm}$, gives only very small gains in efficiency over the simple weighted estimator, $\hat{\lambda}_{1,w}$. It had been hoped that this estimator would achieve some of the expected gains in efficiency of the unweighted MLE while retaining the desirable properties of the p-weighted estimator.

In Table 8.7 we present results for the second eigenvalue, $\hat{\lambda}_2$. These results show quite a different pattern. All four procedures, including the 'SRS' estimator, $\hat{\lambda}_{2,0}$, show little evidence of bias, and now both $\hat{\lambda}_{2,0}$ and $\hat{\lambda}_{2,m}$ show useful gains in efficiency over the probability-weighted estimators, $\hat{\lambda}_{2,w}$ and $\hat{\lambda}_{2,wm}$, for the unequal probability designs D_3 to D_9. The reason for the absence of bias in $\hat{\lambda}_{2,0}$ is that the second principal component, u_2, and the design variable, z, are almost uncorrelated, and so selection on z has no effect on the estimator. Again this is in accord with the theoretical results.

Table 8.7 Unconditional means and standard deviations of the estimators of λ_2 ($= 1.53$) over 1000 replications for each design

	Means				Standard deviations			
Design	$\hat{\lambda}_{2,0}$	$\hat{\lambda}_{2,m}$	$\hat{\lambda}_{2,w}$	$\hat{\lambda}_{2,wm}$	$\hat{\lambda}_{2,0}$	$\hat{\lambda}_{2,m}$	$\hat{\lambda}_{2,w}$	$\hat{\lambda}_{2,wm}$
D_1	1.56	1.55	1.55	1.55	0.14	0.14	0.14	0.14
D_2	1.55	1.55	1.55	1.55	0.16	0.15	0.15	0.15
D_3	1.57	1.56	1.56	1.56	0.15	0.15	0.15	0.15
D_4	1.56	1.56	1.54	1.54	0.15	0.15	0.18	0.18
D_5	1.56	1.56	1.52	1.53	0.15	0.15	0.22	0.22
D_6	1.54	1.56	1.46	1.47	0.14	0.15	0.32	0.33
D_7	1.56	1.55	1.54	1.54	0.15	0.15	0.16	0.16
D_8	1.56	1.55	1.52	1.52	0.14	0.14	0.32	0.32
D_9	1.56	1.54	1.52	1.52	0.15	0.14	0.42	0.42

It is more difficult to present results for the eigenvectors $\hat{\gamma}_1$ and $\hat{\gamma}_2$ in a meaningful way. Following Skinner *et al.* (1986) we define the normalized mean of the sampling distribution of $\hat{\gamma}_i$ by

$$\bar{\gamma}_i^N = \bar{\gamma}_i(\bar{\gamma}_i'\bar{\gamma}_i)^{-1/2}, \qquad \text{where } \bar{\gamma}_i = \sum_{j=1}^{r} \hat{\gamma}_{ij}/r,$$

where $\hat{\gamma}_{ij}$ is the value of $\hat{\gamma}_i$ for the jth replication, and $r = 1000$ is the number of replications for each design. The discrepancy between the mean estimated vector, $\bar{\gamma}_i^N$, and the true vector, γ_i, is then measured by the Euclidean distance

$$d(\bar{\gamma}_i^N, \gamma_i) = (\bar{\gamma}_i^N - \gamma_i)'(\bar{\gamma}_i^N - \gamma_i).$$

In Table 8.8 we present these distances for the first two eigenvectors.

The results are similar to those for the eigenvalues. The 'simple random sampling' estimator, $\hat{\gamma}_{1,0}$, shows large biases for the unequally weighted designs, whereas for $\hat{\gamma}_{2,0}$ there are very small biases. The other three procedures all perform well with the probability-weighted estimators, $\hat{\gamma}_{i,w}$ and $\hat{\gamma}_{i,wm}$, showing some unexpectedly good results for some of the extreme designs.

Our overall conclusion is that this simulation study has confirmed the theoretical results. For samples of size 200 the approximations implicit in the asymptotic results of Section 8.3.2 work very well. If the analyst believes that the assumptions of linearity and homoscedasticity are satisfied then the MLEs $(\hat{\lambda}_{i,m}, \hat{\gamma}_{i,m})$ should be employed because of their potential gains in efficiency. If the assumptions are not satisfied then the position is quite different, as we shall see in Section 8.6.

Table 8.8 Euclidean distance between γ_i and normalized means of its estimators over 1000 replications for each design ($i = 1, 2$)

	Euclidean distance, $d(\bar{\gamma}_1^N, \gamma_1)$				Euclidean distance, $d(\bar{\gamma}_2^N, \gamma_2)$			
Design	$\hat{\gamma}_{1,0}$	$\hat{\gamma}_{1,m}$	$\hat{\gamma}_{1,w}$	$\hat{\gamma}_{1,wm}$	$\hat{\gamma}_{2,0}$	$\hat{\gamma}_{2,m}$	$\hat{\gamma}_{2,w}$	$\hat{\gamma}_{2,wm}$
D_1	0.012	0.012	0.012	0.012	0.028	0.028	0.028	0.028
D_2	0.014	0.014	0.014	0.014	0.027	0.027	0.027	0.027
D_3	0.008	0.011	0.012	0.013	0.020	0.020	0.023	0.023
D_4	0.046	0.012	0.018	0.018	0.032	0.021	0.025	0.025
D_5	0.045	0.023	0.016	0.018	0.020	0.026	0.028	0.028
D_6	0.143	0.034	0.033	0.035	0.033	0.043	0.030	0.027
D_7	0.060	0.014	0.017	0.017	0.023	0.018	0.027	0.027
D_8	0.102	0.026	0.013	0.013	0.034	0.024	0.013	0.013
D_9	0.108	0.029	0.010	0.012	0.034	0.026	0.022	0.021

Conditional Results

In Tables 8.1 and 8.2 we presented the conditional expectations of the four estimators of $\boldsymbol{\mu}_y$ and $\boldsymbol{\Sigma}_{yy}$ by averaging over repeated realizations of the dependent variables $\mathbf{y}$ given z for a fixed sample, s. We have argued that this distribution is relevant for inference because it is the distribution of samples like the one actually drawn. Holt *et al.* (1980) carried out a simulation study based on this conditional distribution in the context of regression analysis which was reported in Chapter 7. In this study we use the randomization sampling distribution described in Section 8.5.1 and we approximate the conditional distribution by averaging samples with similar values of some sample statistic of the z's.

Since the expectations of the estimators of $\boldsymbol{\Sigma}_{yy}$, conditional on the sample s and on $\mathbf{z}$, depend on the sample only through the values of v_{zzs} (or v^*_{zzs}) under the linear homoscedastic model, suitable statistics might be

$$\Delta_{zz} = \frac{v_{zzs} - v_{zzU}}{v_{zzU}} \quad \text{and} \quad \Delta^*_{zz} = \frac{v^*_{zzs} - v_{zzU}}{v_{zzU}},$$

depending on whether or not the estimator is weighted.

In Figures 8.1–8.4 we show plots of the conditional means for each of the four estimators of λ_1. The results for all nine designs are shown on each plot.

These plots show that both $\hat{\lambda}_{1,0}$ and $\hat{\lambda}_{1,w}$ are conditionally biased when Δ_{zz}(or

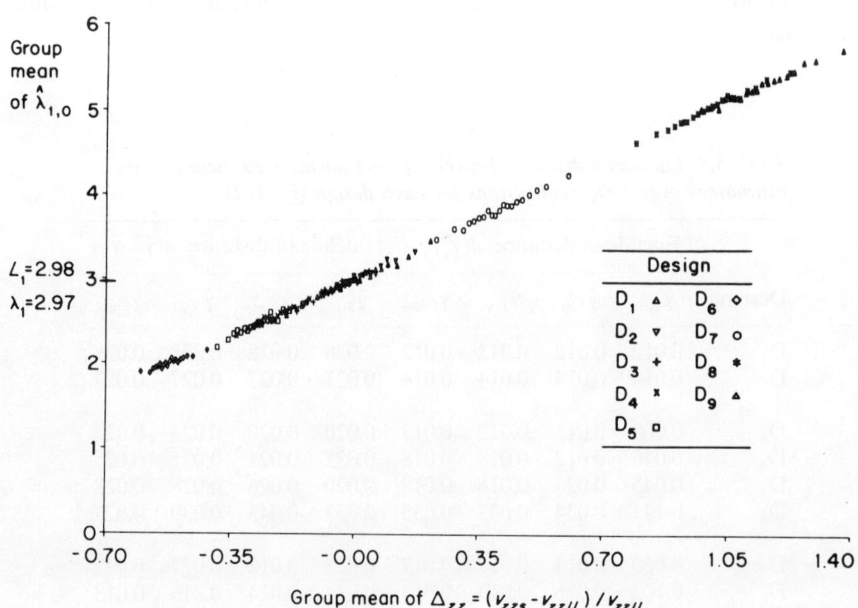

Figure 8.1 Simulated data: plot of group mean of $\hat{\lambda}_{1,0}$ against group mean of Δ_{zz}.

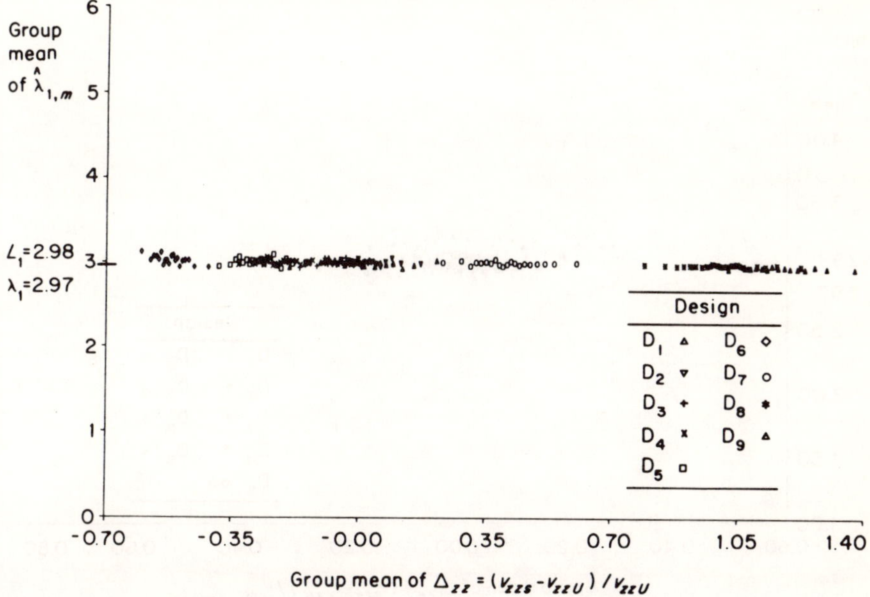

Figure 8.2 Simulated data: plot of group mean of $\hat{\lambda}_{1,m}$ against group mean of Δ_{zz}.

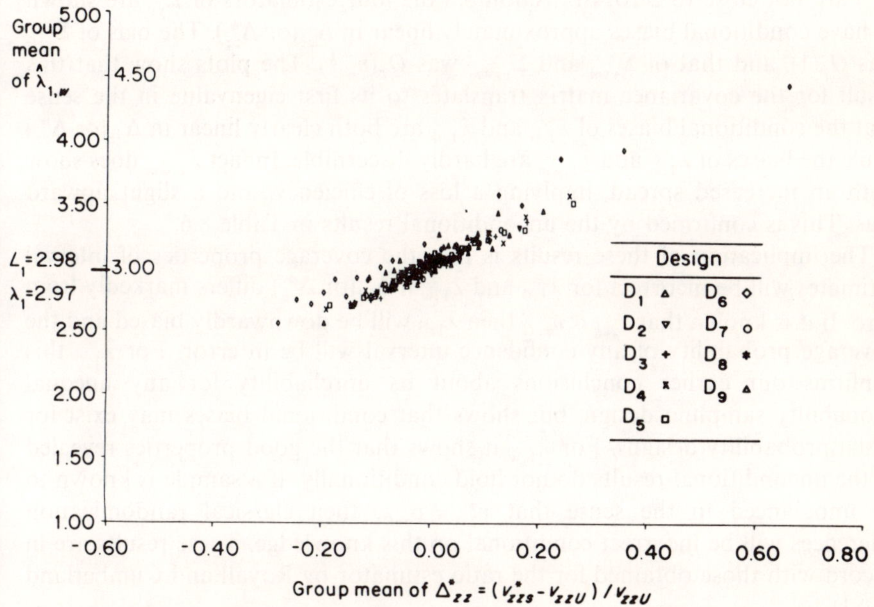

Figure 8.3 Simulated data: plot of group mean of $\hat{\lambda}_{1,w}$ against group mean of Δ^*_{zz}.

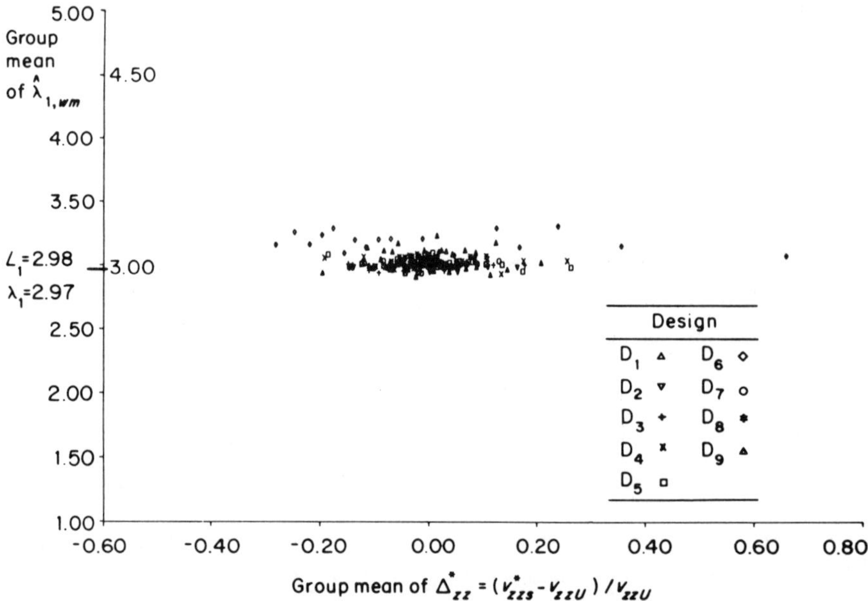

Figure 8.4 Simulated data: plot of group mean of $\hat{\lambda}_{1,wm}$ against group mean of Δ^*_{zz}.

Δ^*_{zz}) are not close to zero. In Section 8.3 the four estimators of Σ_{yy} are shown to have conditional biases approximately linear in Δ_{zz}(or Δ^*_{zz}). The bias of $\hat{\Sigma}_{yyO}$ was $O_p(1)$, and that of $\hat{\Sigma}_{yym}$ and $\hat{\Sigma}_{yywm}$ was $O_p(n_o^{-1})$. The plots show that this result for the covariance matrix translates to its first eigenvalue in the sense that the conditional biases of $\hat{\lambda}_{1,0}$ and $\hat{\lambda}_{1,w}$ are both clearly linear in Δ_{zz}(or Δ^*_{zz}) while the biases of $\hat{\lambda}_{1,m}$ and $\hat{\lambda}_{1,wm}$ are hardly discernible. In fact $\hat{\lambda}_{1,wm}$ does show both an increased spread, implying a loss of efficiency, and a slight upward bias. This is confirmed by the unconditional results in Table 8.6.

The implication of these results is that the coverage properties of interval estimates will be incorrect for $\hat{\lambda}_{1,0}$ and $\hat{\lambda}_{1,w}$ if Δ_{zz}(or Δ^*_{zz}) differs markedly from zero. If it is known that $v_{zzs} \ll v_{zzU}$ then $\hat{\lambda}_{1,0}$ will be downwardly biased and the coverage probability of any confidence interval will be in error. For $\lambda_{1,0}$ this confirms our earlier conclusions about its unreliability for any unequal probability sampling design, but shows that conditional biases may exist for equal probability designs. For $\hat{\lambda}_{1,w}$ it shows that the good properties revealed in the unconditional results do not hold conditionally. If a sample is known to be unbalanced in the sense that $v^*_{zzs} \neq v_{zzU}$, then classical randomization inferences will be incorrect conditional on this knowledge. These results are in accord with those obtained for the ratio estimator by Royall and Cumberland (1981).

8.5.3 Regression Analysis

To see if estimators of other functions of the covariance matrix behaved in a similar way to estimators of principal components we also estimated simple linear regression coefficients. For illustrative purposes we present the results for the regression of expenditure on food on total household expenditure, with expenditure on housing being used as the single design variable (all variables in logarithmic form). For each sample four estimators, $\hat{\beta}_0$, $\hat{\beta}_m$, $\hat{\beta}_w$, and $\hat{\beta}_{wm}$, of the true superpopulation regression slope β $(=0.71)$, or of the finite population regression slope $B_U(=0.72)$, were calculated using equation (8.4) and the appropriate estimator of $\boldsymbol{\Sigma}_{yy}$ from Section 8.2.

The results presented in Table 8.9 show that unconditionally, over all 1000 samples, $\hat{\beta}_0$, the ordinary least squares estimator, shows significant biases when the design is not self-weighting. The other three estimators are all approximately unbiased for all designs with the adjusted estimator, $\hat{\beta}_m$, giving some gains in efficiency for designs D_4 to D_9.

The results for the conditional distributions are presented in Figures 8.5–8.8. They show clearly that both $\hat{\beta}_0$ and $\hat{\beta}_w$ have significant conditional biases while the two adjusted estimators, $\hat{\beta}_m$ and $\hat{\beta}_{wm}$, are approximately conditionally unbiased. These results are very similar to those obtained for the first eigenvalue and show that when the design variable is correlated with the survey variables some form of regression adjustment is needed to remove the conditional biases of both $\hat{\beta}_0$ and $\hat{\beta}_w$.

***Table 8.9** Unconditional means and standard deviations of the estimators of $\boldsymbol{\beta}$ $(=\mathbf{0.71})$ over 1000 replications for each design*

	Means				Standard deviations			
Design	$\hat{\beta}_0$	$\hat{\beta}_m$	$\hat{\beta}_w$	$\hat{\beta}_{wm}$	$\hat{\beta}_0$	$\hat{\beta}_m$	$\hat{\beta}_w$	$\hat{\beta}_{wm}$
D_1	0.721	0.721	0.721	0.721	0.041	0.041	0.041	0.041
D_2	0.721	0.721	0.721	0.721	0.041	0.041	0.041	0.041
D_3	0.725	0.722	0.719	0.719	0.041	0.041	0.043	0.043
D_4	0.735	0.725	0.722	0.722	0.041	0.041	0.054	0.054
D_5	0.737	0.724	0.722	0.722	0.042	0.042	0.063	0.062
D_6	0.746	0.721	0.720	0.719	0.041	0.044	0.110	0.109
D_7	0.702	0.719	0.723	0.723	0.039	0.039	0.043	0.043
D_8	0.677	0.711	0.716	0.716	0.036	0.036	0.085	0.085
D_9	0.673	0.710	0.719	0.719	0.035	0.037	0.123	0.123

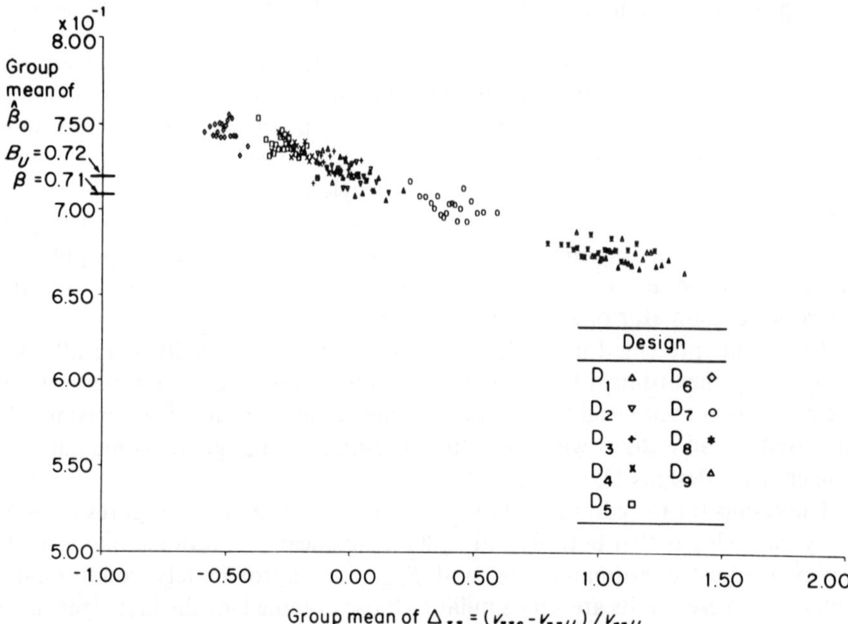

Figure 8.5 Simulated data: plot of group mean of $\hat{\beta}_0$ against group mean of Δ_{zz}.

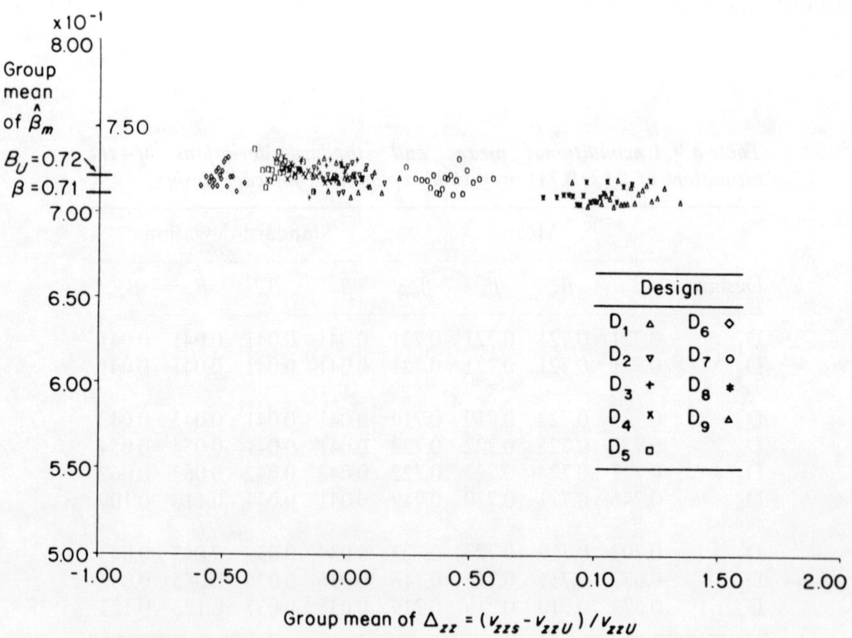

Figure 8.6 Simulated data: plot of group mean of $\hat{\beta}_m$ against group mean of Δ_{zz}.

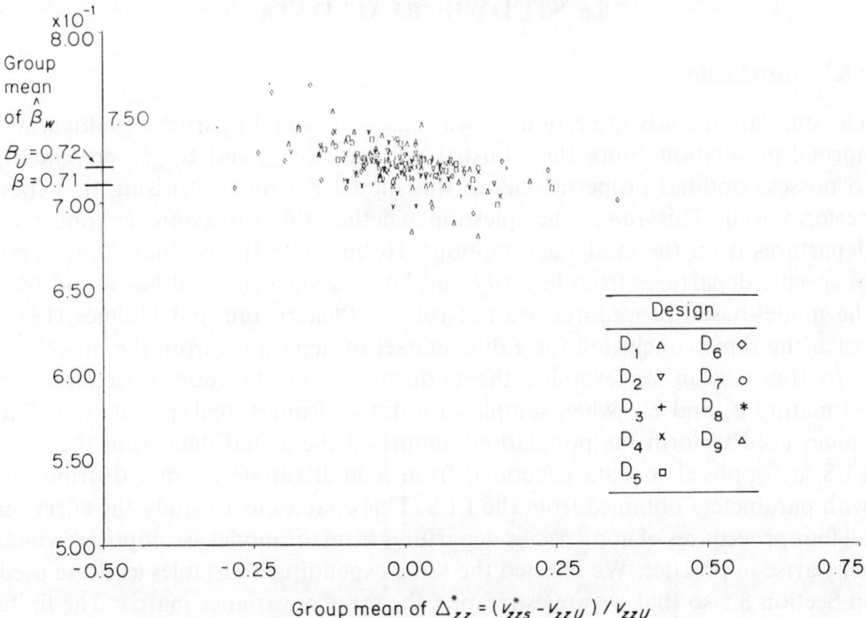

Figure 8.7 Simulated data: plot of group mean of $\hat{\beta}_w$ against group mean of Δ^*_{zz}.

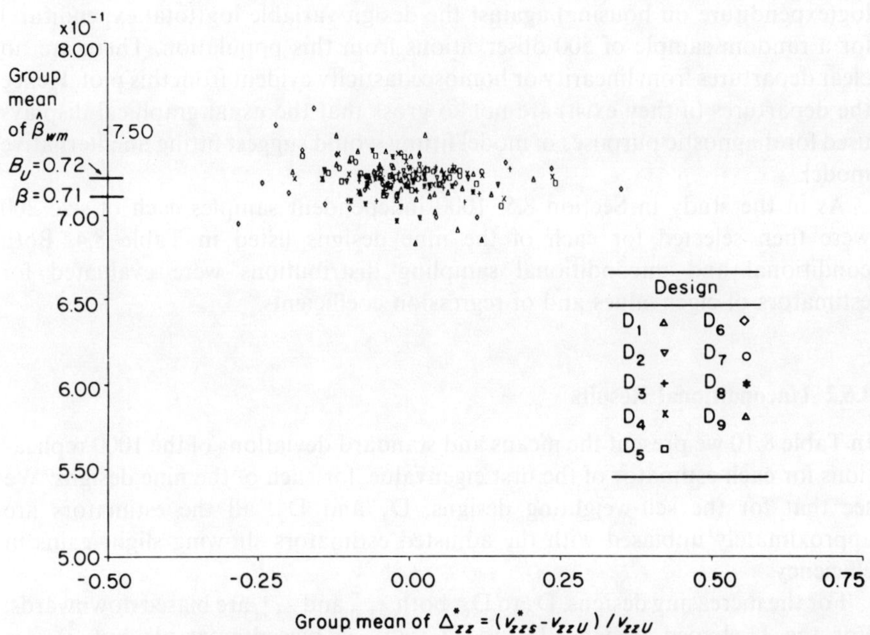

Figure 8.8 Simulated data: plot of group mean of $\hat{\beta}_{wm}$ against group mean of Δ^*_{zz}.

8.6 STUDY II: 'REAL' DATA

8.6.1 Introduction

The simulation study in Section 8.5 was based on sampling from a multivariate normal population. Since the adjusted estimators, $\hat{\boldsymbol{\mu}}_{ym}$ and $\hat{\boldsymbol{\Sigma}}_{yym}$, were chosen to possess optimal properties under this model it is not surprising that they perform well. This raises the question whether this procedure is robust to departures from the model assumptions. Holmes (1987) has studied the effect of specific departures from linearity and homoscedasticity and has found that the model-based procedures are not robust. Pfeffermann and Holmes (1985) reach the same conclusion for a different set of departures from the model.

In this section we examine the performance of the four procedures for estimating $\boldsymbol{\mu}_y$ and $\boldsymbol{\Sigma}_{yy}$ when samples are drawn from a 'real' population. The values used to form the population comprised the actual data from the 1975 FES, as opposed to data generated from a multivariate normal distribution with parameters obtained from the FES. This enables us to study the effect on all four procedures of non-specific departures from the model assumptions which may arise in practice. We selected the same expenditure variables as those used in Section 8.5 so that we are estimating the same covariance matrix. The finite population was divided into five strata of approximately equal size according to the design variable. As an example, Figure 8.9 shows the scatter plot of log(expenditure on housing) against the design variable log(total expenditure) for a random sample of 500 observations from this population. There are no clear departures from linearity or homoscedasticity evident from this plot. Hence the departures (if they exist) are not so gross that the usual graphical displays used for diagnostic purposes in model fitting would suggest fitting an alternative model.

As in the study in Section 8.5, 1000 independent samples each of size 200 were then selected for each of the nine designs listed in Table 8.4. Both conditional and unconditional sampling distributions were evaluated for estimators of eigenvalues and of regression coefficients.

8.6.2 Unconditional Results

In Table 8.10 we present the means and standard deviations of the 1000 replications for each estimator of the first eigenvalue, for each of the nine designs. We see that for the self-weighting designs, D_1 and D_2, all the estimators are approximately unbiased with the adjusted estimators showing slight gains in efficiency.

For the increasing designs, D_3 to D_6, both $\hat{\lambda}_{1,0}$ and $\hat{\lambda}_{1,m}$ are biased downwards. For the U-shaped designs, D_7 to D_9, $\hat{\lambda}_{1,0}$ is biased upwards but $\hat{\lambda}_{1,m}$ is approximately unbiased. The SRS estimator $\hat{\lambda}_{1,0}$ is totally unreliable and cannot

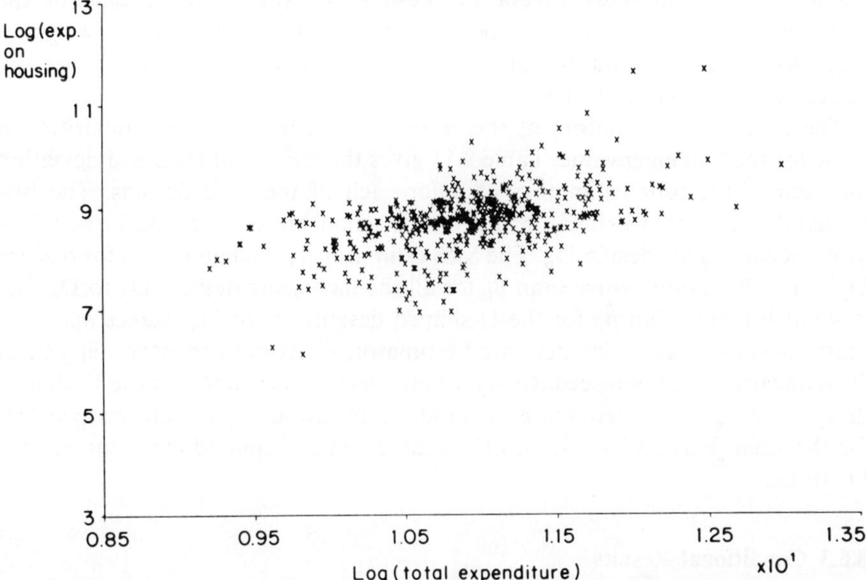

Figure 8.9 Plot of log (expenditure on housing) v. log (total expenditure) for a random sample of size 500.

be recommended. The model-based estimator, $\hat{\lambda}_{1,m}$, has a greatly reduced bias compared to $\hat{\lambda}_{1,0}$ for every design. However, the bias is still substantial for the increasing designs, ranging from 4.4% for design D_3 to 25.4% for design D_6. The two probability-weighted estimators, on the other hand, are approximately

Table 8.10 Unconditional means and standard deviations of the estimators of L_1 (= 2.97) over 1000 replications for each design

	Means				Standard deviations			
Design	$\hat{\lambda}_{1,0}$	$\hat{\lambda}_{1,m}$	$\hat{\lambda}_{1,w}$	$\hat{\lambda}_{1,wm}$	$\hat{\lambda}_{1,0}$	$\hat{\lambda}_{1,m}$	$\hat{\lambda}_{1,w}$	$\hat{\lambda}_{1,wm}$
D_1	2.99	2.99	2.97	2.99	0.30	0.25	0.30	0.25
D_2	3.00	2.99	2.99	2.99	0.28	0.25	0.28	0.25
D_3	2.70	2.83	2.99	3.00	0.24	0.24	0.30	0.28
D_4	2.25	2.64	3.02	3.02	0.21	0.24	0.46	0.39
D_5	2.10	2.51	3.00	3.00	0.21	0.25	0.49	0.43
D_6	1.68	2.21	3.25	3.23	0.17	0.25	1.09	0.96
D_7	3.86	3.00	2.98	2.98	0.30	0.21	0.23	0.22
D_8	5.15	3.05	3.01	3.01	0.36	0.18	0.35	0.34
D_9	5.41	3.06	3.05	3.06	0.37	0.18	0.53	0.52

unbiased unconditionally except for design D_6, the most extreme of the increasing designs. Design D_6 appears to be so extreme that even p-weighted estimators, which should be approximately unbiased unconditionally for any design, show evidence of bias.

The results for estimators of the regression coefficient show similarities to those for the first eigenvalue. Table 8.11 gives the mean and standard deviation for each of the four estimators of β for each of the nine designs. The two probability-weighted estimators have only small biases with the largest bias again occurring for design D_6. The SRS estimator $\hat{\beta}_0$ is again biased for designs D_3 to D_9. $\hat{\beta}_m$ is now worse than $\hat{\beta}_0$ for all the increasing designs D_3 to D_6, but it is much better than $\hat{\beta}_0$ for the U-shaped designs D_7 to D_9. Hence again we must conclude that the model-based estimator, $\hat{\beta}_m$, is not robust to departures from linearity and homoscedasticity for all selection schemes. For the U-shaped designs, however, the reduction in standard deviation more than compensates for the small bias and so $\hat{\beta}_m$ has the smallest mean-squared error for designs D_7 to D_9.

8.6.3 Conditional Results

When an estimator is biased unconditionally it must also show some conditional biases. But as we have seen in Section 8.5, even when an estimator is unbiased unconditionally it may still show significant conditional biases. In Figures 8.10–8.13 we show the plots of the conditional means of the first eigenvalue presented in the same way as Figures 8.1–8.4. Both $\hat{\lambda}_{1,0}$ and $\hat{\lambda}_{1,w}$ show a significant bias linear in Δ_{zz} (Δ^*_{zz}). This is consistent with the results for the

Table 8.11 Unconditional means and standard deviations of the estimators of B_U (= 0.71) over 1000 replications for each design

	Means				Standard deviations			
Design	$\hat{\beta}_0$	$\hat{\beta}_m$	$\hat{\beta}_w$	$\hat{\beta}_{wm}$	$\hat{\beta}_0$	$\hat{\beta}_m$	$\hat{\beta}_w$	$\hat{\beta}_{wm}$
D_1	0.714	0.713	0.714	0.713	0.047	0.047	0.047	0.047
D_2	0.714	0.713	0.714	0.713	0.048	0.047	0.048	0.047
D_3	0.701	0.694	0.712	0.711	0.051	0.060	0.049	0.049
D_4	0.693	0.668	0.711	0.708	0.056	0.066	0.063	0.063
D_5	0.669	0.645	0.706	0.703	0.058	0.066	0.065	0.066
D_6	0.656	0.608	0.699	0.691	0.063	0.063	0.111	0.116
D_7	0.686	0.704	0.711	0.710	0.046	0.046	0.051	0.060
D_8	0.660	0.698	0.701	0.700	0.042	0.044	0.088	0.087
D_9	0.658	0.701	0.712	0.712	0.040	0.043	0.123	0.123

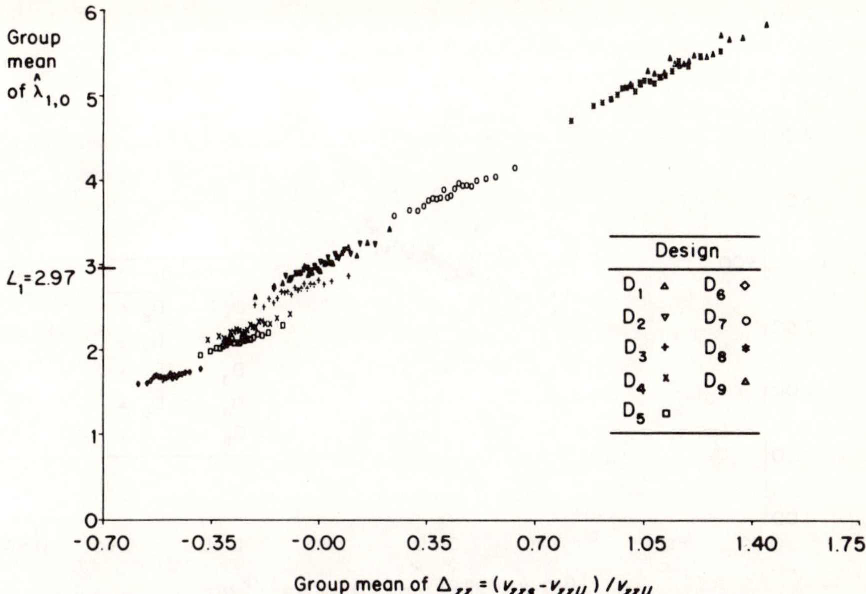

Figure 8.10 Real data: plot of group mean of $\hat{\lambda}_{1,0}$ against group mean of Δ_{zz}.

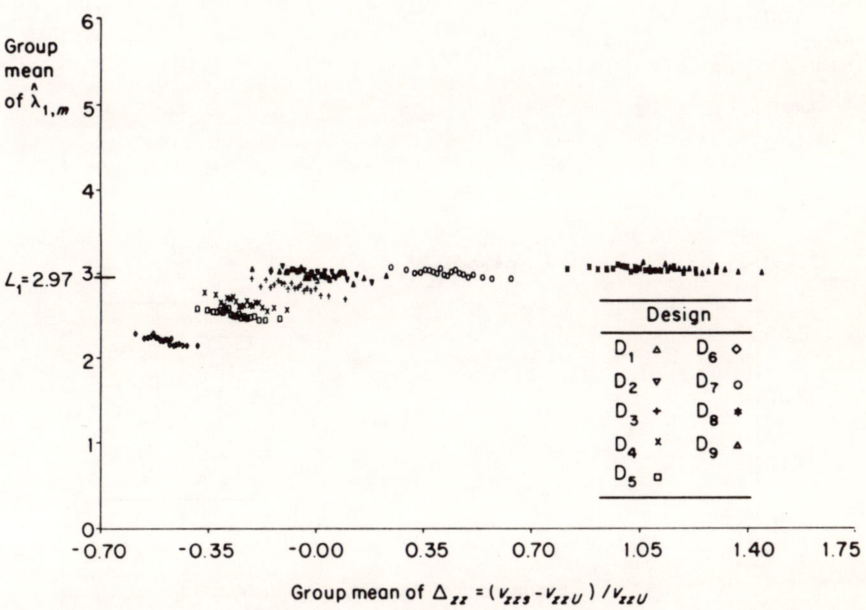

Figure 8.11 Real data: plot of group mean of $\hat{\lambda}_{1,m}$ against group mean of Δ_{zz}.

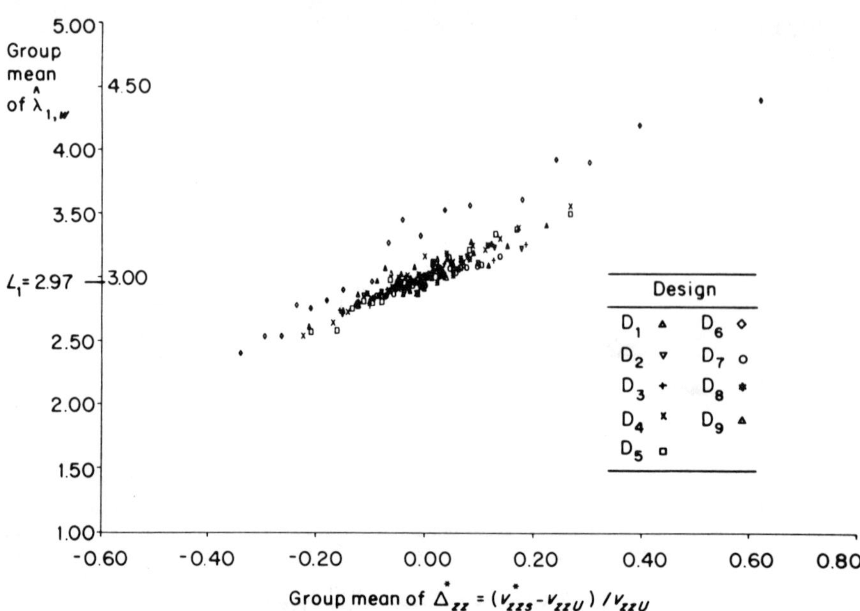

Figure 8.12 Real data: plot of group mean of $\hat{\lambda}_{1,w}$ against group mean of Δ^*_{zz}.

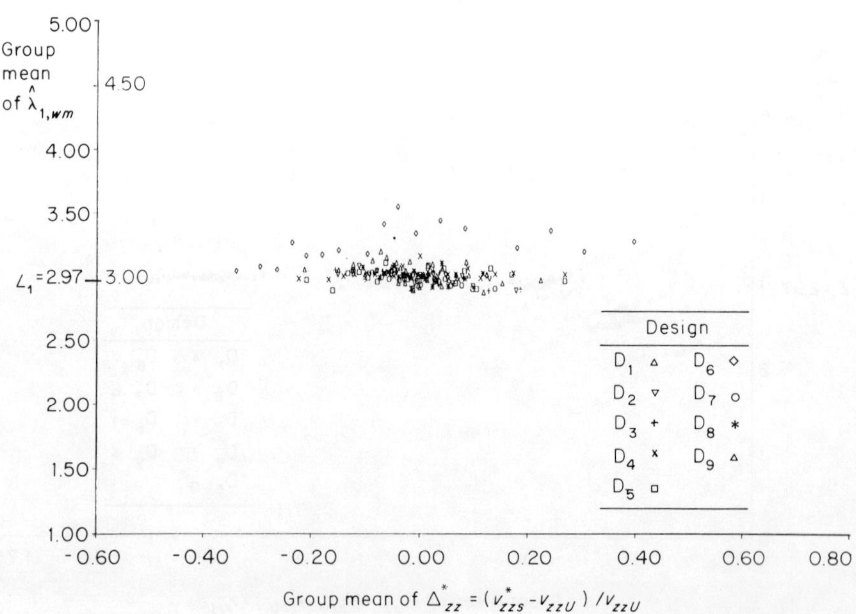

Figure 8.13 Real data: plot of group mean of $\hat{\lambda}_{1,wm}$ against group mean of Δ^*_{zz}.

multivariate normal distribution. The results for $\hat{\lambda}_{1,m}$ are interesting. They show that within a design there is no significant bias which varies with Δ_{zz}; the conditional bias is approximately the same as the unconditional bias within each design. The estimator $\hat{\lambda}_{1,wm}$ shows the best overall performance with no clear evidence of conditional bias. From these results we would conclude that the weighted adjusted estimator, $\hat{\lambda}_{1,wm}$, has the best all-round properties.

The corresponding results for estimation of β are shown in Figures 8.14–8.17. Again both $\hat{\beta}_0$ and $\hat{\beta}_w$ show biases varying with Δ_{zz} (Δ^*_{zz}); the bias of the weighted estimator appears to be less strong than for the eigenvalues. The model-based adjusted estimator, $\hat{\beta}_m$, performs badly for designs D_3 to D_6. There is evidence both of linear effects within designs and of non-linear effects across the increasing designs D_3 to D_6 for both $\hat{\beta}_0$ and $\hat{\beta}_m$. Again the weighted estimators $\hat{\beta}_w$ and $\hat{\beta}_{wm}$ have the best overall properties, neither having large conditional biases.

8.7 CONCLUSIONS

The simulation results in Section 8.5 confirmed the theoretical results of Section 8.3. The unweighted SRS estimator, which is the default option in most computer packages, can be severely biased for designs which are not self-weighting and should not be employed in this situation. Even for self-weighting

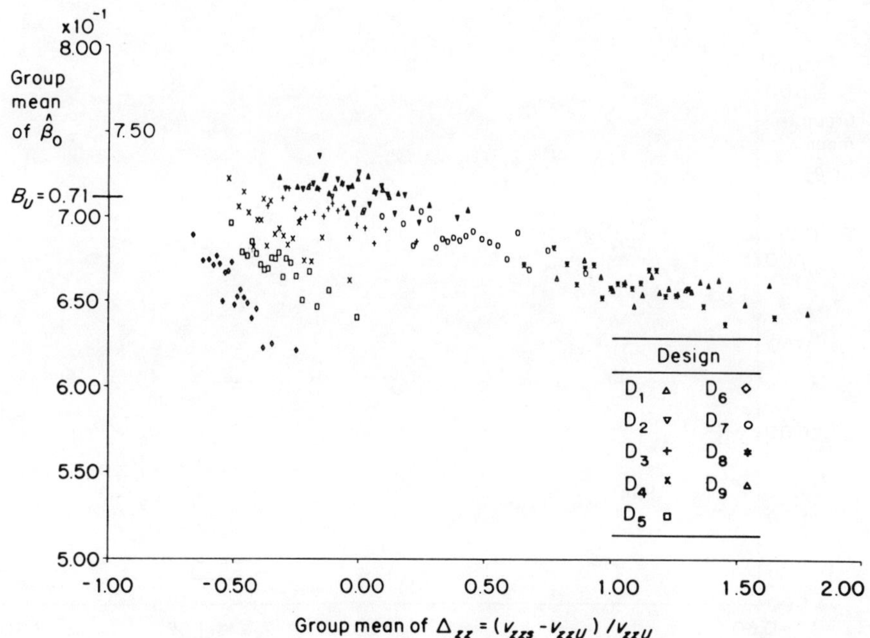

Figure 8.14 Real data: plot of group mean of $\hat{\beta}_0$ against group mean of Δ_{zz}.

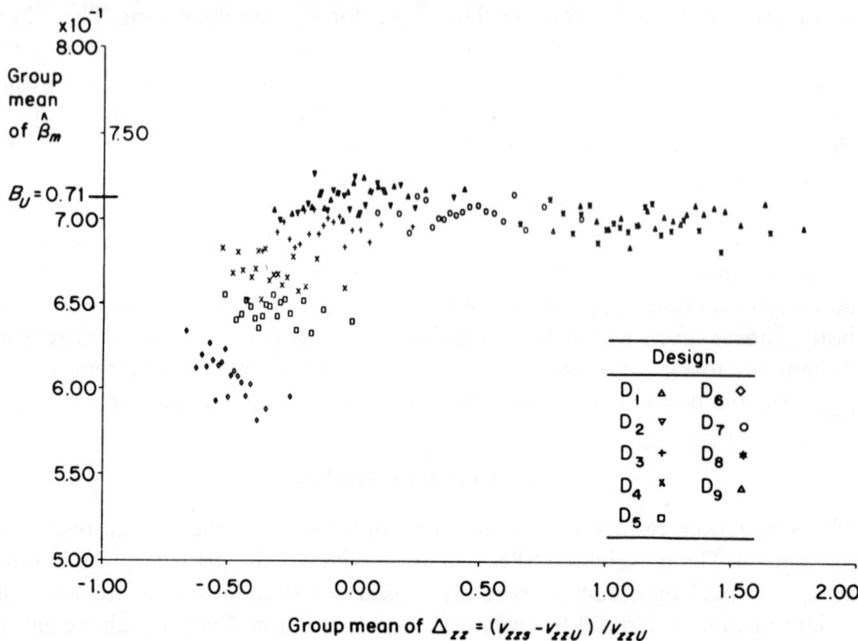

Figure 8.15 Real data: plot of group mean of $\hat{\beta}_m$ against group mean of Δ_{zz}.

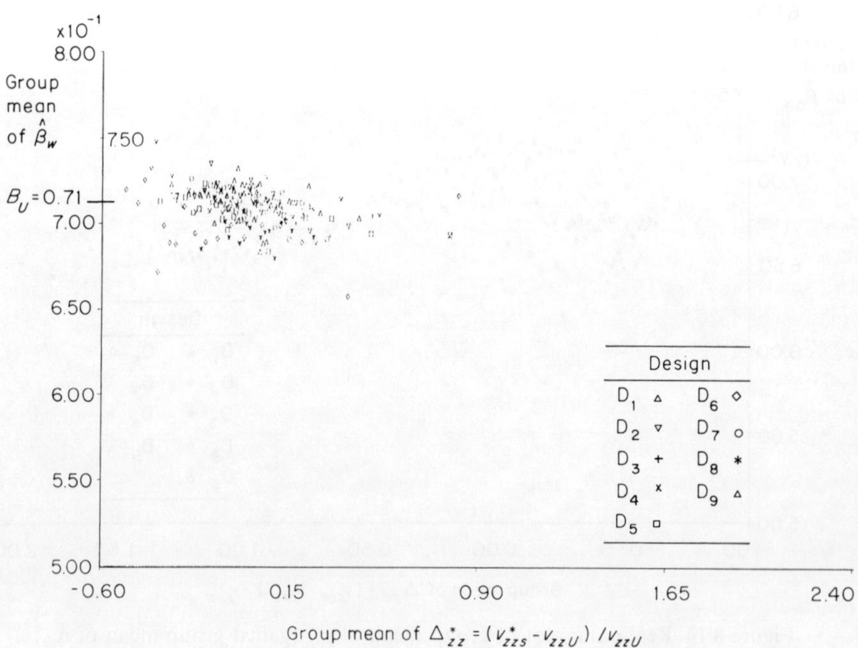

Figure 8.16 Real data: plot of group mean of $\hat{\beta}_w$ against group mean of Δ^*_{zz}.

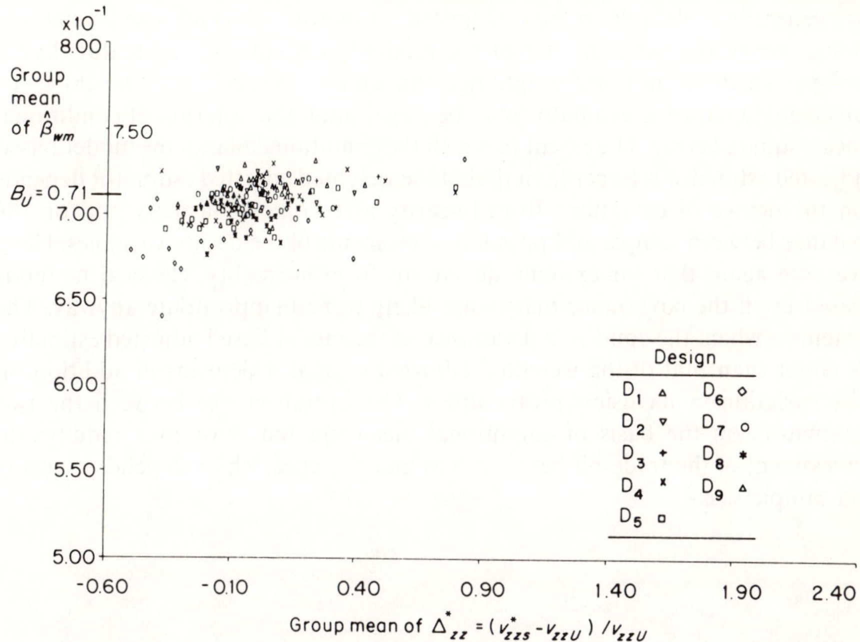

Figure 8.17 Real data: plot of group mean of $\hat{\beta}_{wm}$ against group mean of Δ^*_{zz}.

designs substantial conditional biases can arise if the sample is not balanced. The probability-weighted estimator which is constructed to be approximately unbiased unconditionally for any design can nevertheless have significant conditional biases.

The biases exhibited by these estimators can be substantially reduced by adjusting for differences between the population and sample variances of the design variable. Under ideal conditions of multivariate normality the model-based adjusted estimator works very well, having no noticeable conditional bias and showing substantial gains in efficiency under certain designs. The weighted version of this estimator is also approximately conditionally unbiased but is less efficient for most designs.

The simulations in Section 8.6 carried out with an 'approximately' normal population tell a different story. In particular they demonstrate that the model-based adjusted estimator is both conditionally and unconditionally biased under the increasing designs D_3 to D_6. For the other designs it still works very well and removes the conditional bias which is still present for the unweighted SRS estimators. The probability-weighted estimator of the first eigenvalue still shows a conditional bias but this virtually disappears for the regression coefficient. The weighted adjusted estimator is approximately conditionally unbiased for both estimation of the first eigenvalue and the regression coefficient.

We conclude that the adjusted estimators, whether weighted or model-based, are generally preferable to the unadjusted estimators. Thus for a self-weighting design we recommend the use of the model-based adjusted estimator. For a design which is not self-weighting, the choice between the weighted or unweighted adjusted estimator may be based upon the criterion of conditional mean-squared error. The extent to which the conditional bias of the model-based adjusted estimator is larger than that of the weighted adjusted estimator depends on the degree of deviations from linearity and homoscedasticity and on the balance between sample and population moments of the design variables. Here, we note again that for extreme deviations from normality, classical methods based upon the covariance matrix are likely to be inappropriate anyway. The extent to which the conditional variance of the model-based adjusted estimator is larger than that of the weighted adjusted estimator depends in addition on the variation in inclusion probabilities. The overall choice between the two estimators on the basis of conditional mean-squared error thus requires an assessment of the trade-off between bias and variance, which depends crucially on sample size.

CHAPTER 9

Selection Based on the Response Variable in Logistic Regression

A. J. Scott and C. J. Wild

9.1 INTRODUCTION

In this chapter we look at an important special class of surveys in which the selection probabilities are based directly on the value of the (discrete) response variable. Many surveys, particularly in the health and social sciences, are carried out to help assess the impact of possible explanatory variables, **X**, on a categorical response variable, Y. Roughly speaking, any such survey is an example of a case-control study if the sample is stratified by the value of the response variable. Although case-control studies are widespread, they have received little explicit attention in the sample survey literature. They have received a great deal of attention elsewhere, however, particularly in epidemiology and econometrics. In fact, case-control studies have been described as the 'backbone of epidemiology' in Breslow and Day (1980).

The response variable is often binary in health studies, with $Y = 1$ for an individual who contracts a disease (a 'case') and $Y = 0$ otherwise (a 'control'). However, there may be more categories when, for example, several histologies of a disease are of interest. In a prospective study, a sample of individuals would be chosen (without knowledge of Y) and their responses followed over time. Such studies are time-consuming and expensive; if the disease is rare even a large-scale study may produce very few cases and thus very little information on whether, or to what extent, explanatory variables are risk factors for the disease. In spite of numerous practical difficulties (e.g. with recollection errors and valid randomization), a case-control study can provide an efficient and feasible alternative. A good survey of the medical literature before 1980 is given by Breslow and Day (1980) and there has been an extensive literature since then.

A parallel development has been carried out in econometrics under the heading of choice-based sampling. Economists are often interested in finding out what factors influence individuals to make particular choices. For example, if the mode of transport to work is being investigated, it may be simpler and cheaper to take separate samples of individuals from bus terminals, train stations and car parks than to take a single sample from the whole population of commuters. Manski and McFadden (1981) provide a good introduction to the econometric literature. Examples can be found in many other fields. We have encountered case-control studies of the determinants of unemployment and of causes of business failure, for example.

Since all the developments have taken place in fields where stochastic modelling is second nature, the model-based approach is the conventional one in this situation and, in direct contrast to most other survey work, it is the design-based approach which is the newcomer whose value needs to be demonstrated. For simplicity we concentrate on the binary case where $Y=1$ for a case, and $Y=0$ for a control.

Primary interest centres on modelling

$$\mathscr{P}_1(\mathbf{x}) = P(Y=1|\mathbf{X}=\mathbf{x}),$$

i.e. the probability of being a case given the value $\mathbf{x}$ of the explanatory variables. A common assumption is that $\mathscr{P}_1(\mathbf{x})$ takes a logistic form:

$$\mathscr{P}_1(\mathbf{x}) = \exp(\mathbf{x}'\boldsymbol{\beta})/[1+\exp(\mathbf{x}'\boldsymbol{\beta})], \tag{9.1}$$

or equivalently

$$\log[\mathscr{P}_1(\mathbf{x})/\mathscr{P}_0(\mathbf{x})] = \mathbf{x}'\boldsymbol{\beta},$$

where

$$\mathscr{P}_0(\mathbf{x}) = 1 - \mathscr{P}_1(\mathbf{x}) = P(Y=0|\mathbf{X}=\mathbf{x}).$$

In a basic prospective study, a simple random sample (SRS) of n_o individuals is drawn from the population of N_o individuals. Assuming the population values $y_1, \ldots, y_{N_o}$ are generated independently through equation (9.1), the maximum likelihood estimator (MLE) $\hat{\boldsymbol{\beta}}$ of $\boldsymbol{\beta}$ and associated ML inference procedures are obtained using standard methods as implemented, for example, in GLIM (Payne, 1985), SPSS-X (1986, Chapter 31) or SAS (1985, p. 191).

In a case-control study, however, the sample is drawn using a stratified design with two strata, one consisting of the cases ($y_t=1$), and the other of the controls ($y_t=0$), generally with unequal sampling fractions in these two strata. From the arguments given in Chapter 6, we would therefore expect this selection scheme to induce bias into the conventional estimator $\hat{\boldsymbol{\beta}}$. In fact Prentice and Pyke (1979) have demonstrated the remarkable property of the logistic model that if the model includes a constant term, so $\boldsymbol{\beta}=(\beta_0, \boldsymbol{\beta}_1')'$, $\mathbf{x}=(1,\mathbf{x}_1')'$, $\mathbf{x}'\boldsymbol{\beta}=\beta_0+\mathbf{x}_1'\boldsymbol{\beta}_1$ and $\hat{\boldsymbol{\beta}}=(\hat{\beta}_0, \hat{\boldsymbol{\beta}}_1')$, then although $\hat{\beta}_0$ does become biased (and inconsistent) for β_0, $\hat{\boldsymbol{\beta}}_1$ remains the MLE of $\boldsymbol{\beta}_1$ (and is consistent for $\boldsymbol{\beta}_1$), and furthermore the standard errors of $\hat{\boldsymbol{\beta}}_1$, estimated under the SRS assumption, remain valid. In epidemiology, at least, $\boldsymbol{\beta}_1$ is the parameter of primary importance; when cases are rare in the population $\boldsymbol{\beta}_1$ measures the effect of the explanatory variables on the relative risks of individuals becoming cases. Prentice and Pyke's results are discussed further in Section 9.3, where a simple model-based adjustment of $\hat{\beta}_0$ is also given to correct for bias. This adjustment only requires knowledge of the relative sampling fractions.

Most medical case-control studies incorporate additional stratification on other variables such as race, sex, hospital, etc. Following the arguments in Part C, it may be sensible to disaggregate the model to account for this population structure. For example, the logistic model might be extended to include a separate constant for each of these primary strata so that we have in stratum h

$$P(Y=1\,|\,\mathbf{X}=\mathbf{x}) = \exp(\beta_{oh} + \mathbf{x}'_1\boldsymbol{\beta}_1)/[1+\exp(\beta_{oh}+\mathbf{x}'_1\boldsymbol{\beta}_1)]. \tag{9.2}$$

This model and the corresponding extensions of the Prentice–Pyke results are discussed in Section 9.5.

A fundamental assumption of the MLEs and the Prentice–Pyke results is that the logistic model is correct. This assumption may be weakened using the *pseudo MLE* approach described in Section 3.4.4, where model (9.1) or (9.2) is used only to suggest useful population parameters as targets for inference (Binder, 1983).

If the population values $(y_t, \mathbf{x}'_t)$, $t = 1, \ldots, N_o$, were all known and model (9.1) held then the likelihood would be

$$L(\boldsymbol{\beta}) = \prod_{t=1}^{N_O} p_1(\mathbf{x}_t;\boldsymbol{\beta})^{y_t} p_0(\mathbf{x}_t;\boldsymbol{\beta})^{1-y_t},$$

where

$$p_0(\mathbf{x};\boldsymbol{\beta}) = [1+\exp(\mathbf{x}'\boldsymbol{\beta})]^{-1}, \qquad p_1(\mathbf{x};\boldsymbol{\beta}) = 1 - p_0(\mathbf{x};\boldsymbol{\beta}).$$

Straightforward differentiation shows that the MLE of $\boldsymbol{\beta}$ solves the equations

$$N_o^{-1}\sum_{t=1}^{N_O} \mathbf{u}_t(\mathbf{b}) = \mathbf{0}$$

where

$$\mathbf{u}_t(\mathbf{b}) = [y_t - p_1(\mathbf{x}_t;\mathbf{b})]\mathbf{x}_t. \tag{9.3}$$

As $N_o \to \infty$, the MLE converges to $\mathbf{B}$ the solution of the equations

$$E_{Y\mathbf{X}}[\{Y - p_1(\mathbf{X};\mathbf{b})\}\mathbf{X}] = \mathbf{0}, \tag{9.4}$$

where $E_{Y\mathbf{X}}$ is the expectation with respect to the joint distribution of Y and $\mathbf{X}$. These equations are a special case of equations (3.45) used to define the target parameters in pseudo MLE.

Using the facts that $E_{Y\mathbf{X}} = E_{\mathbf{X}}E_{Y|\mathbf{X}}$ and $E_{Y|\mathbf{X}}(Y) = \mathscr{P}_1(\mathbf{X})$ we may rewrite equation (9.4) as

$$E_{\mathbf{X}}[\{\mathscr{P}_1(\mathbf{X}) - p_1(\mathbf{X};\mathbf{b})\}\mathbf{X}] = \mathbf{0}. \tag{9.5}$$

If the true model is logistic, i.e. $\mathscr{P}_1(\mathbf{X}) = p_1(\mathbf{X};\boldsymbol{\beta})$, it is clear that equation (9.5) is solved by $\mathbf{b} = \boldsymbol{\beta}$ and hence $\mathbf{B} = \boldsymbol{\beta}$. If the true model is not logistic then $\mathbf{B}$ may still be interpreted as defining that logistic model $p_1(\mathbf{X}; \mathbf{B})$ which 'best approximates' the true model $\mathscr{P}_1(\mathbf{X})$. In the pseudo ML approach the aim is to estimate $\mathbf{B}$ without making any assumption about the form of $\mathscr{P}_1(\mathbf{X})$. Of course, $\mathbf{B}$ would be of dubious relevance if $\mathscr{P}_1(\mathbf{X})$ cannot be approximated reasonably well by the logistic form.

To understand in what sense $p_1(\mathbf{X}; \mathbf{B})$ is a good approximation to $\mathscr{P}_1(\mathbf{X})$, define the *approximation error* at $\mathbf{x}$ as

$$\varepsilon(\mathbf{x}) = \mathscr{P}_1(\mathbf{x}) - p_1(\mathbf{x}; \mathbf{B}).$$

Then, letting $\mathbf{X} = (1, \mathbf{X}_1')'$, equation (9.5) may be rewritten as

$$E_{\mathbf{X}}[\varepsilon(\mathbf{X})] = 0, \qquad \mathrm{Cov}_{\mathbf{X}}[\varepsilon(\mathbf{X}), \mathbf{X}_1] = \mathbf{0},$$

i.e. $\mathbf{B}$ defines that logistic model for which the approximation error has zero mean and is uncorrelated with every variable in $\mathbf{X}_1$.

Example 1

Suppose X_1 is scalar and has a standard normal distribution in the population. Suppose the true model is not logistic but is given by

$$\log[\mathscr{P}_1(\mathbf{x})/\mathscr{P}_0(\mathbf{x})] = -5.8 + 4.0x_1 - 0.6x_1^2.$$

The choice of this model is discussed further in Section 9.4. It may be calculated that the solution of equation (9.5) is $\mathbf{B} = (-4.9, 2.4)$ in this case and so the best approximating logistic model is

$$\log[p_1(\mathbf{x}; \mathbf{B})/p_0(\mathbf{x}; \mathbf{B})] = -4.9 + 2.4x_1.$$

The true model and its approximation are displayed in Figure 9.1. Note how the choice of best approximating model depends on the distribution of X_1 in the given population. Greater importance is given to reducing the approximation error in regions where X_1 has high probability. Thus in Figure 9.1 the

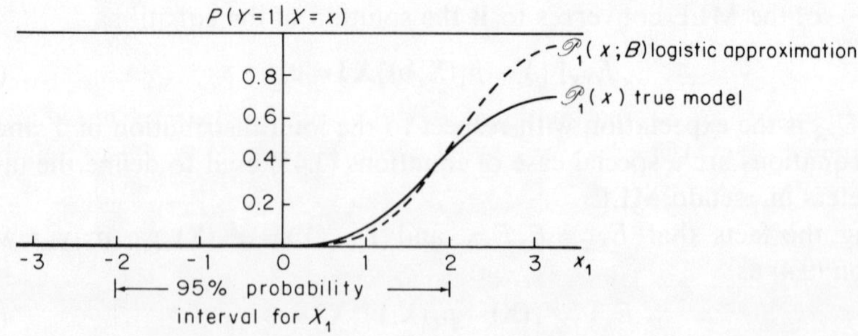

Figure 9.1 True model and its logistic approximation for Example 1.

approximation error is relatively small when $x_1 \in (-1.96, 1.96)$, which happens with probability 0.95, but can become relatively large when x_1 becomes large. Indeed as $x_1 \to \infty$ $\mathscr{P}_1(\mathbf{x}) \to 0$, whereas $p_1(\mathbf{x}; \mathbf{B}) \to 1$.

In Section 9.2 a design-based approach is described in which the parameter **B**, which solves equation (9.4), is estimated using weights inversely proportional to the sampling fractions in the two strata. This approach gives estimators which are design-unbiased for **B** under case-control selection even if the logistic model does not hold. This property is not shared by the model-based estimator of Section 9.3. However, as we would expect, the model-based estimator turns out to be more efficient when the logistic model does hold exactly. The loss of efficiency from using the design-based estimators is often fairly small but it can be large with a very disproportionate allocation of resources between cases and controls.

In theory, the advantages of the design-based approach should become apparent when the model is not followed exactly. In Section 9.4 we present an empirical study taken from Scott and Wild (1986). Clearly **B** would not be a sensible target for inference if the logistic model were grossly inadequate. We look at cases where the population is generated by a quadratic logistic regression model and the quadratic is large enough to be detected by the standard likelihood-ratio test with the given sample sizes about 50% of the time. The results are very similar to those for linear regression presented in Chapters 7 and 8. If we are willing to assume that we really are interested in **B**, then the design-based approach is affected very little by the presence of the quadratic term but it leads to an appreciable bias in the MLE. One disturbing feature is that we appear to need a relatively large sample size before the asymptotic theory gives reasonable approximations for the design-based estimator; the weighting by inverse selection probabilities leads to a smaller effective sample size.

Some large-scale case-control studies have much more complex schemes for selecting cases and controls and here there is no real alternative to the design-based approach. For example, the US National Center for Health Statistics makes available to researchers results from the 1980 National Natality Study, based on a sample of 9941 of the 3 612 258 live births in the USA, and from the corresponding Foetal Mortality Survey, based on a sample of 6386 of the 19 202 foetal deaths of at least 28 weeks' gestation. Both surveys involve complex sample designs and estimation procedures and there is no real possibility of developing exact MLEs, while estimation using pseudo MLE is relatively straightforward.

9.2 DESIGN-BASED ESTIMATORS

We start with the simplest case of an unstratified case-control study and a binary dependent variable, Y. Suppose we have a $p \times 1$ vector, $\mathbf{X}$, of explanatory variables with $P(Y=1|\mathbf{X}=\mathbf{x})=\mathscr{P}_1(\mathbf{x})$ and $P(Y=0|\mathbf{X}=\mathbf{x})=\mathscr{P}_0(\mathbf{x})$. For simplicity, assume that $\mathbf{X}$ is drawn from an (essentially) infinite population with joint density function $f(\mathbf{x})$. In an unstratified case-control study we draw a random sample of n_1 observations (cases), say $\mathbf{x}_{11},\ldots,\mathbf{x}_{1n_1}$, from the conditional distribution of $\mathbf{X}$ given $Y=1$ and an independent random sample of n_0 observations (controls), say $\mathbf{x}_{01},\ldots,\mathbf{x}_{0n_0}$, from the conditional distribution of $\mathbf{X}$, given $Y=0$. Thus, using Bayes's theorem, each $\mathbf{X}_{lt}(l=0,\,1;\,t=1,\ldots,n_l)$ has density

$$f_l(\mathbf{x})=\mathscr{P}_l(\mathbf{x})f(\mathbf{x})/W_l$$

where $W_1=\int\mathscr{P}_1(\mathbf{x})f(\mathbf{x})\mathrm{d}\mathbf{x}$ is the proportion of cases, and $W_0=1-W_1$ the proportion of controls, in the whole population.

The target parameter is taken to be $\mathbf{B}$ the solution of equation (9.4). The left-hand side of equation (9.4) may be rewritten as

$$\begin{aligned}\mathbf{U}(\mathbf{b})&=E_{Y\mathbf{X}}[\{Yp_0(\mathbf{X};\mathbf{b})-(1-Y)p_1(\mathbf{X};\mathbf{b})\}\mathbf{X}]\\&=W_1E_1\{\mathbf{X}p_0(\mathbf{X};\mathbf{b})\}-W_0E_0\{\mathbf{X}p_1(\mathbf{X};\mathbf{b})\},\end{aligned}\tag{9.6}$$

where E_l denotes expectation with respect to the conditional distribution of $\mathbf{X}$ given $Y=l(l=0,1)$.

For any fixed $\mathbf{b}, \mathbf{U}(\mathbf{b})$ is a linear function of means of the two populations (cases and controls) and so can be estimated from case-control data with the corresponding function of the sample means, viz.

$$\hat{\mathbf{U}}(\mathbf{b})=W_1\sum_{t=1}^{n_1}\frac{\mathbf{x}_{1t}p_0(\mathbf{x}_{1t};\mathbf{b})}{n_1}-W_0\sum_{t=1}^{n_0}\frac{\mathbf{x}_{0t}p_1(\mathbf{x}_{0t};\mathbf{b})}{n_0}.$$

The design-based estimator, $\mathbf{b}_d$ say, of $\mathbf{B}$ is given by the solution of

$$\hat{\mathbf{U}}(\mathbf{b})=\mathbf{0}.\tag{9.7}$$

(If $\mathbf{X}$ can take values only in a finite set, then this is equivalent to estimating the population total in each cell and then fitting a logistic model to the estimated totals as in Rao and Scott (1987) or Roberts *et al.* (1987).)

Properties of $\mathbf{b}_d$ can be derived as in Binder (1983) and Scott and Wild (1986). Using the weak law of large numbers, it may be shown that $\mathbf{b}_d$ converges to $\mathbf{B}$ in probability as $n_0,\, n_1\to\infty$.

It may be shown further that if $n=n_0+n_1$ and $n_l/n\to\rho_l$ then $\sqrt{n}(\mathbf{b}_d-\mathbf{B})$ is asymptotically normal with mean $\mathbf{0}$ and covariance matrix

$$\mathbf{V}_d=\mathscr{I}(\mathbf{B})^{-1}(W_0^2\mathbf{V}_0/\rho_0+W_1^2\mathbf{V}_1/\rho_1)\mathscr{I}(\mathbf{B})^{-1},$$

where

$$\begin{aligned}\mathscr{I}(\mathbf{B})&=E\{\mathbf{X}\mathbf{X}'p_0(\mathbf{X})p_1(\mathbf{X})\},\\\mathbf{V}_l&=E_l[\mathbf{X}\mathbf{X}'\{1-p_l(\mathbf{X})\}^2]-[E_l\{\mathbf{X}(1-p_l(\mathbf{X}))\}][E_l\{\mathbf{X}(1-p_l(\mathbf{X}))\}]',\end{aligned}\tag{9.8}$$

and

$$p_l(\mathbf{X}) = p_l(\mathbf{X}; \mathbf{B}), \qquad l = 0, 1.$$

A consistent estimator of $\mathbf{V}_d$ is given by

$$\hat{\mathbf{V}}_d = n\mathbf{A}_n^{-1}(W_0^2\hat{\mathbf{V}}_0/n_0 + W_1^2\hat{\mathbf{V}}_1/n_1)\mathbf{A}_n^{-1},$$

with

$$\mathbf{A}_n = W_0 n_0^{-1} \sum_1^{n_0} \mathbf{x}_{0t}\mathbf{x}'_{0t} p_0(\mathbf{x}_{0t}) p_1(\mathbf{x}_{0t}) + W_1 n_1^{-1} \sum_1^{n_1} \mathbf{x}_{1t}\mathbf{x}'_{1t} p_0(\mathbf{x}_{1t}) p_1(\mathbf{x}_{1t})$$

and

$$(n_l - 1)\hat{\mathbf{V}}_l = \sum_1^{n_l} \mathbf{x}_{lt}\mathbf{x}'_{lt}\{1 - p_l(\mathbf{x}_{lt}; \mathbf{b}_d)\}^2 - n_l\left[\sum_1^{n_l} \frac{\mathbf{x}_{lt}\{1 - p_l(\mathbf{x}_{lt}; \mathbf{b}_d)\}}{n_l}\right]\left[\sum_1^{n_l} \frac{\mathbf{x}_{lt}\{1 - p_l(\mathbf{x}_{lt}; \mathbf{b}_d)\}}{n_l}\right]'$$

for $l = 0, 1$. This enables us to set approximate confidence intervals for individual parameters, B_i or for linear combinations, $\sum_1^p c_i B_i$.

If

$$\mathbf{x} = \begin{pmatrix} 1 \\ \mathbf{x}_1 \end{pmatrix} \quad \text{and} \quad \boldsymbol{\beta} = \begin{pmatrix} \beta_0 \\ \boldsymbol{\beta}_1 \end{pmatrix},$$

so that the first component of $\boldsymbol{\beta}$ represents a constant term, we can simplify the expression for $\mathbf{V}_d$ under the assumption that the model is logistic. In this case $E_l[\mathbf{X}\{1 - p_l(\mathbf{X})\}] = E\{\mathbf{X}p_0(\mathbf{X})p_1(\mathbf{X})\}/W_l$ so that

$$\mathscr{I}^{-1}E_l\{\mathbf{X}(1 - p_l(\mathbf{X}))\} = \mathbf{e}/W_l,$$

where

$$\mathbf{e} = \begin{pmatrix} 1 \\ \mathbf{0} \end{pmatrix}, \quad \text{using the result that} \quad \begin{pmatrix} a & \mathbf{b}' \\ \mathbf{b} & \mathbf{C} \end{pmatrix}^{-1} \begin{pmatrix} a \\ \mathbf{b} \end{pmatrix} = \mathbf{e}.$$

Thus we can write

$$\mathbf{V}_d = \mathscr{I}(\boldsymbol{\beta})^{-1} E\left[\mathbf{X}\mathbf{X}' p_0(\mathbf{X}) p_1(\mathbf{X}) \left\{\frac{W_0 p_1(\mathbf{X})}{\rho_0} + \frac{W_1 p_0(\mathbf{X})}{\rho_1}\right\}\right] \mathscr{I}(\boldsymbol{\beta})^{-1} - \mathbf{e}\mathbf{e}'/\rho_0\rho_1. \tag{9.9}$$

In particular, for proportional allocation when $\rho_0 = W_0$ and $\rho_1 = W_1$, $\mathbf{V}_d$ reduces to $\mathscr{I}^{-1} - \mathbf{e}\mathbf{e}'/W_0 W_1$. Note that $\mathscr{I}^{-1}$ is the asymptotic covariance matrix of $\sqrt{n}(\hat{\boldsymbol{\beta}} - \boldsymbol{\beta})$, where $\hat{\boldsymbol{\beta}}$ is the MLE for a random sample of n observations from the whole population. Thus, asymptotically at least, all the reduction in variance from proportionally allocated stratified sampling is concentrated in the constant term. The stratification into cases and controls has no effect at all (again asymptotically) on the precision of the estimators of the other coefficients which are almost always of primary interest in practice.

Although under proportional allocation there is no increase in precision for the parameters of interest in case-control sampling over SRS, with other allocations there can be very substantial gains. To give some indication of values in practice we consider a small real population made up of the 439 patients who had isolated aortic valve replacements at Greenlane Hospital between 1976 and 1979. The cases are patients who died within 30 days of the operation. Apart from the constant term, there is a single (discrete) explanatory variable, representing the New York Heart Association Index of Severity, which takes values 1, 2, 3, 4, 5, with population frequencies 0.14, 0.36, 0.34, 0.11 and 0.05 respectively. A logistic model with $B_0 = -5.1$ and $B_1 = 0.80$, leading to $W_0 = 0.94$ and $W_1 = 0.06$, fits well. Substituting these values in expression (9.9) we find that the asymptotic variance of $\sqrt{n}(b_{d1} - B_1)$ (assuming the logistic model) is equal to $(1.98/\rho_0) + (0.85/\rho_1)$. For proportional allocation with $\rho_0 = 0.94$, $\rho_1 = 0.06$ (and hence also for random sampling) this takes the value 16.10, while for optimal allocation ($\rho_0 = 0.61$, $\rho_1 = 0.39$) the value is 5.42, a reduction of approximately 66% in the variance. Equal allocation between cases and controls, which is a common strategy in case-control studies, produces a value of 5.66, which is only slightly larger than the optimal value.

In many situations information on controls is much more expensive to obtain than information on cases, which is already available in hospital records, cancer registries, etc. We can use expression (9.9) to get a reasonably efficient allocation for estimating a particular coefficient, or combination of coefficients, provided we have a rough idea of the population density $f(\mathbf{x})$ and the value of $\mathbf{B}$ and the relative sampling costs. Two special cases are worth noting. If $p_0(\mathbf{x})$ and $p_1(\mathbf{x})$ are constant (i.e. if the risk is not affected at all by changes in the explanatory variables) then it follows from equation (9.9) that with equal costs for a fixed total sample size, equal allocation with $n_0 = n_1$ is always optimal. We might expect, therefore, that equal allocation will be very efficient when we are trying to detect rather small effects. It also follows easily from equation (9.9) that, if $p_1(\mathbf{x}) \leqslant kp_0(\mathbf{x})$ across the support of $f(\mathbf{x})$, then $\rho_0 \leqslant \sqrt{(kW_0/W_1)}\rho_1$ for the optimal allocation. Thus if cases are fairly rare, so that k is small, the best allocation will include a disproportionate number of cases in the study, even if we ignore the relative sampling costs.

9.3 MODEL-BASED ESTIMATORS

If the combined sample of $n = n_0 + n_1$ observations had come from a random sample drawn from the whole population and we assumed the logistic model, then the MLE, $\hat{\boldsymbol{\beta}}$, would be the solution of

$$\sum_{t=1}^{n} \mathbf{u}_t(\boldsymbol{\beta}) = \mathbf{0} \tag{9.10}$$

where $\mathbf{u}_t(\boldsymbol{\beta})$ is given in equation (9.3). The asymptotic covariance matrix of

$\sqrt{n}(\hat{\boldsymbol{\beta}}-\boldsymbol{\beta})$ would be $\mathscr{I}(\boldsymbol{\beta})^{-1}$, where $\mathscr{I}$ is defined as in equation (9.8), and the sample information matrix,

$$\hat{\mathscr{I}} = \sum_{l=0}^{1}\sum_{t=1}^{n_l} \mathbf{x}_{lt}\mathbf{x}'_{lt}p_0(\mathbf{x}_{lt};\hat{\boldsymbol{\beta}})p_1(\mathbf{x}_{lt};\hat{\boldsymbol{\beta}})/n, \tag{9.11}$$

would give a consistent estimator of $\mathscr{I}$. (See Cox, 1970, for details.)

We can use exactly the same methods as in the previous section to investigate the behaviour of the random sampling estimator, $\hat{\boldsymbol{\beta}}$, under case-control sampling. If $n_l/n \rightarrow \rho_l$ $(l=0,1)$ as $n \rightarrow \infty$ then it follows that $\hat{\boldsymbol{\beta}}$ converges in probability to $\tilde{\boldsymbol{\beta}}$, the solution of

$$\rho_0 E_0\{\mathbf{X}p_1(\mathbf{X};\tilde{\boldsymbol{\beta}})\} = \rho_1 E_1\{\mathbf{X}p_0(\mathbf{X},\tilde{\boldsymbol{\beta}})\}. \tag{9.12}$$

With proportional allocation $(\rho_l = W_l)$, $\tilde{\boldsymbol{\beta}}$ is identical to $\mathbf{B}$ but in general the value of $\tilde{\boldsymbol{\beta}}$ will depend on the limit of n_0/n_1. When the logistic model is true, however, and

$$\mathbf{x} = \begin{pmatrix} 1 \\ \mathbf{x}_1 \end{pmatrix}$$

(so that the first component of $\boldsymbol{\beta}$ represents a constant term) then a comparison of equation (9.12) with equation (9.6) shows that

$$\tilde{\boldsymbol{\beta}} = \boldsymbol{\beta} + k\mathbf{e}, \quad \text{where } \mathbf{e} = \begin{pmatrix} 1 \\ \mathbf{0} \end{pmatrix}$$

as before and $k = \log(\rho_1 W_0/\rho_0 W_1)$. Thus only the constant term is affected by the change in sampling scheme when the logistic model is true. The model-based estimator, $\mathbf{b}_m$, is given by $\hat{\boldsymbol{\beta}} - k\mathbf{e}$. Prentice and Pyke (1979), following work by Anderson (1972) for the discrete case, show that $\mathbf{b}_m$ can be regarded as the MLE of $\boldsymbol{\beta}$. However, the usual theory does not apply and properties have to be derived from first principles. The asymptotic covariance matrix of $\sqrt{n}(\mathbf{b}_m - \boldsymbol{\beta})$ can be expressed in the form

$$\mathbf{V}_m = \mathbf{G}^{-1} - \mathbf{e}\mathbf{e}'/\rho_0\rho_1,$$

where

$$\mathbf{G} = E\left[\mathbf{X}\mathbf{X}'p_0(\mathbf{X};\tilde{\boldsymbol{\beta}})p_1(\mathbf{X};\tilde{\boldsymbol{\beta}})\left\{\frac{\rho_0}{W_0}p_0(\mathbf{X};\boldsymbol{\beta}) + \frac{\rho_1}{W_1}p_1(\mathbf{X};\boldsymbol{\beta})\right\}\right]. \tag{9.13}$$

The sample information matrix $\hat{\mathscr{I}}$ is a consistent estimator of $\mathbf{G}$, so we end up with the important result of Prentice and Pyke (1979) that, with the appropriate adjustment to the constant term and its variance, the standard ML inferences for SRS remain valid for the case-control sampling scheme, provided the logistic model is valid. The adjustments needed to correct the inferences for the constant term are simple if W_0 and W_1 are known, but one of the big advantages of the model-based approach over the design-based approach is that we do not need to know W_0 and W_1 in order to be able to estimate the parameters of primary interest.

We can make a direct analytic comparison of the efficiencies of the design-based and model-based estimators when the logistic model is valid using expressions (9.9) and (9.13). In the case of proportional allocation the two estimators are identical, of course, and the efficiencies are also equal when $p_0(\mathbf{x})$ and $p_1(\mathbf{x})$ are constant. In other cases the unweighted estimator is more efficient, as the following results show.

THEOREM. *If the logistic model is valid and a constant term is included in the model, then* $\mathbf{V}_d - \mathbf{V}_m$ *is non-negative definite.*

Proof. From equation (9.6), we can express $\mathbf{V}_m$ as

$$\mathbf{V}_m = \mathscr{I}(\boldsymbol{\beta})^{-1} E\{\mathbf{X}\mathbf{X}' p_0(\mathbf{X}) p_1(\mathbf{X}) a(\mathbf{X})\} \mathscr{I}(\boldsymbol{\beta})^{-1}$$

where

$$a(\mathbf{X}) = \frac{W_1/\rho_1 + W_0 \exp(\mathbf{X}'\boldsymbol{\beta})/\rho_0}{1 + \exp(\mathbf{X}'\boldsymbol{\beta})}.$$

From equation (9.13), after some manipulation (and recalling that $\tilde{\boldsymbol{\beta}} = \boldsymbol{\beta} + k\mathbf{e}$), we can express $\mathbf{V}_d$ as

$$\mathbf{V}_d = \left[E\left\{ \mathbf{X}\mathbf{X}' \frac{p_0(\mathbf{X}) p_1(\mathbf{X})}{a(\mathbf{X})} \right\} \right]^{-1}.$$

If we define

$$\mathbf{U} = \mathbf{X}\{p_0(\mathbf{X}) p_1(\mathbf{X}) a(\mathbf{X})\}^{1/2},$$

and

$$\mathbf{V} = \mathbf{X}\{p_0(\mathbf{X}) p_1(\mathbf{X}) / a(\mathbf{X})\}^{1/2},$$

then

$$\mathscr{I}(\boldsymbol{\beta})(\mathbf{V}_d - \mathbf{V}_m)\mathscr{I}(\boldsymbol{\beta}) = E(\mathbf{U}\mathbf{U}') - E(\mathbf{U}\mathbf{V}')E(\mathbf{U}\mathbf{U}')^{-1}E(\mathbf{V}\mathbf{U}').$$

This is the residual sum of squares matrix when $\mathbf{U}$ is regressed on $\mathbf{V}$ and hence is non-negative definite. The theorem then follows immediately.

It follows that the variance of any linear combination of estimated coefficients is smaller with the model-based estimators than their design-based equivalents. For example, in the population of aortic valve replacement patients described in Section 9.2, the asymptotic variance of $\sqrt{n}(\hat{\beta}_1 - \beta_1\}$ is equal to 4.81 when $\rho_0 = 0.61$ (as against 5.42 for the weighted estimator) and 4.70 when $\rho_0 = 0.50$ (compared to 5.66 for the weighted estimator). The optimal allocation in this example is $\rho_0 = 0.525$ where the variance is 4.69 (5.56 for the weighted estimator). We see that the weighted estimator is, indeed, less efficient, although the difference is not very large for moderate values of ρ_0. The loss of efficiency increases as ρ_0 gets smaller. For example, with $\rho_0 = 0.1$ $V_m = 12.20$ while $V_d = 20.740$.

9.4 SIMULATION RESULTS

The comparison of the model-based and design-based estimators in the previous section is under the assumption that the logistic model holds exactly. If this is not the case then $\tilde{\boldsymbol{\beta}}$, the solution of equation (9.12), is not equal to $\mathbf{B}$ unless $\rho_l = W_l (l = 0, 1)$. Thus if estimation of $\mathbf{B}$ is the desired goal, we have to add the squared bias to the variance of $\mathbf{b}_m$ when we compare its performance with that of $\mathbf{b}_d$ as an estimator of $\mathbf{B}$. Obviously $\mathbf{b}_m$ will have the larger mean-squared error (MSE) if the true model is far enough from the logistic, but $\mathbf{B}$ itself is of little interest if the logistic approximation is too crude. In our simulations, therefore, we have attempted to compare the performance of $\mathbf{b}_m$ and $\mathbf{b}_d$ in cases where the logistic model is reasonable but not perfect. A further reason why simulation results are desirable is that large-sample results for the logistic are notoriously misleading in small samples so it is important to augment the asymptotic results of the previous section with some small-sample comparisons.

The values in Table 9.1 are typical of the results that we have obtained so far. The distribution of the single explanatory variable is taken to be standard normal in all three cases and the results are based on 500 repetitions of each experiment. Parameter values were chosen so that the proportion of cases in the population was always 5% and a reasonable degree of separation was obtained between the two conditional densities $f(x \mid y = 0)$ and $f(x \mid y = 1)$. A value of $B_1 \approx 2.5$ achieves the latter aim. In Table 9.1 'St. dev.' denotes the mean of the estimated standard deviations and hence is proportional to the average width of the confidence intervals. In Table 9.1(a), where the postulated logistic model is exactly true, the model-based estimator performs well and the coverage frequency of the corresponding confidence intervals is close to the nominal value. We see from the MSE term that weighting involves a considerable loss of information. Moreover, it seems that larger samples are required before the asymptotic theory provides reasonable approximations in the design-based approach. In effect, the design-based approach scales down the number of cases, leading to a smaller effective sample size. If we obtain $\hat{\boldsymbol{\beta}}$ and $\mathbf{b}_m$ by iteratively reweighted least squares, we see from the final weights that the design-based approach tends to draw more information from regions where observations are relatively scarce. (See Smith, 1981 for analogous results in linear regression.)

In Table 9.1(b) and (c) we see the effect of ignoring a small quadratic term in the model. The coefficient of the extra term is negative in Table 9.1(b) (see Example 1 in Section 9.1), positive in Table 9.1(c), and is chosen so that it will be detected by the usual likelihood ratio test about 50% of the time when $n_0 = n_1 = 200$. (We have taken this as indicating a 'reasonable but not perfect'

***Table 9.1** Comparison of the model-based (b_{m1}) and design-based (b_{d1}) estimators of B_1 (reproduced with permission from Scott and Wild 1986)*

(a) Model: $\log \frac{p_1(\mathbf{x})}{p_0(\mathbf{x})} = -5.5 + 2.75x_1 \quad (B_0 = -5.5, B_1 = 2.75)$

	$n_0 = n_1 = 100$		$n_0 = n_1 = 200$	
	b_{d1}	b_{m1}	b_{d1}	b_{m1}
Bias	0.23	0.11	0.11	0.05
MSE	0.51	0.22	0.20	0.10
St. dev.	0.48	0.41	0.36	0.28
Coverage of 95% intervals	0.85	0.94	0.89	0.93
Coverage of 99% intervals	0.93	0.99	0.96	0.99

(b) Model: $\log \frac{p_1(\mathbf{x})}{p_0(\mathbf{x})} = -5.8 + 4.0x_1 - 0.6x_1^2$

$(B_0 = -4.9, B_1 = 2.4)$

	$n_0 = n_1 = 100$		$n_0 = n_1 = 200$	
	b_{d1}	b_{m1}	b_{d1}	b_{m1}
Bias	0.22	0.51	0.11	0.46
MSE	0.49	0.47	0.21	0.30
St. dev.	0.47	0.42	0.36	0.29
Coverage of 95% intervals	0.85	0.85	0.88	0.70
Coverage of 99% intervals	0.93	0.98	0.95	0.92

(c) Model: $\log \frac{p_1(\mathbf{x})}{p_0(\mathbf{x})} = -5.1 + 2.0x_1 + 0.3x_1^2$

$(B_0 = -5.4, B_1 = 2.8)$

	$n_0 = n_1 = 100$		$n_0 = n_1 = 200$	
	b_{d1}	b_{m1}	b_{d1}	b_{m1}
Bias	0.14	−0.20	0.01	−0.26
MSE	0.35	0.21	0.12	0.14
St. dev.	0.45	0.21	0.33	0.24
Coverage of 95% intervals	0.89	0.85	0.92	0.75
Coverage of 99% intervals	0.97	0.94	0.98	0.87

fit.) The performance of the design-based estimator is affected very little by the extra term. The model-based unweighted estimator still has the smaller mean-squared error when $n_0 = n_1 = 100$, but this is reversed when $n_0 = n_1 = 200$ and the confidence intervals for the unweighted estimator perform very badly with this amount of contamination. However, the model-based intervals do cover $\tilde{\beta}$ at essentialy the nominal ratio. Here $\tilde{\beta}$ differs from B by about 12%, but with larger samples (large enough for the weighted asymptotics to be adequate, say) the difference between $\tilde{\beta}$ and B will be much smaller at the point where the non-linearity can be detected only 50% of the time.

9.5 STRATIFIED CASE-CONTROL STUDIES

Most real case-control studies incorporate stratification on other variables such as age and sex. A common assumption is that the relative risk is constant across strata but the absolute levels of risk vary from stratum to stratum. This leads to a model of the form

$$P(Y = 0|\mathbf{x}) = [1 + \exp(\beta_{0h} + \mathbf{x}_1'\boldsymbol{\beta}_1)]^{-1} \tag{9.14}$$

for stratum h ($h = 1, \ldots, L$).

Consider the simplest situation in which independent samples of cases and controls are drawn within each stratum. Let W_{1h} and W_{0h} denote the population proportion of cases and controls, and n_{1h} and n_{0h} the corresponding sample sizes in the hth stratum ($h = 1, \ldots, L$). The results in Sections 9.2 and 9.3 generalize in a straightforward way with

$$\mathbf{x} = \begin{pmatrix} 1 \\ \mathbf{x}_1 \end{pmatrix}$$

replaced by

$$\mathbf{x} = \begin{pmatrix} \mathbf{x}_0 \\ \mathbf{x}_1 \end{pmatrix}$$

where $\mathbf{x}_0$ is a vector of dummy variables indicating stratum membership. The MLE $\hat{\boldsymbol{\beta}}$, that would be appropriate if we had observed the whole population, would again converge to $\mathbf{B}$, the solution of equation (9.4).

The design-based estimator of $\mathbf{B}$, say $\mathbf{b}_{sd}$, is obtained using the stratified equivalent of $\hat{\mathbf{U}}(\mathbf{b})$ in equation (9.7), viz.

$$\hat{\mathbf{U}}_s(\mathbf{b}) = \sum_{h=1}^{L} \left\{ W_{1h} \sum_{t=1}^{n_{1h}} \frac{\mathbf{x}_{h1t} p_0(\mathbf{x}_{h1t}; \mathbf{b})}{n_{1h}} - W_{0h} \sum_{t=1}^{n_{0h}} \frac{\mathbf{x}_{h0t} p_1(\mathbf{x}_{h0t}; \mathbf{b})}{n_{0h}} \right\}.$$

We deal only with situations in which there are relatively large samples in each stratum and consider asymptotic behaviour when $n_{lh}/n_h \to \rho_{lh}$ as $n_h = n_{0h} + n_{1h} \to \infty$. As in Section 9.3, $\sqrt{n}(\mathbf{b}_{sd} - \mathbf{B})$ is asymptotically normal with mean zero and covariance matrix

$$\mathbf{V}_{sd} = \mathscr{I}(\mathbf{B})^{-1}\left\{\sum_h (W_{0h}^2\mathbf{V}_{0h}/\rho_{0h} + W_{1h}^2\mathbf{V}_{1h}/\rho_{1h})\right\}\mathscr{I}(\mathbf{B})^{-1},$$

where $\mathbf{V}_{lh}$ is defined in the same way as $\mathbf{V}_l$ for the units in stratum h ($l = 0, 1$; $h = 1, \ldots, L$).

The model-based estimator is again obtained by a simple modification of the MLE for SRS. Write

$$\boldsymbol{\beta} = \begin{pmatrix} \boldsymbol{\beta}_0 \\ \boldsymbol{\beta}_1 \end{pmatrix},$$

where $\boldsymbol{\beta}_0$ is the coefficient vector corresponding to $\mathbf{x}_0$, the vector of stratum indicators. Then, if we assume that model (9.14) is valid we find that $\hat{\boldsymbol{\beta}}$ is asymptotically normal with mean $\tilde{\boldsymbol{\beta}}$, where $\tilde{\boldsymbol{\beta}}_1 = \boldsymbol{\beta}_1$ and

$$\tilde{\beta}_{0h} = \beta_{0h} + \log\left(\frac{\rho_{1h}W_{0h}}{\rho_{0h}W_{1h}}\right).$$

The asymptotic covariance matrix has the form

$$(n\mathscr{I})^{-1} - \begin{pmatrix} \mathbf{D} & \mathbf{0}' \\ \mathbf{0} & 0 \end{pmatrix},$$

where $\mathbf{D} = \text{diag}(d_1, \ldots, d_L)$ with $d_h = (n_h\rho_{0h}\rho_{1h})^{-1}$. Moreover, the sample information matrix for SRS gives a consistent estimator of $\mathscr{I}$. Thus inferences about $\boldsymbol{\beta}_1$ can again be made using the unmodified output from a standard logistic regression program assuming SRS. The model-based estimator of $\boldsymbol{\beta}$ is obtained by making the appropriate adjustment of $\hat{\boldsymbol{\beta}}_0$. This adjustment requires knowledge of the relative sampling fractions of cases and controls within each stratum. The estimator of $\boldsymbol{\beta}_1$ requires no such information, and one of the major attractions of the model-based estimator is that estimates of relative risk can be obtained without knowledge of the relative sampling fractions. This knowledge is often not available in medical studies.

Comparisons between the two approaches in the stratified case are similar to those for the unstratified case. The model-based estimator is more efficient if the model is true, although the difference in efficiency is not large except when there is extreme imbalance in the sampling fractions. If the model is not exactly true and we are willing to accept $\mathbf{B}$ as a reasonable target for inference, then the bias in the model-based estimator eventually outweighs its smaller variance. Again, however, we need a large sample size within each stratum for the large-sample theory to provide a reasonable approximation to the design-based estimator's performance.

To give some indication of the relative efficiency in a realistic situation we calculated the asymptotic variances of both estimators for the stratified population (modified slightly to make a linear model for the stratum constants fit exactly) given in Fears and Brown (1986). There are twelve strata, corresponding to geographical regions, and a single explanatory variable, a binary index of susceptibility. The variance of the model-based estimator is 0.1118 while the variance of the design-based estimator is 0.1295. Thus the relative efficiency of the design-based approach in this study is 86%.

9.6 CONCLUSION

The results in the preceding sections indicate that the model-based approach can lead to a substantial increase in efficiency over the design-based approach when fitting logistic regression models to case-control data if the model fits well and the allocation of observations between cases and controls is extreme. The gains are fairly modest, however, with the sort of allocations typically found in practice.

When the model does not fit so well, and it is agreed that the estimation of $\mathbf{B}$, defined as the solution to equation (9.4), is a reasonable goal, then the bias in the model-based estimator eventually outweighs its smaller variance as the sample size increases. Under these circumstances it seems reasonable to advocate using the design-based estimator. It remains to be seen whether or not epidemiologists can be persuaded that $\mathbf{B}$ is the natural target for inferences. In many cases the fitted model will be used to predict likely future cases, and it seems plausible that the best discriminator would be obtained using a model that fits well in the region where there is a substantial overlap between cases and controls. With an equal number of cases and controls this might well correspond to using $\tilde{\boldsymbol{\beta}}$, the limit of the model-based estimator. Further work on this would certainly be worth while.

All this work is based on stratified random sampling of cases and controls. As studies become larger, more complex sampling schemes are needed. There seems no real possibility of obtaining exact MLEs in these cases, although weighted least squares may eventually provide estimators with the same asymptotic efficiency (see Scott *et al.*, 1989). At present the only alternative is the pseudo MLE approach which, in principle at least, is straightforward for any design for which we have consistent estimators of population totals along with the appropriate variance estimators. The results in previous sections are fairly reassuring, since the design-based estimators combine robustness with reasonable efficiency in most practical situations.

PART C

Disaggregated Analysis: Modelling Structured Populations

CHAPTER 10

Introduction to Part C

D. Holt

10.1 AGGREGATION VERSUS DISAGGREGATION

In previous chapters we have accepted the objectives of the analysis as a set of prespecified parameters to be estimated or particular hypotheses to be tested. Part A has been concerned with standard procedures such as ordinary least squares (OLS) regression analysis and χ^2 procedures for contingency table analysis. Part B has been concerned with the effects of selection on point estimation. Throughout both parts there is a sense in which the structure of the survey data has been viewed as a nuisance: an unwanted complication which invalidates standard procedures but leaves the basic objectives unaltered. In this final part of the book this will not be the case. We change the emphasis to bring the population structure more explicitly into the analysis procedure. The approach taken is a model-building one where a statistical model is formulated to describe the relationship between the variables of interest. The complexity of the population structure is used as evidence that simple models and standard procedures will be inadequate in general. Hence the models chosen are made more elaborate to take account of the population structure and this can change the targets of the inference. Specifically the model may be so modified that old parameters are abandoned and new ones introduced in an adaptive process which is much more dependent on the survey data. Thus cluster effects are not regarded as a nuisance that stands between the data and well-established procedures, but as an intergral part of the population structure that should be properly modelled and can contribute to our understanding of the relationships between variables.

The methods described in Part A could be called aggregate or marginal analyses since the parameters of interest are averaged over some aspects of the population structure. There is a fundamental issue, for which no clear resolution exists, as to whether and to what extent this approach is generally appropriate for survey data. We will use the example of a two-way contingency table to illustrate the issue.

Table 10.1 Two-way contingency table

n_{11}	n_{12}	n_{13}	$n_{1.}$
n_{21}	n_{22}	n_{23}	$n_{2.}$
$n_{.1}$	$n_{.2}$	$n_{.3}$	$n_{..}$

We assume that two categorical variables are measured for each survey respondent and these are used to construct a two-way table of counts n_{ij} (Table 10.1).

The data are collapsed across aspects of the population structure such as strata and clusters, but assumed to be generated from an underlying table of probabilities μ_{ij}; $i = 1, 2; j = 1, 2, 3$. If the table is a cross-classification of two response variables then $\sum_i \sum_j \mu_{ij} = 1$, whereas if the marginal row totals are fixed and the conditional probabilities within each row are of interest then $\sum_j \mu_{ij} = 1$; $i = 1, 2$. We assume that the principal objectives are taken to be the estimation of the cell probabilities $\{\mu_{ij}\}$ or some function of these such as the logarithms or logits. Of primary interest also will be models which impose constraints on $\{\mu_{ij}\}$ such as the independence hypothesis $\mu_{ij} = \mu_{i.}\mu_{.j}$ where $\mu_{i.}$ and $\mu_{.j}$ are the marginal probabilities. The essential assumption is that the parameters $\{\mu_{ij}\}$, possibly constrained, adequately describe the relationship between the variables and these are therefore the targets of inference.

We have deliberately not stated whether the values $\{\mu_{ij}\}$ are considered to be model parameters which characterize the data-generation process or finite population proportions. In this discussion we wish to refer to both possibilities although there are important conceptual differences between the two approaches. In the model-based approach to survey analysis, the parameters $\{\mu_{ij}\}$ relate to the model and capture the variation in the underlying cell probabilities. The finite population is regarded as a realization of size N generated as a stochastic process under the appropriate model assumptions. In the framework provided by the randomization distribution of the survey design, $\{\mu_{ij}\}$ are regarded as an arbitrary set of finite population proportions. In this latter case the idea that these should exhibit some additional underlying structure, for example $\mu_{ij} = \mu_{i.}\mu_{.j}$ is difficult to fit within the framework since such constraints will not be satisfied exactly by finite population values. Nevertheless, survey researchers do fit such models even though they sometimes use a finite population framework for the properties of estimators and test statistics.

The analysis of the two-way table is described as aggregate or marginal since it has been formed by collapsing the data across many aspects of the population structure. If we consider a social survey of households, for example, we may recognize that there are systematic differences between regions and between metropolitan, urban and rural households. These differences will be reflected in the survey data and are almost invariably reflected also in the survey design with stratification by region and type of location for the households. Similarly, within strata, the survey population will exhibit a hierarchical structure with higher level units (e.g. local authority areas) containing smaller units (e.g. electoral wards) and these in turn containing households. In general, households in the same hierarchical unit (cluster) exhibit positive intra-cluster correlation; the values of the survey variables tend to be more alike within clusters than would occur by a random grouping. This is a feature of the population structure and will be due in part to the influence of each household on its neighbours and in part to the fact that households in the same cluster will be exposed to common influences. The model specification encapsulated $\{\mu_{ij}\}$ does not reflect these aspects of the population structure.

Suppose, for example, that the population is not homogeneous but can be described by a two-stage structure with differences between regions which have been used as strata and with differences between clusters of households within regions which have been used as primary sampling units.

By differences between strata we mean that a separate two-way table could be formed for each stratum, h (Table 10.2):

Table 10.2 Two-way table of means for the hth stratum

μ_{11h}	μ_{12h}	μ_{13h}	$\mu_{1.h}$
μ_{21h}	μ_{22h}	μ_{23h}	$\mu_{2.h}$
$\mu_{.1h}$	$\mu_{.2h}$	$\mu_{.3h}$	$\mu_{..h}$

where μ_{ijh} are the cell probabilities for the i,jth cell and the hth stratum $h = 1, \ldots, H$. These would exhibit differences between strata in the sense that the conditional probabilities $\mu_{ij|h} = \mu_{ijh}/\mu_{..h} \neq \mu_{ij}$ for some h.

Cluster effects are exhibited by between cluster differences in the cell proportions which are larger than one would expect on the basis of the multinomial variation associated with selected clusters formed at random.

If there were no stratum or cluster effects then the population would be homogeneous as far as the variables of interest were concerned and the objectives of the aggregate analysis would be undisputed. However, if stratum or cluster effects exist then the population is not homogeneous and this calls into question whether the aggregate analysis, collapsed across strata and clusters, is appropriate. The aggregate analysis implies that we are interested in parameters to model the marginal proportions $\{\mu_{ij}\}$ when the existence of important population structure suggests that the observations contributing to each cell of the table are not well represented by all respondents having a common probability μ_{ij} of being classified into the i,jth cell for example.

Consider stratification only, in which case

$$\mu_{ij} = \sum_h \mu_{ij|h}\mu_{..h}.$$

The observations in the i, jth cell are a mixture of individuals with probabilities $\mu_{ij|h}$ conditional on belonging to the hth stratum with mixing probabilities $\{\mu_{..h}\}$. At this stage the analyst has two choices:

1. To maintain that the original objectives of inference about $\{\mu_{ij}\}$ are appropriate but to recognize that the data are not independent, identically distributed random variables which needs to be taken into account in the analysis. In model terms the structural part of the model is maintained but the error assumptions are modified to reflect the properties of the observations. This is the approach in Part A.
2. To recognize that the existence of stratum effects calls into question the relevance of the parameters $\{\mu_{ij}\}$. This suggests that the structural part of the model needs to be modified, perhaps by changing the target of inference to $\{\mu_{ij|h}\}$. The natural approach would be to disaggregate the two-way table into a three-way table with layers of the table being determined by stratum membership and to extend the analysis to a higher dimension contingency table.

In most circumstances where stratification effects exist, the model-building approach suggests that the aggregate analysis is inappropriate and strategy (2) should be adopted. The key issue is whether there are subgroups for which the interaction structure between the first two variables is different and if there are such subgroups an elaboration of the model by introducing an extra dimension to the contingency table is appropriate. Whether or not the subgroups were used in the survey design as the basis for stratification is not the issue.

There are, however, situations which may occur where stratum effects exist and yet the objectives of the aggregate analysis should be maintained. Suppose for example that measurements are made upon individuals at the time they enter a programme or system, such as attainment tests for children entering education. We suppose further that the allocation of individuals to clusters is correlated in some way with the measurements made, as when the allocation of children to schools or classes depends on the test scores attained. Finally, suppose that a sample of clusters and individuals is now selected and the individual data recorded including specifically the children's original attainment test scores as the variable of interest. In this case it could be argued that the population structure has been created after the variables of interest were attained and that the stratum and cluster effects which may occur are simply an impediment to the real target of inference. In this case it is the marginal distribution of the original measured variables which is required and we wish to make estimates and test hypotheses in this marginal distribution despite the population structure. Here the aggregated analysis is appropriate and this type of situation is the subject-matter of Parts A and B of the book.

However, it is certainly not the case that the appropriate target of inference is clear in all cases. Consider the following situation which arose in a survey of marathon runners (Barrell *et al.* 1987). A sample of marathon runners had been selected with much higher selection probabilities for faster runners. This was achieved by creating four finish time strata with disproportionate allocation to each. Questions were asked to investigate the factors that had been major influences on the runners' original decision to take up distance running, and secondly on the decision to enter the runners' first marathon event. The factors were reasons such as 'personal challenge'; 'to improve fitness'; 'the influence of family and friends' and so on. The finish time used for selection was the result of a recent marathon and reflected the ability and standard of fitness of the runner at the time of the survey rather than when the original decisions were made. This finish time reflects current commitment, amount of training in each individual training session and the consistency of training over a period of months or years. Nevertheless there were clear differences in the reasons given for the original decisions to run and to enter a marathon between those who were in the faster time strata and those who were slower. Clearly the reasons for the original decisions were not homogeneous and the finish time stratification captured some of the heterogeneity. One argument is that current fitness and speed is something which has occurred after the original decisions were made and as such has no part to play in the analysis. Hence the marginal parameters $\{\mu_{ij}\}$ are the inference targets. However, it could also be argued that the original motivation for runners was heterogeneous and that this is also reflected in the current fitness and speed of the runners. Whilst current speed cannot cause previous motivation it might be argued that it is a surrogate variable for characteristics which the runners had at the time of making the decisions. In these circumstances, perhaps a disaggregated analysis can be justified and the inference targets would be the conditional parameters $\{\mu_{ij|h}\}$. The description of the results would be of the kind 'of those who subsequently ran at under 2 hours 45 minutes the interaction between reasons for starting to run and for running a first marathon were…'. Both approaches to the analysis seem to have some validity and the choice is not clear cut. Of course, with this example there is a third interpretation of the results which is that the reasons given for beginning to run and of first running a marathon are not the original reasons but are the reasons given *post hoc* at the time of data collection. In this case the performance has been achieved before the *ex post* justification is required and as such the conditional analysis is the appropriate one. This third interpretation should not detract from the fundamental issues which can occur as to the appropriate targets of the analysis.

The same issues arise for clusters as for strata. The existence of between-cluster variation for the cell probabilities implies that the clusters do not have identical interaction structure. Once again the question occurs as to whether we should simply collapse the data across clusters or seek to model, and hopefully to understand, the between-cluster differences. The stratification approach is to create a separate layer of the table for each stratum, but this is not a viable option for clusters. First of all there is the simple practical point that the sample size in each cluster is generally small, often smaller than the size of the table that would be created for each cluster. Secondly there is the more fundamental point that data are obtained from every stratum in the population but from only a sample of clusters. Inference is required which goes beyond the sampled clusters to the entire population, and one approach is to treat cluster effects as a random process rather than adding fixed parameters as is implied for strata. We expand on this point in the next section. We note that some designs use a large amount of stratification with only one or two primary sampling units per stratum. Thus smallness of the sample size within each stratum and the difficulty of interpreting results for a very large number of strata becomes a problem. This is even more of a difficulty for contingency table analysis compared with, for example, regression analysis since the number of parameters to be estimated can increase quickly. Hence the point at which the within-stratum sample size is inadequate for the number of parameters to be estimated is reached sooner. At some point it becomes better to treat strata (or substrata within the principal stratification) in the same way as we describe in the next section for clusters.

There may be a tendency to interpret the previous comments to suggest that if an aggregate analysis is required then the contents of Parts A and B are relevant, whereas Part C will provide an alternative approach: a disaggregated analysis which will overcome the problems described in previous chapters. This is not the case.

As statistical models are refined to take account of the population structure it is generally true that some of the previous issues are diminished—but not removed. When disaggregated target parameters are identified, for example, the residual cluster effects will be reduced because some of the between-cluster variation will be accounted for in the more elaborate model specification. Thus the factors that dominated Part A will not be as strong. For practical considerations, the process of disaggregation cannot go to an extreme. The statistical models will be chosen to incorporate systematic differences between strata and clusters, for example, but even so there will be residual cluster effects and the issues in Part A will still be relevant. The methods described in this final part of the book will automatically make due allowance for these residual effects and standard errors will properly reflect the data structure. Similarly, test procedures will reflect the modifications or alternative methods described in Part A.

Using a disaggregated analysis also tends to reduce the selection effects for point estimation referred to in Part B. However, this is not always the case and at some points reference will be made to *p*-weighting procedures as a protection against these effects. It would be possible to include the adjustment methods for point estimation, described in Part B, as part of these final chapters. We have not done this so as not to complicate the presentation. However, it should be borne in mind that the methods to be described could be used in conjunction with the adjustment methods described in Part B.

10.2 MODELLING THE STRUCTURE OF THE SURVEY POPULATION

We will use the example of regression analysis to illustrate the ideas behind the modelling process.

For stratification the elaboration of the regression model is relatively straightforward. The modification for strata implies additional terms to allow for a separate intercept for each stratum. Similarly, one or more of the regression variables may require different regression coefficients within each stratum. The natural approach is to introduce a set of fixed-effects terms for stratum membership and to explore the possibility of interaction between these and the explanatory variables in order to introduce different regression slopes in different strata. Thus consider the case of simple linear regression and two strata with marginal model:

$$E(Y_i) = \beta_0 + \beta_1 x_i. \tag{10.1}$$

We can introduce a single dummy variable Z to identify stratum membership for each respondent and explore models of the kind

$$E(Y_i) = \beta_0 + \beta_1 x_i + \beta_2 z_i + \beta_3 x_i z_i. \tag{10.2}$$

If $\beta_3 = 0$ the stratum effect is a change in the intercept from β_0 when $z_i = 0$ to $\beta_0 + \beta_2$ when $z_i = 1$. If $\beta_3 \neq 0$ there is a different slope in each stratum also.

Before describing the elaboration of the regression model to take account of cluster effects, it should be noted that the term 'intra-cluster correlation' (and for that matter 'design effect') has come to be used rather loosely, almost as a euphemism for some cluster effect without being too specific as to its precise nature. If we consider the case of multiple regression, for example, with p explanatory variables, $\mathbf{x}_i' = (x_{1i}, \ldots, x_{pi})$, there are several different intra-cluster correlation coefficients that might be relevant. Each of the explanatory variables x_{ji} in $\mathbf{x}_i$ may separately exhibit an intra-cluster correlation ρ_{x_j} of different values and this in turn may result in the dependent variable, Y, exhibiting an intra-cluster correlation ρ_y. As an added complication the independent variables may be correlated as may the cluster effects for the separate $\mathbf{x}$ variables. Since for the regression analysis we condition on the observed values of the independent variables, the error terms $\varepsilon_i = \mathbf{Y}_i - \mathbf{x}_i'\boldsymbol{\beta}$ may also exhibit intra-cluster correlation $\rho_{y|\mathbf{x}}$ (see Scott and Holt, 1982; Wu *et al.*, 1988). Similarly, the partial regression coefficients between Y and each of the $\mathbf{x}$ variables may vary between clusters.

For the example of the two-way table in Table 10.1 the term 'intra-cluster correlation' is taken to mean that the proportion of individuals falling into a particular category or cell varies between clusters in a way which exhibits greater variation than would be expected from binomial variation alone. As in the regression example this may relate to one or both of the marginal classification without necessarily affecting the conditional probability of classification within each row or column of the table. More generally intra-cluster correlation may be exhibited in each cell of the table even when viewed conditionally on the marginal totals in each cluster.

A complete specification of the population structure holds many possibilities and the way in which the basic IID assumptions are violated will have a strong bearing on the actual properties of the standard procedures.

We return to the example of the marginal simple linear regression, written as

$$Y_{ct} = \beta_0 + \beta_1 x_{ct} + \varepsilon_{ct}, \tag{10.3}$$

where now the double subscripts c, t denote the tth respondent from the cth cluster. For survey data we often observe the fact that $\{\varepsilon_{ct}\}$ do not satisfy the IID assumption. On closer examination we may find that both the intercept β_0 and the regression parameter β_1 appear to vary between clusters. A scatterplot of the clusters might look as in Figure 10.1. (*N.B.* Each ellipse indicates the scatter of individual data points for one cluster.)

One possibility is that β_0 and β_1 in equation (10.3) be replaced by random

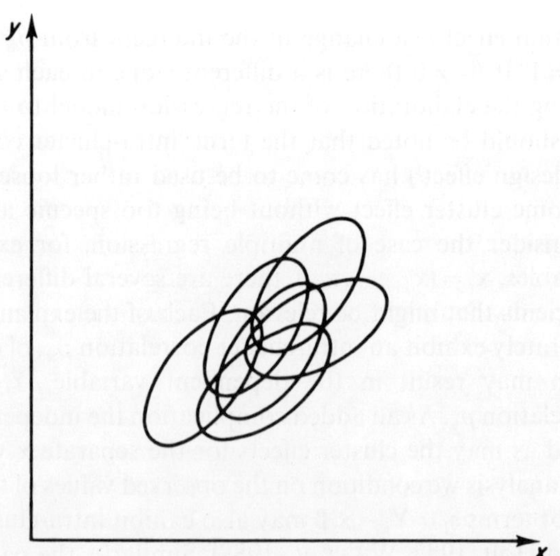

Figure 10.1 Clustered survey data diagram.

coefficients that depend upon the cluster membership:

$$
\begin{aligned}
Y_{ct} &= \beta_{0c} + \beta_{1c} x_{ct} + \varepsilon_{ct} \\
\beta_{0c} &= \beta_0 + \eta_{0c} \qquad (10.4) \\
\beta_{1c} &= \beta_1 + \eta_{1c}
\end{aligned}
$$

where β_0 and β_1 are fixed but unknown coefficients and ε and η are random variables.

We assume

$$
\begin{aligned}
E(\varepsilon_{ct}) &= E(\eta_{0c}) = E(\eta_{1c}) = 0, \\
\operatorname{var}(\varepsilon_{ct}) &= \sigma^2, \operatorname{var}(\eta_{0c}) = \sigma_0^2, \operatorname{var}(\eta_{1c}) = \sigma_1^2, \\
\operatorname{cov}(\varepsilon_{ct}, \eta_{0c'}) &= \operatorname{cov}(\varepsilon_{ct}, \eta_{1c'}) = 0. \\
\operatorname{cov}(\varepsilon_{ct}, \varepsilon_{c't'}) &= 0, \qquad c \neq c' \text{ or } t \neq t',
\end{aligned}
$$

and

$$
\begin{aligned}
\operatorname{cov}(\eta_{0c}, \eta_{1c'}) &= \sigma_{01} \qquad c = c', \\
&= 0 \qquad c \neq c'. \qquad (10.5)
\end{aligned}
$$

This elaboration of the original model permits the inclusion of random cluster effects into the model, and using appropriate methods will permit efficient estimation of parameters, interval estimates and tests of hypotheses. These will properly reflect the population structure and overcome the invalidity of procedures such as OLS which make no allowance for the cluster effects. However, our understanding of the relationship between the variables is not much advanced because we have been able to do no more than attribute the cluster effects to random variation between clusters (the terms η_{0c} and η_{1c}). The size of terms such as $\sigma_0^2, \sigma_1^2, \sigma_{01}$ and σ^2 will warn us of the between-cluster variation which exists but do little to explain it.

The model-building approach goes further to seek additional elaboration of the model to 'explain' the variation between clusters. This may to some extent be captured by variables measured or constructed at the cluster level. For example, in an epidemiological survey variables such as water hardness or levels of air pollution which are taken to be common to all cluster members may be measured for each cluster. Other variables, constructed at the cluster level may be based upon the individual characteristics of the cluster members. For example, the proportion of households within each cluster containing people over 70 years of age or with income below a prespecified amount. In this case the proportion may be separately measured for the cluster, for example from census data, or may only be available by aggregating the sample observations from the cluster. These cluster level variables are sometimes referred to as ecological or contextual factors respectively (Blalock, 1985). We note that the constructed cluster level variables need not only be means but may be other measures such as cluster standard deviations or percentiles.

A cluster level variable, say a_c is introduced as a way of reducing the unexplained variation σ_0^2 and σ_1^2 and also perhaps as a way of reducing the covariance σ_{01}. Thus the model is further elaborated to

$$\begin{aligned} Y_{ct} &= \beta_{0c} + \beta_{1c} x_{ct} + \varepsilon_{ct}, \\ \beta_{0c} &= \gamma_{00} + \gamma_{01} a_c + \eta_{0c}, \\ \beta_{1c} &= \gamma_{10} + \gamma_{11} a_c + \eta_{1c}. \end{aligned} \tag{10.6}$$

In practice either γ_{01} or γ_{11} may be zero and decisions such as this will be part of the model-building process based on the survey data.

The new model will alter the basic objectives of the estimation process since interest now centres on the new parametrization $(\gamma_{00}, \gamma_{01}, \gamma_{10}, \gamma_{11}, \sigma_0^2, \sigma_1^2, \sigma^2, \sigma_{01})$ and interval estimates and hypotheses to be tested will be adjusted accordingly. Thus the cluster level information is used to elaborate the model and to deepen our understanding of the relationships between the original variables (Y, X) and the cluster level variable a_c. To some extent the variation in β_{0c} and β_{1c} between clusters is 'explained' by a_c.

This simple example illustrates the basic objectives of the model-building approach. Several levels of clustering may be allowed and a number of variables included at each level. A more rigorous development is given in the following chapters.

Methods are described for model building, estimation and hypothesis testing. In Chapter 11 a careful development of the multilevel model formulation is given and the analyses of two data sets are described to illustrate the approach. The basic statistical techniques of estimation and testing in mixed models are described more fully in Chapter 12. In one important sense the models described in Chapter 12 are more restrictive than those described in the other chapters, including this introduction, since in Chapter 12 the assumption is made that the cluster effect random variables are uncorrelated even within the same cluster (i.e. $\sigma_{01} = 0$ in equation (10.5)). However, the models and methods can be extended to include the more general assumptions. In Chapters 11 and 13 we adopt a pure model-building approach in which the sample design plays no part in the analysis, except in so far as it coincides with the population structure which is being modelled. This does not mean that the sample design is unimportant, since, for example, the number of clusters and the number of observations per cluster have a direct effect on the standard errors of parameter estimates, etc. In Chapter 12, which is restricted to regression models, there is more of a synthesis between the modelling approach and aspects of the survey design and in particular the probabilities of selection. More extensive use of weighted estimation is made. The basic reasons are an attempt to protect against selection effects and to ensure a degree of robustness to model misspecification. The weighted estimators are also simpler since estimates of variance components are not needed. A comparison of Chapters 11 and 12 demonstrates how essentially the same methods can be applied in a purely model-based approach or in a synthesis between models and the survey design. For the former approach it is the population structure not the formal sampling mechanism which determines the choice of units at each level, but in many situations survey design will mirror the population structure and the two approaches will coincide to a large extent. The question of selection effects and the arguments for using *p*-weighted estimation to protect against model misspecification have already been discussed in Part B.

As well as estimating the model parameters a second objective is sometimes to predict the (random) coefficients for individual clusters or linear combinations of these across a set of clusters. This illustrates how the previous discussion on aggregated and disaggregated analyses has been somewhat oversimplified since a linear combination of regression coefficients across clusters could be regarded as an aggregated analysis. However, it probably helps to think of this objective as a disaggregated analysis since the statistical model is elaborated to allow for differences between strata and clusters.

The fact that a linear combination of coefficients is required can be embodied within the disaggregated approach as is described in Chapter 12. Chapter 13 formulates a model for logistic regression for contingency table analysis. The methods of variance estimation are different to the two previous chapters and the concern with small numbers of observations per cluster is more explicit. Nevertheless the basic approach is similar to the random coefficient regression models of the previous chapters and correlated cluster effects are permitted. The other development is in the explicit attempt to decompose the cluster effects into meaningful components and to interpret them in terms of the underlying social processes generating them.

CHAPTER 11

Multilevel and Multivariate Models in Survey Analysis

H. Goldstein and R. Silver

11.1 INTRODUCTION

Recent developments in the theory of linear model estimation (Aitkin and Longford, 1986; Goldstein, 1986a) have made possible the specification, efficient estimation and testing of models fitted to data obtained from nested or hierarchical structures. A good example of a hierarchical structure is an educational system where students are 'clustered' or grouped within classes, classes are grouped within schools, schools within local boards or authorities and so forth. Other kinds of data too can be viewed in this manner, most notably repeated measurement longitudinal data, which are an example of two-level data with the first or lower level comprising the measurement occasions 'within' individual subjects, and the second or higher level comprising the subjects themselves (see Goldstein, 1986b). Of course, the individual subjects may themselves be grouped within classes, schools, etc., so giving rise to a three-level or higher-level structure. It is important to emphasize that these hierarchies are intrinsic properties of the systems being studied, and that the use of statistical models to describe these structures is motivated by the structures themselves, independently of any sampling procedure which generates the data. Thus, a sample may be drawn from a school population by a simple random procedure, but it will still generally need to be modelled with due attention paid to the structure of the population itself.

From this viewpoint a multi-stage sampling procedure may be important for providing a valid and efficient analysis, not merely to reduce costs. Thus, if we wish to obtain stable estimates of within-school variation we need sufficiently large numbers of children within individual schools. A simple random sample which produced, say, just one child per school on average would be unsuitable. Further discussion of this point is given in Chapter 12.

The same considerations apply to most social data. Societies tend to have inbuilt hierarchies, for example, individuals within households which are grouped within localities and so on. In this chapter our concern is with a model-based approach with inferences to an assumed superpopulation, and our models are devised explicitly to incorporate the hierarchical structure of the population. In particular, these models can incorporate measurements made, say, at the level of the cluster, and if clusters are geographical areas, we can think of using variables such as the average social composition of the area or its general amenities. Thus the emphasis is on describing the between-unit variability at each level of hierarchical aggregation. This contrasts with the more usual emphasis in survey analysis where this variation is regarded as constituting a 'nuisance' since interest centres on relationships amongst the lowest level units—typically individual subjects.

Because so-called 'multilevel' models have been applied most extensively to educational data, the principal exposition will be in terms of educational variables. Nevertheless, as our second example illustrates, the methods of analysis will apply to data collected from other hierarchical systems, and to social surveys in particular. To introduce the basic idea of multilevel models, consider a simple two-level structure with students grouped within classrooms. Suppose there is a response variable, say a mathematics achievement test score (y) measured on each student, which we wish to relate to the gender (x_1) and social background (x_2) of each student, and to the average social background of all the students in each class (a_1) and years of mathematics teaching experience of the classroom teacher (a_2). Note that there are two explanatory variables measured at the student level, and two at the classroom level, one of which is a characteristic of the teacher and one is a so-called 'contextual' variable, based on an aggregated characteristic of all the students in the class. Note that the values of the aggregated variables may be available even though the sample itself does not include all the children in the class.

The next section deals with the basic statistical model. The approach is informal, rather than statistically rigorous, and uses a simple notation and terminology. The full statistical details of how to obtain generalized least squares and maximum likelihood estimates can be found in Goldstein (1986a). We deal first with the case of a continuously distributed response variable, and in the following section with discrete response variables.

11.2 THE TWO-LEVEL MODEL

The 'fixed' part of the model mentioned above can be written as

$$y_{ct} = \beta_0 x_0 + \beta_1 x_{1ct} + \beta_2 x_{2ct} + \gamma_{10} a_{1c} + \gamma_{20} a_{2c}, \tag{11.1}$$

where the c subscript indexes classes and the t subscript indexes students within classes. The coefficients $\beta_0, \beta_1, \beta_2, \gamma_{10}$ and γ_{20} are those which we want to estimate. In this example, the variables x_1 and x_2 are categorical and are defined using 'indicator' variables to denote the levels or categories, and the variables a_1 and a_2 are basically continuous so that γ_{10} and γ_{20} can be interpreted in the usual way as regression coefficients. The variable x_0 is in fact a constant which is set to 1.0 so that β_0 becomes the 'overall constant' or 'intercept' term in the model.

Turning to the 'residual' terms, these are random variables with an assumed mean of zero. We start by defining two of them, motivated as follows. Suppose β_0 varies randomly across individuals and classrooms. That is, for a fixed set of values for x_1, x_2, a_1, a_2, assuming we know (or that we have estimates of) $\beta_1, \beta_2, \gamma_{10}, \gamma_{20}$ then we rewrite it with extra subscripts as β_{0ct} which varies from student to student with a mean value for the cth class of β_{0c}. We write:

$$\beta_{0c} = \beta_0 + \eta_{0c} \tag{11.2}$$

and

$$\beta_{0ct} = \beta_0 + \eta_{0c} + \varepsilon_{ct}, \tag{11.3}$$

where η_{0c} is the deviation of the cth class from the overall mean and where ε_{ct} is the deviation of the tth student from the mean of the cth class, with variances σ_0^2 and σ^2 respectively. The η_c are mutually independent and so are the ε_{ct} and they are independent of each other.

Thus, remembering that the variable $x_0 = 1$ we can write:

$$\begin{aligned} y_{ct} &= \beta_{0ct} + \beta_1 x_{1ct} + \beta_2 x_{2ct} + \gamma_{10} a_{1c} + \gamma_{20} a_{2c} \\ &= \beta_0 x_0 + \sum_{k=1}^{2} \beta_k x_{kct} + \sum_{k=1}^{2} \gamma_{k0} a_{kc} + (\eta_{0c} + \varepsilon_{ct}). \end{aligned} \tag{11.4}$$

The term is brackets is the random part of the model, and we need to estimate the two parameters associated with it, namely σ_0^2, σ^2.

It will be seen that equation (11.4) is the usual form for the linear model, but with the additional random term η_{0c}. It is the presence of two random terms that requires special estimation procedures and means that ordinary regression techniques cannot be applied (unless of course σ_0^2 is zero or negligibly small).

We can extend this model to include 'interactions' between level 1 and level 2 variables by allowing, say, β_1 to be a function of level 2 variables, and writing:

$$\beta_{1ct} = \sum_{k=1}^{2} \gamma_{k1} a_{kc} + \eta_{1c} + \varepsilon_{1ct}. \tag{11.5}$$

This allows β_1 to vary at both level 1 and level 2, and in general any coefficient can vary randomly at any level.

As an example, suppose the variable x_{1ct} is coded as (0, 1), so that its coefficient measures the gender difference, and let this coefficient vary randomly across classes. This means that the gender difference is greater in some classes than others, which seems a reasonable assumption. In this case, if the variance of β_{1c} is σ_1^2 with mean β_1, an extra term is added to the random part of the model and we now have the variance of y_{ct} as

$$\sigma_0^2 + \sigma_1^2 x_{1ct}^2 + 2\sigma_{01} x_{1ct} + \sigma^2, \tag{11.6}$$

where we allow η_{0c} and β_{1c} to be correlated with covariance σ_{01}. Alternatively or in addition, we can allow β_1 to vary randomly over students within classes, and this leads to a model where the level 1 contribution to the variance for males is

$$\sigma_0^2 + \sigma_m^2$$

and for females

$$\sigma_0^2 + \sigma_f^2.$$

It is clear that we can accumulate a large number of random parameters (variances and covariances) by allowing further coefficients to be random. It is also possible to have a random coefficient whose mean value is 'constrained' to be zero, so that the effect of the explanatory variable is seen only in the random part of the model. A fuller study of such possibilities can be found in Goldstein (1986b, 1987b).

Returning to the basic two-level model of equation (11.4) we see that it implies a positive correlation between the responses of any two students in the same class, but a zero correlation between the responses of any two students chosen from different classes. Thus, the covariance of y_{ct} and $y_{ct'}$ is the covariance between $(\eta_{0c} + \varepsilon_{ct})$ and $(\eta_{0c} + \varepsilon_{ct'})$ and since ε_{ct} and $\varepsilon_{ct'}$ are assumed to be independent, this covariance then becomes simply σ_0^2. The variance of y_{ct} or $y_{ct'}$ conditional on the fixed part of the model, is $\sigma_0^2 + \sigma^2$ and so the correlation between the responses is

$$\tau = \sigma_0^2(\sigma_0^2 + \sigma^2)^{-1}. \tag{11.7}$$

This correlation is the usual 'intraclass' correlation and measures the degree of similarity of students within classrooms, or alternatively how well the response variable y_{ct} is 'clustered' by classrooms. The larger the value of this correlation the greater the clustering and the more important it is to use a fully efficient estimation procedure. In general this correlation will be referred to as the 'intra-unit' correlation since we can determine correlations for each higher level of a model.

One particular feature which gives these models a considerable flexibility is that we can extend the idea of repeated measurements as part of a two-level structure to a specification of multivariate data. This is done by designating level 2 to be that of the student denoted now by c, with the variates considered as repeated within-student measurements at level 1, and defined by suitable indicator variables.

Suppose each student c has four measurements $y_{c1}, \ldots, y_{c4}$. These could be different test scores, or alternate forms of the same test. We can write a simple basic two-level model for this as follows:

$$y_{ct} = \beta_{1c}x_{1ct} + \beta_{2c}x_{2ct} + \beta_{3c}x_{3ct} + \beta_{4c}x_{4ct}, \tag{11.8}$$

where t ranges from 1 to 4 and where x_1, x_2, x_3, x_4, are indicator variables (e.g., $x_{1ct} = 1$ if $t = 1$ and 0 otherwise).

We also suppose that the coefficients are random variables at the student level with variances and covariances which are to be estimated. In equation (11.8) this yields ten random parameters. The result of such an analysis will be an estimate of the 4×4 covariance matrix together with the means of the four measurements. These are simply the estimates of the coefficient means. Of course, if every student had all four measurements, these estimates would be the same as those obtained by the usual procedure for calculating the means and covariance matrix for a set of variables. The flexibility arises because the two-level model provides efficient estimates where some of the measurements are missing, and also allows nesting of students within higher levels of a hierarchy. In particular, we can see that 'multiple matrix sampling' or 'rotated form' designs, where not every student responds to the same combination of tests of forms, are special cases of data which are 'missing' by design. All such designs can in principle be regarded as special cases of the general multilevel model, and hence analysed by a suitable procedure. Any mixture of rotated designs, longitudinal repeated measurements and multiple levels of institutional organization, can be analysed within the same model structure in an integrated and efficient manner. In Section 11.5 we illustrate some of these possibilities with an example.

11.3 ESTIMATION IN THE MULTILEVEL MODEL

There are several methods available for estimating the fixed and random parameters in a multilevel model. The iterative generalized least squares (IGLS) approach used in the examples in this chapter, is outlined in Chapter 12. Under the assumption of multivariate normality the resulting estimates are also maximum likelihood. Goldstein (1986a, 1987a) shows how efficient computer algorithms can be constructed for the general case by making use of the known structure of the data, and how significance tests and confidence intervals can be derived. A trial version of a program is available from the authors.

A further useful extension of the purely hierarchical model is given in Goldstein (1987b, Chapter 7). This allows units to be cross-classified by two or more random factors. Thus the level 2 units could be the cells of a cross-classification of schools by neighbourhoods, with each student belonging to one cell. In sample surveys individuals could be classified by geographical area and workplace. In longitudinal surveys, where migration takes place between measuring occasions, individuals can be described by a cross-classification of their residence neighbourhood at the different occasions.

An important aspect of any analysis is a study of model adequacy. Procedures for the hierarchical fitting of parameters in both the fixed and random parts are available, analogously to those for ordinary linear models. At each level we can also construct residual plots. In the ordinary linear model there is only a single (level 1) residual, but in the models considered here we may have several (correlated) residuals at any level and bivariate residual plots are very useful. In the case of the basic model of equation (11.4) we obtain estimates of the residuals, namely:

$$\hat{\eta}_c, \hat{\varepsilon}_{ct}$$

which are typically referred to as 'shrunken' estimates. Very often these are themselves the focus of interest, as estimates of level 2 (school) 'effects' and from a Bayesian viewpoint would be termed 'posterior' means (Aitkin and Longford, 1986). A detailed discussion of the estimation and interpretation of such residuals is given by Goldstein (1987b, Chapters 2, 3).

11.4 LOGLINEAR MULTILEVEL MODELS FOR PROPORTIONS

In this section we describe the analysis of multilevel models where the response variable is a proportion with a logit transformation. We show how this provides a generalization of the usual single-level loglinear model to the multilevel case. Further details are given in Goldstein (1987b).

Consider a two-level model where, within each level 2 unit (cluster), the level 1 units are classified into I response categories. These may themselves have a structure, for example a two-way cross-classification corresponding to a contingency table, which then could be further modelled in terms of constants fitted to the table margins. For simplicity we shall suppose that we have a one-way classification of categories.

Since the response proportions for each cluster add to one we consider only $I-1$ of them. As in Section 11.2, we define a set of $I-1$ dummy variables (x) corresponding to the $I-1$ response proportions. Thus we have a multivariate model with a response vector of length $I-1$. As in equation (11.8), the dummy variable coefficients are assumed to vary between clusters, that is, they are random at level 2. For a given cluster the $I-1$ responses themselves have a covariance matrix, constituting the level 1 variation, which if the proportions have a multinomial distribution is given by

$$\begin{aligned} \operatorname{var}(p_{ci}|c) &= \mu_{ci}(1-\mu_{ci})n_c^{-1}, \\ \operatorname{cov}(p_{ci}, p_{cj}|c) &= -\mu_{ci}\mu_{cj}n_c^{-1}, \end{aligned} \tag{11.9}$$

where c refers to cluster, i and j to response categories, μ_{ci} is the expected proportion and n_c is the cluster size. In general, however, the μ_{ci} are unknown, and although we could obtain estimates based on the samples for each cluster, these will tend to be unstable unless cluster sizes are very large. Hence we replace μ_{ci} in effect by an average value by writting:

$$\text{var}(p_{ci}|c) = \sigma_i^2 n_c^{-1},$$
$$\text{cov}(p_{ci}p_{cj}|c) = \sigma_{ij} n_c^{-1}, \tag{11.10}$$

where the σ_i^2, and σ_{ij} are to be estimated.

An alternative procedure is outlined in Chapter 13. This assumes a multinomial distribution within each cluster and replaces the unknown parameters in equation (11.10) by the corresponding linear or loglinear functions of the (estimated) proportions. As in equation (11.10) the same values are assumed for each cluster. An estimate of the between-cluster (level 2) covariance matrix is then obtained by differencing. We note that where the distribution is not strictly multinomial the variances and covariances will still be proportional to n_c^{-1} and this suggests that equation (11.10) provides a robust procedure.

In the simple model where the μ_{ci} have the same level 2 variance, we have:

$$\text{var}(p_{ci}) = \sigma^2 + \sigma_i^2 n_c^{-1}. \tag{11.11}$$

We can specify the covariance structure as follows. For the level 1 variation we define a set of dummy variables taking the values 0 or, $n_c^{-0.5}$ analogously to the dummy variables described above. The coefficients of these dummy variables are random at level 2 so giving the covariance structure (11.10). For the between-cluster variation we have a single constant term with a coefficient random at level 2. In the next section we provide an example of this model.

For the logit and loglinear cases we define a multivariate logit as:

$$\log(p_{ci}/p_{cI}) \tag{11.12}$$

which reduces to the ordinary logit when $I = 2$. Wong and Mason (1985) give a maximum likelihood procedure for the ordinary logit two-level model. In the general case the choice of denominator as the last proportion is made for convenience.

For the multivariate logit transformation the variances and covariances are also inversely proportional to n_c and so the same procedures for specifying the covariance structures can be used as in the untransformed case. This multivariate logit model is equivalent, in the case of a single-level model, to the corresponding loglinear model based on the underlying frequencies, and so gives us the corresponding loglinear multilevel model for frequency tables.

One difficulty with the use of logit functions occurs when some of the observed proportions are zero or one. In this case we could adopt the common procedure of adding or subtracting 0.25 as appropriate to the relevant frequencies. An iterative version of this can be based upon the EM algorithm (Dempster, *et al.*, 1977), whereby the parameter estimates obtained are used to predict the (0, 1) frequencies and these predicted values treated as if they were observed values for the next cycle of iterations. The process would be repeated until convergence. An alternative is to collapse relevant cells across clusters as described in Chapter 13.

It is clear that even where the number of response categories is small, there are, potentially, a large number of random parameters at level 2. In the example in Chapter 13 the full set of variances and covariances are fitted and the structure of the resulting level 2 correlation matrix is examined. The model formulation in this section allows a hierarchically structured design for the level 2 random parameters, since the level 2 and level 1 random parameters are separately specified. Thus, we might first allow the coefficients for each of the (one-way) main effects to vary randomly at level 2, then see whether adding interactions improved the model fit, and so forth. In this way one may be able to arrive at a parsimonious model which can be interpreted and which describes the data adequately.

11.5 EXAMPLES

The data for the first example are taken from the second International Educational Achievement (IEA) survey test score results at the eighth grade or year of compulsory schooling. Two populations are used, one from Japan and one from British Columbia in Canada. In each case all the students within one class in each study school were measured. There are 187 and 47 schools respectively. For each student there is a core 'post-test' of geometry items taken towards the end of the school year. In addition each student responded to an 'alternative form' post-test. In fact there were four alternative forms of post-test of which each student took only one. Preliminary analyses revealed no significant differences between the alternative forms and we make no distinction in the analyses presented here. We treat the data as if every student answered the core post-test and the alternative form. Further discussion of this point and an extended model structure to allow for alternative forms of the post-test is given in Goldstein (1987b).

Two classroom-level variables; (a_1) the years of experience of the class mathematics teacher and (a_2) the cumulative percentage of the post-test items covered in the curriculum were measured. This second variable is often referred to as 'opportunity to learn' (OTL). In addition each student achieved a pretest score taken at the start of the school year.

The fact that only one class of students is measured within each school means that the school and class effects are confounded. We shall refer to this level of clustering as the school effect although the two explanatory variables a_1 and a_2 actually relate to the particular class sampled.

If we were to analyse only the core post-test scores (putting to one side the alternative form) then a two-level model would suffice where level 2 was the school and level 1 the individual student. To model overall ability the pretest score would not be used and thus we could write:

$$y_{ct} = \sum_{k=1}^{2} \gamma_{k0} a_{kc} + \beta_{ct},$$

$$\beta_{ct} = \beta + \eta_{0c} + \varepsilon_{ct}, \tag{11.13}$$

where $k = 1, 2$ refers to the two explanatory variables, c refers to school and t to the student. Here $\text{var}(\eta_{0c}) = \sigma_0^2$ and $\text{var}(\varepsilon_{ct}) = \sigma_\varepsilon^2$.

If we introduce into equation (11.13) the individual pretest score x_{ct} with a corresponding regression parameter then the emphasis is altered to modelling the change in performance for each child with the pretest score treated as a student level covariate. The results to be presented will cover both cases of including and excluding the pretest score.

The existence of an alternative form post-test introduces an added complexity which may be incorporated by extending the model to three levels. An additional subscript $j = 1, 2$ identifies the core post-test ($j = 1$) and the alternative form ($j = 2$). The three-level model becomes

$$y_{ctj} = \sum_k \gamma_{k0j} a_{kc} + \beta_{ctj}, \tag{11.14}$$

where

$$\beta_{ctj} = \beta_j + \eta_{0c} + \varepsilon_{ctj}$$

with

$$\text{var}(\eta_{0c}) = \sigma_0^2,$$
$$\text{var}(\varepsilon_{ct1}) = \sigma_{\varepsilon.1}^2,$$
$$\text{var}(\varepsilon_{ct2}) = \sigma_{\varepsilon.2}^2,$$

and

$$\text{cov}(\varepsilon_{ct1}, \varepsilon_{ct2}) = \sigma_{\varepsilon.12}^2$$

As in equation (11.13), the pretest score x_{ct} may be included if the change in attainment is the objective.

The formulation allows for the variances of ε_{ctj} to be unequal for $j = 1, 2$ and for the two terms to be correlated. Thus the intra-school correlation for the two forms of the post-test may be different since although σ_0^2 is common, the individual level components of variance $\sigma_{\varepsilon.1}^2$ and $\sigma_{\varepsilon.2}^2$ may differ.

The scores are standardized on a scale which gives the percentage of items correctly answered. The results of the analysis are given in Tables 11.1 and 11.2.

We consider first the case when the pretest score is included in the model. As might be expected, in both countries there is a fairly strong relationship with the pretest, as judged by the coefficient of the pretest score. There is a difference between the core and alternate form tests for British Columbia which appears as an interaction with the pretest score. The only other variable showing any relationship with progress between pre- and post-tests is OTL in the case of

Table 11.1 Summary statistics for British Columbia and Japan: Grade 8 geometry

Variable	Pretest	Post-test	Yrs exp.	OTL
British Columbia				
Mean	88.1	54.0	8.2	50.1
SD	43.3	24.3	5.9	19.6
Japan				
Mean	106.9	64.2	6.2	50.6
SD	37.2	21.6	4.6	14.1

N.B Number of students = 253 (B. Columbia), 671 (Japan),
Number of schools = 47 (B. Columbia), 187 (Japan).

British Columbia. The coefficients for OTL are fairly similar in both countries (0.07 and 0.11) but the standard error for the Japanese coefficient is no smaller despite being based on a larger number of schools. The reason is seen in Table 11.1 where the standard deviation for OTL is larger in British Columbia. Since the OTL values are more dispersed the standard error of the regression coefficient will be smaller and this compensates for fewer schools in the sample. Turning to the fixed-effect terms when the pretest is omitted (i.e. for a purely cross-sectional analysis) the contribution of the alternate form score is less than that of the core in both countries. There is also a suggestion in British Columbia that the years of teaching experience is negatively related to test score.

A persistent problem in cross-cultural studies of educational achievement is to obtain consistent and comparable interpretations of common test scores when the curricula, examination systems and so forth differ. This difficulty applies particularly to the fixed-effect parameters. When we look at the random parameters, however, we are studying how the variation is distributed over the different levels of the educational system and it may be that such comparisons between systems will be more meaningful. This will not remove the difficulty that a common test may be more relevant to some educational systems than to others, so that it seems generally preferable to design and use different assessment measures, each appropriate to its own system. The same argument would apply to studies of a single system subject to curriculum change where comparisons across long time periods were required.

Bearing in mind these caveats, we see that the intra-school correlations for Japan are much smaller than those for British Columbia. After accounting for the pretest, the intra-school correlations in Japan become zero, whereas those for British Columbia are still substantial. Thus Japanese schools appear to have

Table 11.2 Grade 8 geometry for British Columbia and Japan

	British Columbia	Japan
Post-test related to class-level variables		
Fixed		
Constant	50.8	58.2
Alt. minus core	−12.5 (1.4)	−5.1 (0.7)
Years exp. maths	−0.61 (0.28)	0.16 (0.17)
OTL	0.17 (0.09)	0.10 (0.06)
Random		
Between schools	70.0 (13.9)	15.7 (8.2)
Variance (core)	550.7 (30.3)	437.7 (24.0)
Variance (alt.)	353.9 (24.0)	443.0 (20.9)
Covariance	203.9 (23.2)	265.9 (17.3)
Correlation	0.46	0.60
Intra-school (core)	0.11	0.04
Intra-school (alt.)	0.17	0.03
Post-test related to class-level variables and pretest		
Fixed		
Constant	16.0	34.1
Alt. minus core	−1.7 (3.1)	−3.8 (2.5)
Pretest	0.40 (0.03)	0.24 (0.02)
Pretest × (alt. minus core)	−0.12 (0.03)	−0.01 (0.02)
Years exp. maths	−0.27 (0.17)	0.19 (0.19)
OTL	0.11 (0.05)	0.07 (0.06)
Random		
Between schools	21.1 (10.5)	0.0
Variance (core)	319.8 (29.8)	532.3 (21.2)
Variance (alt.)	257.1 (23.2)	378.0 (20.6)
Covariance	55.9 (18.9)	223.3 (16.8)
Correlation	0.20	0.50
Intra-school (core)	0.09	0.0
Intra-school (alt.)	0.08	0.0

greater homogeneity of achievement than those in British Columbia. It is also interesting to note that the correlation between core and alternate forms is only moderate and appears to be stronger in Japan. The variances $\sigma_{\varepsilon.2}^2$ are generally smaller than the corresponding terms $\sigma_{\varepsilon.1}^2$ for the core pretest which suggests that the alternative form may be preferred in terms of reliability. It should be remembered, however, that we have made no adjustment for measurement error in the pretest score, and this might alter the inferences we make if measurement error were taken into account (see Goldstein, 1986a). In general, a major aim of these kinds of analysis will be to see how far the between-school variation

***Table 11.3* Support for law on racial discrimination**

		Numbers of individuals responding (row percentages in parentheses)		
		1984		
		Yes	No	Total
1983	Yes	253 (79.1)	67 (20.9)	320 (100)
	No	71 (47.3)	79 (52.7)	150 (100)
	Total	324	146	470

can be reduced by the further fitting of explanatory variables, at either the school or student level.

The data for the second example are taken from a survey of social attitudes (Jowell and Airey, 1985). The response is the proportion of individuals in 1984 who said that they supported the law in the United Kingdom which outlawed racial discrimination in housing, employment, etc. There is one dichotomous explanatory variable, namely the (two-category) response to the same question asked of the same individuals one year previously in 1983.

Table 11.3 shows the overall numbers responding in each category. The data is based on a sample of 49 randomly selected clusters, and in one sense this example, too, could be regarded as a three-level structure. There are two responses for each individual who are in turn nested within the 49 clusters. However, for data of this kind we may consider the observed proportions within each cluster as the random variables to be analysed rather than the individual-level responses. We formulate a two-level model involving clusters (level 2) and the categories for the contingency table within each cluster are level 1. Thus for each cluster there are four categories defined by the responses in 1983 and 1984. In each cluster we observe the proportions falling into each cell of the table. We analyse three of these proportions directly or calculate a set of three logits by using one cell as a baseline response. In either case we refer to this variable as y_{ct} for the tth proportion (or logit) in the cth cluster.

Table 11.4 contains a set of dummy variables which are used in various model formulations for the fixed and random parts of models for these data. We first consider the model reported under (A) in Table 11.5, which is an analysis of

Table 11.4 Explanatory design variables for the* c*th level 2 unit

	1983 = yes 1984 = yes	1983 = yes 1984 = no	1983 = no 1984 = yes
x_1	1	0	0
x_2	0	1	0
x_3	0	0	1
x_4	1	1	0
x_5	1	0	1
x_6	$n_c^{-0.5}$	0	0
x_7	0	$n_c^{-0.5}$	0
x_8	0	0	$n_c^{-0.5}$
x_9	1	1	1

Table 11.5 A two-level multivariate analysis of proportions. Social attitudes data

Explanatory variable	Analysis A	B	C	D
Fixed				
x_1	—	1.77 (0.23)	—	0.54 (0.03)
x_2	—	—	—	0.14 (0.02)
x_3	—	—	—	0.15 (0.02)
x_4	0.63 (0.12)	−0.27 (0.17)	0.20 (0.01)	—
x_5	0.67 (0.12)	−0.23 (0.17)	0.21 (0.01)	—
Random				
Level 1				
σ_6^2	2.15	3.09	0.51	0.34
σ_7^2	13.1	6.49	0.16	0.13
σ_8^2	13.2	6.94	0.19	0.16
σ_{67}	−2.57	−1.69	−0.19	−0.13
σ_{68}	−2.18	1.23	−0.20	−0.13
σ_{78}	−4.41	−1.99	0.008	−0.02
Level 2				
σ_9^2	0.77 (0.56)	0.66 (0.47)	0.0001 (0.007)	0.0001 (0.007)

Note: Analyses A and B use the logits as responses: analyses C and D use the proportions.

the logits of the proportions.

$$\begin{aligned} y_{ct} &= \beta_{0ct} + \beta_4 x_{4ct} + \beta_5 x_{5ct}, \\ \beta_{0ct} &= \beta_0 + \eta_9 x_{9ct} + \eta_6 x_{6ct} + \eta_7 x_{7ct} + \eta_8 x_{8ct}, \\ \text{var}(\eta_i) &= \sigma_i^2 \qquad i = 6\ldots 9, \\ \text{cov}(\eta_i, \eta_{i'}) &= \sigma_{ii'} \qquad i, i' = 6, 7, 8, \\ \text{cov}(\eta_i, \eta_9) &= 0 \qquad i = 6, 7, 8. \end{aligned} \tag{11.15}$$

In the fixed part of the model the dummy variables x_4 and x_5 represent the two main effects for responses in 1983 and 1984. In the random part of the model the dummy variable x_9 is the explanatory variable defining the between-cluster variance and x_6, x_7 and x_8 define the level 1 variances and covariances. Note that the variables x_6, x_7 and x_8 include the factor $n_c^{-0.5}$ to reflect the fact that observed proportions are based on a sample of size n_c in the cth cluster.

In this description we have taken the four cell proportions together and used dummy variables to define the model structure of main effects and interactions. An alternative approach would have been to analyse the conditional proportions of responses in 1983. A comparison between this approach and that described above is analagous to the use of logistic or loglinear models for contingency tables. The conditional approach could be adopted using the same sorts of models as described above and a discussion of this is given in Goldstein (1987b).

In column B of Table 11.4 an extra dummy variables x_1, is included in the fixed part of the model to allow for interaction between the responses in 1983 and 1984.

A comparison of columns A and B shows, as we would expect, that the interaction term is very important with standard error much smaller than the coefficient. In the random part of the model we observe that the components of variance reduce with the introduction of the extra explanatory variable. This is often the case.

Columns C and D of Table 11.4 contain the corresponding analyses for the cell proportions rather than the logits. In column D, for illustrative purposes a different parametrization of the three fixed terms has been used but still accounts for all three available degrees of freedom. The results in columns C and D give the same general picture as those in columns A and B, although because of the change in scale from the logistic to the ordinary proportions the actual numerical values are different. As before, the interaction term is needed in the fixed part of the model and the random components are generally smaller in column D.

The variances and covariances between x_6, x_7 and x_8 generally take the signs and approximate values that we would expect under the within-cluster multinomial sampling assumption. This approach is described in Chapter 13. Ths cluster component σ_9^2 is very small. In this analysis the level 1 random parameter estimates have little substantive interest attached to them, apart from acting as a check on the adequacy of a multinomial distribution. They may be regarded as nuisance parameters. Also, their standard errors are complicated and have not been calculated.

For more complex models we can obtain an approximate chi-squared goodness-of-fit test based on the observed and predicted frequencies, the latter being derived from the predicted proportions. For degrees of freedom we would use the total number of frequencies less the number of fitted model parameters.

11.6 DISCUSSION

In our examples we have shown how multilevel modelling can explore the contribution of measurements made at different levels of a hierarchical system to both the fixed and random parts of the model. In many cases, primary interest will centre on the random variation rather than the fixed coefficients. This offers researchers a new range of techniques for exploring and understanding the effects of hierarchical systems in terms of the relative heterogeneity of units. The introduction of variables measured on higher level units allows us to attempt to 'explain' such variation. In social surveys we can introduce 'ecological' variables at the cluster level, and this opens up many new analysis possibilities.

11.7 ACKNOWLEDGEMENTS

Our grateful thanks are due to the following for their help and comments during the preparation of this paper. Leigh Burstein, Skip Kifer, Rod McDonald, Les McLean and Richard Wolfe. We are particularly grateful to Richard Wolfe for his help with the analysis of the mathematics attainment data and for discussions of models.

CHAPTER 12

Regression Models for Stratified Multi-Stage Cluster Samples

D. Pfeffermann and L. LaVange

12.1 INTRODUCTION

Survey data are very often the outcome of a stratified multi-stage cluster-sampling design. The US NHANES survey which will be described in Section 12.6 is a typical example of the use of such a design in a large-scale survey, consisting of several stages of cluster sampling within superstrata. When fitting regression models to such data, the population grouping has to be taken into consideration unless it can be assumed that the relationship between the dependent variable and the regressors is the same across the groups.

In this chapter we introduce the idea of using a two-stage model to analyse survey data. We use as first-stage units design clusters which define a division of the population into subgroups which are homogeneous with respect to socioeconomic, demographic and environmental characteristics. When collecting data from institutions, as for example in Chapter 11 where there were schools, classes and students, a three- or four-stage model may be appropriate. However, for household surveys we suggest that most between-group variation can often be captured by a two-stage model even if the survey design is multi-stage. Defining the first-stage units as design clusters (not necessarily primary sampling units) is reasonable from a subject-matter point of view and has the further advantage that external information which is often available for these clusters can be employed in the inference process. The external information includes the sampling inclusion probabilities and a variety of cluster characteristics like median income, type of urbanization, etc., which is recorded for example in censuses and used in part for the design.

The purpose of this chapter is to discuss ways of fitting regression models which permit different coefficients in different design clusters. An important feature of the proposed inference procedure is the use of the sampling clusters at various stages of the analysis. In Chapters 10 and 11, the approach has been purely model-based and the selection probabilities have not featured in the analysis. In this chapter the approach is still model-based, but at various points we will choose to use the selection probabilities in the analysis. There are several reasons for this. The estimation methods for a multilevel model are complex and require iterative methods to obtain estimates of variance components as an essential step in the total estimation procedure. This is time-consuming, particularly at the initial model selection stages when a variety of models are being considered and individual explanatory variables are added or dropped as part of the model selection procedure. The probability-weighted estimators, on the other hand, are easy to obtain from standard weighted least-squares programs; they provide consistent estimators of parameters even if they are not as efficient as the model-based estimators. In particular, tests of hypotheses to assess the importance of individual regression terms are available without estimation of the full set of components of variance required by the model-based approach. Another justification is that model misspecification is a potentially serious problem, particularly for the pure model-based approach. The use of p-weighted estimators effectively ensures that the fitted model will reflect the average relationship in the finite population from which the sample is drawn.

Employing the sampling design in conjunction with the external information available for the design clusters is a useful way to deal with 'selection bias' implied by the use of an informative sampling design. This important aspect of the analysis is illustrated at various places in the chapter. The implications of using informative sampling designs have been discussed more generally in Part B of the book. Simplicity and availability of computer programs are major considerations in the present approach and, as will be shown, all the analyses can be carried out, with simple modifications, using common software routines for ordinary regression analysis of survey data.

12.2 THE MODEL

In what follows we use the index c to signify the design clusters with separate regression coefficients (not necessarily primary sampling units) and the index t for defining units within these clusters. The vector of population values of the dependent variable in cluster c is denoted by $\mathbf{y}_{Uc}$ and the corresponding design matrix of the regressor variables by $\mathbf{X}_{Uc}$. The known cluster characteristics are denoted by the vector $\mathbf{a}_c$ of length $p+1$. The model assumed for the population is a two-level model as described in Chapters 10 and 11 and defined as follows:

$$\mathbf{y}_{Uc} = \mathbf{X}_{Uc}\boldsymbol{\beta}_c + \boldsymbol{\varepsilon}_{Uc}; \qquad E(\boldsymbol{\varepsilon}_{Uc}) = \mathbf{0}, \qquad E(\boldsymbol{\varepsilon}_{Uc}\boldsymbol{\varepsilon}_{Uc'}) = \boldsymbol{\sigma}^2\mathbf{I}, \tag{12.1}$$

$$\boldsymbol{\beta}_c = \mathbf{A}_c\boldsymbol{\gamma} + \boldsymbol{\eta}_c; \qquad E(\boldsymbol{\eta}_c) = \mathbf{0}, \qquad E(\boldsymbol{\eta}_c\boldsymbol{\eta}_{c'}) = \boldsymbol{\Delta}, \tag{12.2}$$

where $\boldsymbol{\beta}_c$ is of length $k+1$ and $\boldsymbol{\Delta} = \mathrm{diag}[\boldsymbol{\sigma}_0^2, \boldsymbol{\sigma}_1^2, \ldots, \boldsymbol{\sigma}_k^2]$. It is also assumed that $E(\boldsymbol{\varepsilon}_{Uc}\boldsymbol{\eta}_{c'}) = \mathbf{0}$ and that residuals pertaining to different clusters are independent.

The matrix $\mathbf{A}_c$ is $[(k+1) \times (k+1)(p+1)]$ and has the form $\mathbf{A}_c = \mathbf{I} \otimes \mathbf{a}_{c'}$ where $\otimes$ defines the Kronecker product. The major difference between the above model and those described in Chapters 10 and 11 is the assumption that $\mathbf{\Delta}$ is diagonal. This implies that the cluster level variance components are uncorrelated. This assumption can be changed to allow correlated errors and the methods described can be extended. However, in the empirical results that are given in this chapter the assumption that $\mathbf{\Delta}$ is diagonal is made.

Equation (12.1) postulates ordinary regression relationships within the clusters but permits different vectors of coefficients in different clusters. Equation (12.2) models the variation of the regression coefficients as a function of the cluster characteristics. For a particular coefficient β_{ic}, the corresponding equation is

$$\beta_{ic} = \gamma_{i0} + \gamma_{i1} a_{1c} + \cdots + \gamma_{ip} a_{pc} + \eta_{ic} = \boldsymbol{\gamma}_i' \mathbf{a}_c + \eta_{ic}, \tag{12.3}$$

so that different vectors of coefficients $\boldsymbol{\gamma}_i$ are allowed for different regression coefficients β_{ic}. In particular, by setting some of the coefficients $\{\gamma_{i1}, \ldots, \gamma_{ip}\}$ equal to zero the set of cluster characteristics used to explain the variation of the regression coefficients, β_{ic}, may be different for different values of i. The error terms η_{ic} account for the variation of the coefficients unexplained by the available cluster characteristics. In Section 12.5 we discuss test statistics which can be used for identifying significant cluster characteristics and for testing for the existence of residual variances $\sigma_i^2 = \text{var}(\eta_{ic})$.

The need for equations of the form (12.2) can be justified on four different grounds:

1. The variables $\mathbf{a}_c$ help to explain the cluster level variation of the coefficients and so aid our understanding.
2. In practical applications the sample sizes n_c within the sampled design clusters with different vectors of coefficients can be very small, excluding the possibility of estimating the vectors $\boldsymbol{\beta}_c$ separately.
3. Estimators of the regression coefficients based on the full model defined by equations (12.1) and (12.2) borrow information from one cluster to another which results in increased efficiency under the model.
4. Modelling the variation of the regression coefficients as a function of the cluster characteristics used for the design, eliminates the effects of selecting a non-representative sample of design clusters, at least in so far as this is captured by cluster level variables.

The model defined by equations (12.1) and (12.2) is general enough to include as special cases most of the models proposed in the literature for the regression analysis of grouped populations. Two well-known examples are (i) the random coefficients regression model which postulates that the regression coefficients in the various groups are exchangeable, and (ii) the intra-cluster correlation model which assumes a fixed vector of coefficients across the groups but permits equal, non-zero correlations between observations belonging to the same group. These and other models proposed for the analysis of grouped populations are reviewed in Pfeffermann and Smith (1985). Moreover, all of the analyses described in subsequent sections can be modified straightforwardly to permit non-zero covariances between individual coefficients β_{ic} pertaining to the same cluster c. Consideration of non-diagonal covariance matrices $\mathbf{\Delta}$ is necessary if the model is to be invariant under linear transformations of the regressor variables like change of origin, difference between regressors, etc. (see Aitkin and Longford, 1986). Obviously, any such extension of the original model may involve a substantial increase in the number of unknown model parameters.

12.3 ESTIMATION OF THE FIXED REGRESSION COEFFICIENTS

12.3.1 The Sample

The model defined by equations (12.1) and (12.2) refers to the target population. We assume that $m > 1$ of the design clusters with separate regression coefficients have been selected by a probability sampling scheme with known inclusion probabilities π_c and that the ultimate sample, denoted by s, includes $n_c \geqslant 1$ units from the cth selected cluster, $c = 1, \ldots, m$, drawn by another probability sampling scheme. The overall inclusion probabilities $\pi_{ct} = P(ct \in s)$ are assumed to be known for every unit in the sample. Both sampling schemes may involve a complex, multi-stage selection procedure. The ultimate sample units are distinguished by the double index (ct) so that y_{ct} and $\mathbf{x}'_{ct} = (1, x_{1ct}, \ldots, x_{kct})$ define the observations on the dependent and independent variables recorded for unit (ct). The data observed for the entire sample from cluster c will be denoted correspondingly by $\mathbf{y}_c$ and $\mathbf{X}_c$.

12.3.2 Estimation of γ

We consider the estimation of the γ-coefficients first since these estimators form the basis for the estimation of the regression coefficients $\boldsymbol{\beta}_c$. The model defined by equations (12.1) and (12.2) implies the following relationships obtained by inserting equation (12.2) into equation (12.1):

$$\mathbf{y}_c = \mathbf{X}_c\mathbf{A}_c\boldsymbol{\gamma} + \mathbf{X}_c\boldsymbol{\eta}_c + \boldsymbol{\varepsilon}_c = \mathbf{X}^*_c\boldsymbol{\gamma} + \mathbf{u}_c,$$
$$E(\mathbf{u}_c) = \mathbf{0}; \qquad E(\mathbf{u}_c\mathbf{u}'_c) = \sigma^2\mathbf{I}_{n_c} + \mathbf{X}_c\mathbf{\Delta}\mathbf{X}'_c; \qquad E(\mathbf{u}_c\mathbf{u}'_d) = \mathbf{0} \qquad c \neq d. \tag{12.4}$$

The tth row of the matrix $\mathbf{X}^*_c$ has the form $\mathbf{x}^{*\prime}_{ct} = (\mathbf{a}'_c, x_{1ct}\mathbf{a}'_c, \ldots, x_{kct}\mathbf{a}'_c)$ so that it consists of all cross-products between the values of the regressor variables and the values of the cluster characteristics. The model for the entire sample data can be written compactly as

$$\mathbf{y} = \mathbf{X}^*\boldsymbol{\gamma} + \mathbf{u}; \qquad E(\mathbf{u}) = \mathbf{0}, \qquad E(\mathbf{u}\mathbf{u}') = \boldsymbol{\Sigma}, \tag{12.5}$$

where $\mathbf{y}' = (\mathbf{y}'_1 \cdots \mathbf{y}'_m)$, $\mathbf{X}^{*\prime} = [\mathbf{X}_1^{*\prime}, \ldots, \mathbf{X}_m^{*\prime}]$ and $\boldsymbol{\Sigma}$ is block diagonal with $\boldsymbol{\Sigma}_{cc} = E(\mathbf{u}_c\mathbf{u}'_c)$ as the cth block.

The matrix $\mathbf{X}^*$ would usually be of full rank $p^* = (p+1)(k+1)$, and so for known variances σ^2 and σ_j^2, $j = 0, \ldots, k$, the best linear unbiased estimator (BLUE) of $\boldsymbol{\gamma}$ is the generalized least squares (GLS) estimator,

$$\hat{\boldsymbol{\gamma}}_{\mathrm{GLS}} = (\mathbf{X}^{*\prime}\boldsymbol{\Sigma}^{-1}\mathbf{X}^*)^{-1}\mathbf{X}^{*\prime}\boldsymbol{\Sigma}^{-1}\mathbf{y}. \tag{12.6}$$

In practice, the variances are rarely known and in the next section we discuss the ways of estimating them consistently. Inserting the sample estimators instead of the true variances in equation (12.6) yields the empirical estimator $\hat{\boldsymbol{\gamma}}_{\mathrm{E}}$ which has the same asymptotic properties as the GLS estimator $\hat{\boldsymbol{\gamma}}_{\mathrm{GLS}}$ (Anderson, 1973). Alternativerly, one may consider the ordinary least squares (OLS) estimator $\hat{\boldsymbol{\gamma}}_{\mathrm{OLS}} = (\mathbf{X}^{*\prime}\mathbf{X}^*)^{-1}\mathbf{X}^{*\prime}\mathbf{y}$ which is unbiased and under mild conditions also consistent but not efficient.

A third alternative for unequal probability of selection designs is to pre-multiply each vector observation $(y_{ct}, \mathbf{x}'_{ct})$ by the square root of the inverse of the unit selection probability and then calculate the OLS estimator based on the weighted observations. The resulting estimator is the familiar probability-weighted (PW) estimator which has the form

$$\hat{\boldsymbol{\gamma}}_{\mathrm{PW}} = [\mathbf{X}^{*\prime}\boldsymbol{\Pi}^{-1}\mathbf{X}^*]^{-1}\mathbf{X}^{*\prime}\boldsymbol{\Pi}^{-1}y, \tag{12.7}$$

where $\boldsymbol{\Pi}$ is a diagonal matrix with the inclusion probabilities π_{ct} on the main diagonal. The computation of a PW GLS estimator is somewhat more complicated and is deferred to the Appendix.

The rationale behind the weighting procedure is that under mild conditions on the design, ordinarily satisfied in practice, $\hat{\boldsymbol{\gamma}}_{\mathrm{PW}}$ is approximately design-unbiased and consistent for the 'census estimator' $\gamma_{U,\mathrm{OLS}}$, defined as the least-squares solution based on all the population data (see Section 1.6 for definitions of design-unbiased and consistent). The census estimator is unaffected by the design and so $\hat{\boldsymbol{\gamma}}_{\mathrm{PW}}$ is consistent for $\boldsymbol{\gamma}$. A similar consideration applies to the use of the weighted GLS estimator. Thus the use of the p-weighted estimator protects against selection bias. The properties of such estimators are described in Section 12.8 and have been developed in Part B of the book. A formulation for asymptotic analysis in finite population sampling is given by Isaki and Fuller (1982). The condition of the design and the manner by which the population and the sample have to increase are essentially the same as needed for the consistency of Horvitz–Thompson-type estimators and imply in particular that the number of primary sampling units (PSUs) is allowed to increase.

The computation of $\hat{\boldsymbol{\gamma}}_{\mathrm{PW}}$ does not require the knowledge of the unknown model variances and can be carried out using any software for regression analysis. The design covariance matrix of the estimator can easily be estimated using, for example, the linearization method, see Sections 2.13 and 3.4.

Under mild conditions, the estimated covariance matrix accounts for the total variation of $\hat{\gamma}_{PW}$ around γ, where by the total variation we mean the design variance around the census estimator $\gamma_{U,OLS}$ and the model (ξ) variance of the census estimator. The conclusion from this discussion is that it is possible to test hypotheses on γ before actually estimating the model variances although the test may not be as powerful as the full model-based Wald test (12.16). The implementation of these tests is very simple and may save considerable computer time. Thus, the use of the PW estimator can be motivated on grounds other than protection against selection bias. We elaborate on the rationale and computation of such test statistics in Section 12.5.

12.4 ESTIMATION OF THE UNKNOWN VARIANCES

12.4.1 Preface

The model defined by equations (12.1) and (12.2) is a special case of the 'classical' mixed model of variance components for which the covariance matrix of the error terms is a linear combination of known matrices with unknown coefficients. Variance estimation under these kinds of models has been a major area of research in recent years and methods with known properties and feasible computation algorithms have been developed. The review article by Harville (1977) contains an extensive discussion of the theoretical and computational properties of the various methods.

In the next section we describe a relatively simple way of obtaining maximum likelihood estimates (MLEs) of the unknown variances under the assumption that the error terms in equations (12.1) and (12.2) are normal deviates. As argued by Harville (1977), MLEs derived on the basis of normality have desirable properties even when the form of the distribution is not specified. The method is not new and goes back to Anderson (1973). See also Brown and Burgess (1983) and Goldstein (1986a) for more recent applications.

The prominent feature of the method is that it transforms the problem of variance estimation into a generalized regression estimation problem. This has some important advantages in the present context which are discussed in Section 12.4.3.

12.4.2 Maximum Likelihood Estimation

For units $ct \in s$, define

$$\mathbf{u}_{ct} = (\mathbf{y}_{ct} - \mathbf{x}_{ct}^{*\prime}\gamma) = (\mathbf{y}_{ct} - \mathbf{x}_{ct}'\boldsymbol{\beta}_c) + \mathbf{x}_{ct}'(\boldsymbol{\beta}_c - \mathbf{A}_c\gamma) = \boldsymbol{\varepsilon}_{ct} + \mathbf{x}_{ct}'\boldsymbol{\eta}_c. \qquad (12.8)$$

The following model holds for the squares and cross-products of the error terms u_{ct} within clusters (see comment 2 below).

$$\begin{aligned} E(u_{ct}^2) &= \sigma_0^2 + x_{1ct}^2\sigma_1^2 + \cdots + x_{kct}^2\sigma_k^2 + \sigma^2 = (\mathbf{z}'_{ct}, 1)\boldsymbol{\sigma}^*, \\ E(u_{ct}u_{cl}) &= \sigma_0^2 + (x_{1ct}x_{1cl})\sigma_1^2 + \cdots + (x_{kct}x_{kcl})\sigma_k^2 \\ &= (\mathbf{z}'_{ctl}, 0)\boldsymbol{\sigma}^*, \qquad t \neq l, \end{aligned} \tag{12.9}$$

where $\boldsymbol{\sigma}^* = (\sigma_0^2 \cdots \sigma_k^2, \sigma^2)'$, $\mathbf{z}'_{ct} = (1, x_{1ct}^2, \ldots, x_{kct}^2)$ and

$$\mathbf{z}'_{ctl} = (1, x_{1ct}x_{1cl}, \ldots, x_{kct}x_{kcl}).$$

In addition,

$$\operatorname{cov}(u_{ct}u_{cl}, u_{cp}u_{cq}) = \tilde{\lambda}_{ctp}\tilde{\lambda}_{clq} + \tilde{\lambda}_{ctq}\tilde{\lambda}_{clp}, \tag{12.10}$$

where

$$\tilde{\lambda}_{ctt^*} = \operatorname{cov}(u_{ct}, u_{ct^*}) = \begin{cases} \mathbf{x}'_{ct}\Delta\mathbf{x}_{ct} + \sigma^2, & t = t^*, \\ \mathbf{x}'_{ct}\Delta\mathbf{x}_{ct^*}, & t \neq t^*, \end{cases} \tag{12.11}$$

$$\operatorname{cov}(u_{ct}u_{cl}, u_{c^*t^*}u_{c^*l^*}) = 0, \qquad c \neq c^*.$$

Equations (12.9) and (12.11) follow directly from equation (12.1) and (12.2) (see also equation (12.20) and are not restricted to the normal case. Equation (12.10) translates known properties of the multivariate normal distribution to the present model.

Let $\mathbf{g}_c$ be the vector of squares and cross-products of the errors u_{ct} corresponding to the cluster c, arranged in some convenient order and define $\mathbf{g}' = (\mathbf{g}'_1, \ldots, \mathbf{g}'_m)$. Let $\mathbf{Z}_c$ be the matrix consisting of the rows $(\mathbf{z}'_{ct}, 1)$ and $(\mathbf{z}_{ctl}, 0)$ arranged in the same order as the elements of $\mathbf{g}_c$, and define $\mathbf{Z}' = [\mathbf{Z}'_1, \ldots, \mathbf{Z}'_m]$. It follows from equations (12.9)–(12.11) that the vector $\mathbf{g}$ satisfies the model

$$\mathbf{g} = \mathbf{Z}\boldsymbol{\sigma}^* + \mathbf{e}; \qquad E(\mathbf{e}) = \mathbf{0},\ E(\mathbf{e}\mathbf{e}') = \boldsymbol{\Lambda}, \tag{12.12}$$

where the elements of $\boldsymbol{\Lambda} = \oplus[\boldsymbol{\Lambda}_1, \ldots, \boldsymbol{\Lambda}_m]$ are known functions of the unknown variances $\{\sigma_j^2\}$ and σ^2 as defined by equation (12.10). The notation $\oplus[\mathbf{A}, \mathbf{B}, \mathbf{C}\ldots]$ defines a block diagonal matrix with $\mathbf{A}, \mathbf{B}, \mathbf{C}\ldots$ as the block matrices. If $\boldsymbol{\gamma}$ was known so that the vector $\mathbf{g}$ was observable, and $\boldsymbol{\Lambda}$ was also known then the BLUEs of the unknown variances would have been

$$\tilde{\boldsymbol{\sigma}}^* = [\mathbf{Z}'\boldsymbol{\Lambda}^{-1}\mathbf{Z}]^{-1}\mathbf{Z}'\boldsymbol{\Lambda}^{-1}\mathbf{g}. \tag{12.13}$$

In practice, both $\boldsymbol{\Lambda}$ and $\boldsymbol{\gamma}$ are unknown, and $\boldsymbol{\sigma}^*$ can be estimated by iterating between equation (12.13) and equation (12.6) which defines the BLUE of $\boldsymbol{\gamma}$ for given $\boldsymbol{\Delta}$. The iterations proceed as follows: let $\boldsymbol{\gamma}_{(0)}$ define a first estimate of $\boldsymbol{\gamma}$ obtained by initializing $\boldsymbol{\Sigma}^{(0)} = \mathbf{I}$ and define $\hat{u}_{ct}^{(0)} = (y_{ct} - \mathbf{x}_{ct}^{*\prime}\boldsymbol{\gamma}_{(0)})$. Substituting $\hat{u}_{ct}^{(0)}$

for u_{ct} in the vector $\mathbf{g}$ and setting $\mathbf{\Lambda}^{(0)} = \mathbf{I}$ yields a first set of estimates

$$\boldsymbol{\sigma}^{*\prime}_{(1)} = [\sigma^2_{0(1)}, \ldots, \sigma^2_{k(1)}, \sigma^2_{(1)}].$$

Inserting these estimates in the matrix $\mathbf{\Sigma}$ yields a new estimate $\boldsymbol{\gamma}_{(1)}$, new residuals $\{\hat{u}^{(1)}_{ct}\}$ and hence new estimates $\boldsymbol{\sigma}^*_{(2)}$. The procedure is continued until convergence or until a given number of iterations is completed.

It follows from Anderson (1973) that the set of equations defined by equations (12.6) and (12.13) are equivalent to the likelihood equations obtained by equating the first derivatives of the log likelihood function to zero. Thus, the solutions to these equations are MLEs provided they are non-negative (see next section).

Comments

1. Inspection of the first two equations in (12.9) indicates that the variances σ^2_0 and σ^2 are estimable provided that the sample sizes within some of the clusters are larger than one.
2. The model defined by equation (12.12) does not include cross-products between error terms pertaining to different clusters. This facilitates the computations enormously. The exclusion of these products is permissible because by equation (12.8) they have zero expectations and are uncorrelated with the squares and cross-products included in the vectors $\mathbf{g}_c$.
3. The set of equations (12.6) and (12.13) assume a non-informative sampling design. The computation of a 'PW' analogue of $\tilde{\boldsymbol{\sigma}}^*$ seems too cumbersome due to the complex structure of the matrix $\mathbf{\Lambda}$. In order to deal with an informative design, one may iterate, however, between the PW GLS estimator of $\boldsymbol{\gamma}$ presented in the Appendix and the estimator

$$\tilde{\boldsymbol{\sigma}}^*_{\Pi} = (\mathbf{z}'\tilde{\mathbf{\Pi}}^{-1}\mathbf{\Lambda}_0^{-1}\mathbf{z})^{-1}\mathbf{z}'\tilde{\mathbf{\Pi}}^{-1}\mathbf{\Lambda}_0^{-1}\mathbf{g}$$

obtained by pre-multiplying every vector observation $(u^2_{ct}, \mathbf{z}'_{ct}, 1)$ and $(u_{ct}u_{cl}, \mathbf{z}'_{ctl}, 0)$ in equation (12.12) by the inverse of the square root of the corresponding sample inclusion probability and setting the off-diagonal elements of $\mathbf{\Lambda}$ to zero to give $\mathbf{\Lambda}_0$. The matrix $\tilde{\mathbf{\Pi}}$ is diagonal with the corresponding inclusion probabilities on the main diagonal. The estimators obtained as the solutions to the above modified equations are consistent with respect to the $p\xi$-distribution. Assuming the practical situation of a large number of design clusters with only a few observations in each, it is evident that ignoring the off-diagonal elements of $\mathbf{\Lambda}$ has only a minor effect on the estimation of $\mathbf{\Lambda}$ since residuals u_{ct} pertaining to different clusters are independent anyway.

12.4.3 Advantages of the Regression Method for ML Estimation

In this section we overview several desirable properties of the regression method.

(a) Simplicity

The procedure can be carried out using common statistical software packages with only minor adjustments and there is no need for extra software development. Furthermore, as already mentioned in Section 12.2, the procedure can easily be extended to include more general structures of the matrix $\boldsymbol{\Delta} = \text{var}(\boldsymbol{\beta}_i)$. In fact all that is needed is to redefine the linear model to include the squares and cross-products of the residuals u_{ct} as implied by the specific structure of the matrix $\boldsymbol{\Delta}$.

(b) Non-negativity constraints on the variance estimators

As with the solution of the ordinary likelihood equations, the iterative solution of equations (12.6) and (12.13) may converge to negative variance estimates or may become unstable if an iteration produces negative estimates. The common routine of replacing a negative estimate by zero, either on each iteration where it occurs or in the final iteration, does not in general produce ML estimation.

An important advantage of the regression method for ML estimation is that it offers a simple way of constraining the estimates to the parameter space, that is, to non-negative values. The idea is that instead of iterating between equations (12.6) and (12.13), one iterates between equation (12.6) and the solution to the constrained minimization problem:

$$\min_{\boldsymbol{\sigma}^* \geqslant 0} [(\mathbf{g} - \mathbf{Z}\boldsymbol{\sigma}^*)' \boldsymbol{\Lambda}^{-1} (\mathbf{g} - \mathbf{Z}\boldsymbol{\sigma}^*)]. \tag{12.14}$$

Notice that the estimator defined by equation (12.13) solves the unconstrained minimization problem, that is, without the restriction $\boldsymbol{\sigma}^* \geqslant 0$. Obviously, a positive solution of equation (12.13) solves equation (12.14) also.

For given values of $\boldsymbol{\gamma}$ and $\boldsymbol{\Lambda}$ (as obtained in the previous iteration), the solution of equation (12.14) reduces to the familiar problem of minimizing a quadratic function subject to linear inequality constraints for which standard computer programs are available. An alternative algorithm for solving equation (12.14) in any given iteration, where one or more of the elements of the unconstrained solution is negative, is to search on the boundaries. This algorithm, recommended in Goldstein (1986a) and applied in the empirical study of Section 12.6, consists of fixing in turn each variance σ_j^2 of $\boldsymbol{\sigma}^*$ to zero (i.e. omitting the jth column of $\mathbf{Z}$) and recomputing the unconstrained solution $\hat{\boldsymbol{\sigma}}^*_{(j)}$. The vector $\hat{\boldsymbol{\sigma}}^*_{(j0)}$ which minimizes equation (12.14) out of all non-negative vectors $\boldsymbol{\sigma}^*_{(j)}$ is the global solution to equation (12.14) in that iteration. If none of the vectors $\hat{\boldsymbol{\sigma}}^*_{(j)}$ has all its components in the feasible region, the procedure is repeated with pairs of variances and so forth until an optimal boundary solution is identified.

It should be emphasized that the use of this algorithm does not increase the computation too much because the vector $\mathbf{g}$ has to be calculated only once on each iteration and because the inverses of the reduced matrices $[\mathbf{Z}'_{(j)} \hat{\boldsymbol{\Lambda}}^{-1} \mathbf{Z}_{(j)}]$

needed for the computation of the constrained estimators $\hat{\boldsymbol{\sigma}}^*_{(j)}$ can be obtained from the inverse of the matrix $(\mathbf{Z}'\hat{\boldsymbol{\Lambda}}^{-1}\mathbf{Z})$ used for the calculation of $\hat{\boldsymbol{\sigma}}^*$.

The use of non-negativity constraints in conjunction with the regression method was first proposed by Brown and Burgess (1983). The authors present simulation results for random-effects models which illustrate the superiority of constraining the estimators over the *ad hoc* procedure of replacing negative estimates by zeros in terms of bias and MSE.

(c) Variances of variance estimators

Under the model, a consistent estimate of the covariance matrix of the variance estimators is readily obtained from equation (12.13) as

$$\widehat{\mathrm{Var}}(\hat{\boldsymbol{\sigma}}^*) = (\mathbf{Z}'\hat{\boldsymbol{\Lambda}}^{-1}\mathbf{Z})^{-1}, \tag{12.15}$$

where $\hat{\boldsymbol{\Lambda}}$ defines the estimator of $\boldsymbol{\Lambda}$ as obtained in the final iteration. The consistency of $\mathrm{Var}(\hat{\boldsymbol{\sigma}}^*)$ follows from the consistency of $\hat{\boldsymbol{\Lambda}}$ (Anderson, 1973) which holds simultaneously for all its elements if both the 'within-clusters' sample sizes and the number of clusters are allowed to increase.

The estimators defined by equations (12.13) (or 12.14) and (12.15) are model dependent. In particular, the form of the matrix $\boldsymbol{\Lambda}$ results from the assumption that the error terms $\{u_{ct}\}$ are normal deviates. Harville (1977) argues that estimators of variance components derived under normality are reasonable estimators even when the form of the distribution is not specified. This assertion, however, does not necessarily apply to the estimation of the variances of these estimators. Thus, an alternative and more robust procedure for estimating the covariance matrix of $\hat{\boldsymbol{\sigma}}^*$ is to estimate the design covariance matrix of the variance estimators over all possible sample selections from the finite population. The latter matrix estimates also the total variation over both the model and the randomization distributions. (See the next section for an explanation of this property.)

Estimators of the design variances and covariances can be derived similarly to the estimation of the design covariance matrix of OLS and PW estimators. The computations in the case of the estimator $\hat{\boldsymbol{\sigma}}^*$ involve:

1. Calculation of $\hat{\mathbf{g}}_{\mathrm{E}}$ obtained from $\mathbf{g}$ by replacing the unobserved errors $\{u_{ct}\}$ by the observed residuals $\{\hat{u}_{ct} = y_{ct} - \mathbf{x}^{*\prime}_{ct}\hat{\boldsymbol{\gamma}}_{\mathrm{E}}\}$, where $\hat{\boldsymbol{\gamma}}_{\mathrm{E}}$ denotes the empirical estimator of $\boldsymbol{\gamma}$ as obtained in the final iteration;
2. Premultiplying $\hat{\mathbf{g}}_{\mathrm{E}}$ and $\mathbf{Z}$ by $\hat{\boldsymbol{\Lambda}}^{-1/2}$, where $\hat{\boldsymbol{\Lambda}}$ is the estimate of $\boldsymbol{\Lambda}$ as obtained in the final iteration;
3. Applying the usual formula for the estimation of the design covariance matrix of OLS estimators with $\hat{\boldsymbol{\Lambda}}^{-1/2}\hat{\mathbf{g}}_{\mathrm{E}}$ as the dependent variable and $\hat{\boldsymbol{\Lambda}}^{-1/2}\mathbf{Z}$ as the regressors. See Pfeffermann and LaVange (1986) for the appropriate formulae in the case of the linearization method.

The key assumptions underlying the use of the above procedure are that the asymptotic design covariance matrix of $\hat{\boldsymbol{\sigma}}^*$ is the same as that of $\tilde{\boldsymbol{\sigma}}^*$ which assumes known $\boldsymbol{\gamma}$ and $\boldsymbol{\Lambda}$ (equation 12.13) and that selection of primary sampling units within strata can be considered as being with replacement. The latter property permits the estimation of the covariance matrix by considering only the variance releated to the first stage of selection (see Section 2.13).

12.5 MODEL TESTING

12.5.1 Hypotheses of Interest

Having postulated a general model of the form (12.1) and (12.2), there are two classes of hypotheses one would usually like to test:

(A) Hypotheses related to the $\boldsymbol{\gamma}$ coefficients of the cluster characteristics used to explain the variation of the regression coefficients;
(B) Hypotheses related to the residual variances $\boldsymbol{\sigma}^*$ of the regression coefficients.

It should be noted that hypotheses related to the original regressor variables can be formulated in terms of (A) and (B) above. Thus, the jth regressor, X_j, is 'non-significant' if and only if both $\boldsymbol{\gamma}_j = \mathbf{0}$ (implying $E(\beta_{jc}) = 0$, equation 12.3) and $\sigma_j^2 = \text{var}(\beta_{jc}) = 0$. In the next section we discuss the use of Wald statistics for testing that certain elements of $\boldsymbol{\gamma}$ and/or $\boldsymbol{\sigma}^*$ are zero.

12.5.2 Large-sample Wald Statistics

Initially the use of the Wald (1943) statistic will be illustrated by considering hypotheses related to the vector $\boldsymbol{\gamma}$. Suppose one is interested in testing that a particular set of r cluster characteristics have a non-significant marginal effect when explaining the variation of the regression coefficients. Under the model, this is equivalent to testing that the corresponding elements of $\boldsymbol{\gamma}$ are zero. Any such hypothesis can be formulated as $H_0: \mathbf{C}\boldsymbol{\gamma} = \mathbf{0}$, where $\mathbf{C}$ is an $[r \times (k+1)(p+1)]$ matrix with corresponding single entries of ones in each row and zeros elsewhere. The *model-based Wald statistic* for testing H_0 is

$$X^2_{\mathrm{W,M}} = (\mathbf{C}\hat{\boldsymbol{\gamma}}_{\mathrm{E}})'[\mathbf{C}(\mathbf{X}^{*\prime}\hat{\boldsymbol{\Sigma}}^{-1}\mathbf{X}^*)^{-1}\mathbf{C}']^{-1}(\mathbf{C}\hat{\boldsymbol{\gamma}}_{\mathrm{E}}), \tag{12.16}$$

where $\hat{\boldsymbol{\gamma}}_{\mathrm{E}}$ is the empirical estimator of $\boldsymbol{\gamma}$ obtained by replacing $\boldsymbol{\Sigma}$ by $\hat{\boldsymbol{\Sigma}}$ in equation (12.6). For $\hat{\boldsymbol{\Sigma}}$ close enough to $\boldsymbol{\Sigma}$, the distribution of $X^2_{\mathrm{W,M}}$ under H_0 is approximately chi-square with r degrees of freedom.

The statistic $X^2_{\mathrm{W,M}}$ is model dependent and requires the estimation of $\boldsymbol{\sigma}^*$ which is computationally intensive. Thus, the alternative *design-based Wald* *statistic* for testing H_0 is

$$X^2_{\mathrm{W,D}} = (\mathbf{C}\hat{\boldsymbol{\gamma}}_{\mathrm{PW}})'[\mathbf{C}\{\widehat{\mathrm{Var}}_p(\hat{\boldsymbol{\gamma}}_{\mathrm{PW}})\}\mathbf{C}']^{-1}(\mathbf{C}\hat{\boldsymbol{\gamma}}_{\mathrm{PW}}), \tag{12.17}$$

where $\hat{\boldsymbol{\gamma}}_{\mathrm{PW}}$ is the PW estimator of $\boldsymbol{\gamma}$ defined in equation (12.7) and $\widehat{\mathrm{Var}}_p(\hat{\boldsymbol{\gamma}}_{\mathrm{PW}})$ is its estimated covariance matrix with respect to the p-distribution.

The use of equation (12.7) can be justified by noticing that the estimator $\hat{\boldsymbol{\gamma}}_{\mathrm{PW}}$ is approximately unbiased for $\boldsymbol{\gamma}$ with respect to the $p\xi$-distribution and that the $p\xi$-variance of $\hat{\boldsymbol{\gamma}}_{\mathrm{PW}}$ can be decomposed as

$$\mathrm{Var}_{p\xi}(\hat{\boldsymbol{\gamma}}_{\mathrm{PW}}) = E_\xi[\mathrm{Var}_p(\hat{\boldsymbol{\gamma}}_{\mathrm{PW}})] + \mathrm{Var}_\xi[E_p(\hat{\boldsymbol{\gamma}}_{\mathrm{PW}})] = E_\xi[\mathrm{Var}_p(\hat{\boldsymbol{\gamma}}_{\mathrm{PW}})] + \mathrm{O}\left(\frac{1}{N_o}\right), \tag{12.18}$$

where N_o is the population size. The right-hand-side equality results from the observation that $E_p(\hat{\boldsymbol{\gamma}}_{\mathrm{PW}})$ is a finite population parameter. Now, assuming a multi-stage cluster design, $\mathrm{Var}_p(\hat{\boldsymbol{\gamma}}_{\mathrm{PW}})$ is of order $1/m^*$, where m^* denotes the number of PSUs in the sample. Thus, if only a fraction of the population PSUs are selected, $\mathrm{Var}_p(\hat{\boldsymbol{\gamma}}_{\mathrm{PW}}) \simeq E_\xi[\mathrm{Var}_p(\hat{\boldsymbol{\gamma}}_{\mathrm{PW}})]$ and any consistent estimator of $\mathrm{Var}_p(\hat{\boldsymbol{\gamma}}_{\mathrm{PW}})$ is also consistent for $\mathrm{Var}_{p\xi}(\hat{\boldsymbol{\gamma}}_{\mathrm{PW}})$.

The major advantage of the statistic $X^2_{\mathrm{W,D}}$ is its ease of computation. It is also less 'model dependent' compared to the statistic $X^2_{\mathrm{W,M}}$ and can be used under an informative sampling design. However, when selecting the sample by a stratified multi-stage cluster sampling design with only a few PSUs per stratum, the design variance and covariance estimators may become unstable, hence distorting the nominal chi-square distribution of the statistic. Transformations of the Wald statistic to an approximate F-statistic taking the number of degrees of freedom for variance estimation into account have been proposed by Shah *et al.* (1977) and Fuller (1984) (see Section 3.4.2), but the theory underlying the transformations has to be further established.

The model-based variance estimators measure the variation between the design clusters with separate vectors of coefficients and there are usually many more of these than there are PSUs. Hence, the problem of unstable variance estimators is avoided when using the statistic $X^2_{\mathrm{W,M}}$.

Wald statistics for testing hypotheses on the model variances can be constructed in a similar way. Moreover, in principle, one can test simultaneously that corresponding elements of $\boldsymbol{\gamma}$ and $\boldsymbol{\sigma}^*$ are zero (e.g. that both $\boldsymbol{\gamma}_j = \mathbf{0}$ and $\sigma_j^2 = 0$ implying that the jth regressor is non-significant). In particular, under normality assumptions, $\hat{\boldsymbol{\gamma}}_{\mathrm{E}}$ and $\hat{\mathbf{g}}_{\mathrm{E}}$ are asymptotically independent (Anderson, 1973) implying that $\mathrm{cov}_\xi(\hat{\boldsymbol{\gamma}}_{\mathrm{E}}, \hat{\boldsymbol{\sigma}}^*) \simeq \mathbf{0}$. As for the design-based Wald statistic, the joint design covariance matrix of the vector $(\hat{\boldsymbol{\gamma}}'_{\mathrm{PW}}, \hat{\boldsymbol{\sigma}}^{*\prime})$ can be estimated by combining equations (12.5) and (12.12) so that $\{(\boldsymbol{\Pi}^{-1/2}\mathbf{y})', (\hat{\boldsymbol{\Lambda}}^{-1/2}\hat{\mathbf{g}}_{\mathrm{E}})'\}'$ are the dependent observations and $\oplus\{\boldsymbol{\Pi}^{-1/2}\mathbf{X}^*, \hat{\boldsymbol{\Lambda}}^{-1/2}\mathbf{Z}\}$ is the design matrix and then applying the usual formulae for the estimation of the design covariance matrix of OLS estimators.

12.6 EMPIRICAL RESULTS I: ESTIMATION AND TESTING OF FIXED REGRESSION COEFFICIENTS AND VARIANCE COMPONENTS

12.6.1 The NHANES Design

In this section we report results obtained when applying the methods described in the previous sections to the National Health and Nutrition Examination Survey (NHANES). Further empirical results, illustrating the prediction of the random cluster coefficients, are given in Section 12.7.2. A detailed description of NHANES is given in McDowell *et al.* (1981). The NHANES consists of a stratified multi-stage probability cluster sample of households in the USA. At the first stage, PSUs that coincide with counties or groups of contiguous counties and selected for the National Health Interview Survey were stratified into 64 'superstrata' based on size, income and racial distribution. One PSU was selected from each stratum with probability proportional to size (PPS). In order to over-sample persons with low incomes, enumeration districts (EDs) within the selected PSUs were sorted into poverty and non-poverty strata. Then EDs were selected separately from each stratum using a PPS design with sampling rates within poverty strata being three times larger than the sampling rates in corresponding non-poverty strata. Households within EDs were clustered into segments of eight adjacent addresses, and a systematic sample of segments was selected across all the EDs with no more than one segment per ED. At the final stage, persons were selected one per household, roughly, with young and old age-groups being sampled at a rate of 3:4 and other ages at a rate of 1:4.

12.6.2 Statement of Problem and Regression Variables

The methods discussed in previous sections are illustrated through an analysis of the relationship between blood lead levels and diastolic blood pressure (the dependent variable) using NHANES II data. Several articles appearing recently in the literature have studied this important issue by fitting ordinary multiple regression models to data from NHANES I and NHANES II. Since our main goals in this study are to illustrate methodological issues, we decided to use previously published model equations, thereby taking advantage of the extensive variable selection procedures used to fit these models.

The model chosen as a basis for the study was fitted by Harlan *et al.* (1985) using data for men aged 12–74 years. Harlan *et al.* postulated a uniform regression line throughout the population and used PW estimators and randomization variances for model estimation and diagnostic testings. A detailed description of their analysis can be found in Pirkle *et al.* (1985). The importance and practical implications of the model are likewise discussed in these two articles. See also Shaper and Pocock (1985) who analysed data obtained from the British Regional Heart Study.

The model estimated by Harlan *et al.* is displayed in the bottom row of Table 12.4. It includes, other than the lead variable and an intercept, six regressor variables representing medical control measurements (body mass, haemoglobin and serum zinc) and demographic characteristics (race, age and age square).

12.6.3 Design Clusters and Design Cluster Characteristics

We use as design clusters the EDs which, as described in Section 12.6.1, are the second-stage sampling units in NHANES. Enumeration districts are homogeneous groups of households. Furthermore, data characterizing the EDs and available on the US 1970 census tapes could be merged onto the NHANES II data base. These data form the design cluster characteristics used to explain the variation of the regression vectors of coefficients. They include: a_1—median family income; a_2—proportion of non-whites; a_3—proportion of persons aged 25+ with less than 9 years of school; a_4—living areas (urban outside central city, urban inside central city, rural); a_5—poverty level.

In addition to these characteristics, the ED sample inclusion probabilities are also known. The number of EDs used in the analysis is 1892 with observations on one to eight persons per ED. The total sample size is 2676 persons.

12.6.4 Analysis and Results

Having defined the design clusters as EDs, subsequent analysis included the following main stages.

1. Identification of a subset of ED characteristics which explain significantly the variation of the ED regression coefficients. This stage was implemented utilizing the *p*-weighted estimators $\hat{\gamma}_{PW}$.
2. Estimation of the unknown model variances.
3. Repeat of Stage 1 employing the estimates obtained in Stage 2 for a 'model-dependent' analysis.
4. Repeat of Stage 2 with the newly identified significant ED characteristics.

Next we describe the various stages in more detail.

The selection of an initial set of significant ED characteristics was carried out by forming the design matrix $\mathbf{X}^*$ of equation (12.5) based on all five ED characteristics, computing the *p*-weighted γ-coefficients of the model (12.5) and testing their significance using the design-based linearization variance estimates for the computation of the corresponding design-based Wald statistics.

The analysis of Stage 1 was implemented using the SURREGR software (Holt, 1977). By this program the statistic $X^2_{\mathrm{W,D}}/r$ is assumed to have an F-distribution with r and v degrees of freedom where v equals the number of PSUs minus the number of strata, although several authors have found this approximation to be too conservative—see e.g. Shah *et al.* (1977) and Fuller (1984).

Table 12.1 presents the p-weighted γ-coefficients of the models identified for the regression coefficients. The values in brackets are the estimated randomization standard errors of the coefficients. It should be noted that the models presented in Table 12.1 have been selected by stepwise routines which were necessary due to the large number of regressors used initially to form the matrix $\mathbf{X}^*$. The effect of using stepwise routines is that the nominal P-values are often too small which is in contrast to the effect of transforming the Wald statistic to an F-statistic mentioned above.

The results of Table 12.1 suggest that the intercept and the regression coefficients of 'age', 'age²', 'race' and in particular blood lead indeed vary amongst the EDs and that this variation can be explained, at least in part, by two of the ED characteristics: a_2—proportion of non-whites and a_3—proportion of persons aged 25 + with less than 9 years of school. The regression coefficients of 'body mass', 'haemoglobin' and 'serum zinc', on the other hand, do not vary with any of the ED characteristics. We discuss the interpretation of the γ-parameters at the end of this section. As indicated there, for this example the initial results obtained from a 'design-based inference' change only slightly when employing a model-dependent inference. It is also important to emphasize that estimating the γ-coefficients by OLS gives almost identical results. This can be taken as an indication of the non-informativeness of the design after incorporating the ED characteristics into the model.

Having identified an initial set of significant ED characteristics, the next stage

Table 12.1 p-weighted γ-coefficients and randomization standard errors (in brackets) of models identified for regression coefficients. Design-based analysis

	Regression coefficients ($\boldsymbol{\beta}_c$)							
ED characteristic	Intercept	Age	Age²	Body mass	Race	Lead (log)	Haemoglobin	Serum zinc (log)
Constant	43.49	0.45	−0.004	0.96			1.21	−3.26
	(8.52)	(0.07)	(8.6×10^{-4})	(0.06)			(0.25)	(1.69)
Proportion non-white		0.99	−0.01			−7.24		
		(0.31)	(0.003)			(2.22)		
Proportion <9 years' education	−17.87				8.33	6.30		
	(9.74)				(2.84)	(2.73)		

Table 12.2 Positive variance estimates and standard errors of estimates

Variance component	Estimates	Model SE of estimates	Design SE of estimates
Age^2	1.8×10^{-6}	4.26×10^{-7}	4.4×10^{-7}
Body mass	0.014	0.0065	0.008
Race	34.324	12.5954	12.3954
Res. var.(σ^2)	92.432	4.4446	4.9266

was to calculate the MLEs of the unknown model variances. This stage was implemented by using the procedure described in Section 12.4.2. Table 12.2 shows the positive variance component estimates as obtained in the last iteration, along with their estimated standard errors under both the model and with respect to the randomization distribution. Notice that only four of the variance components estimates are positive but all four seem significant. A variance component equal to zero implies that either the variation of the corresponding regression coefficient among the EDs is fully explained by the ED characteristics or that the coefficient is fixed across the EDs.

Another interesting outcome arising from Table 12.2 is that the non-parametric design-based standard errors are very close to the corresponding model-based standard errors despite the difference in their computation and interpretation. The rationale and relationship between these two sets of estimators are discussed in Section 12.4.2.

In Stage 3 we used the variance estimates to construct the covariance matrix Σ defined by equation (12.5). The analysis of Stage 1 was then repeated by computing the GLS γ-coefficients and testing their significance using the model-dependent Wald statistic $X^2_{W,M}$.

The model-dependent analysis of Stage 3 validated the models exhibited in Table 12.1 in the sense that all corresponding GLS γ-coefficients came out significant. In addition, the ED characteristic of 'living area', not found significant in the preliminary analysis of Stage 1 turned out to be significant in explaining the variation of the intercept, the ED characteristic 'median income' was found significant in explaining the variation of the log lead coefficient and 'proportion of non-whites' was found significant in explaining the variation of the race variable coefficient. (The race variable is an individual-level dummy variable which takes the value of 1 when the person is non-white and the value of 0 otherwise.)

In Stage 4 we re-estimated the unknown model variances and γ-coefficients using again the procedure described in Section 12.4.2, applied this time to the extended models identified in Stage 3. As it turned out, the estimates of the variances came out almost identical to the estimates obtained in Stage 2 which is explained by (or rather illustrates) the very low marginal contribution of

***Table 12.3** GLS γ-coefficients and standard errors (in brackets) of models identified for regression coefficients. Model-based analysis*

	Regression coefficients ($\boldsymbol{\beta}_c$)							
Ed characteristic	Intercept	Age	Age^2	Body mass	Race	Lead (log)	Haemog-lobin	Serum zinc (log)
Constant	46.68	0.45	−0.004	0.91			0.96	−4.06
	(6.06)	(0.074)	(9×10^{-4})	(0.05)			(0.19)	(1.30)
Proportion non-white		1.09	−0.001		−6.22	−6.08		
		(0.29)	(0.003)		(2.90)	(1.96)		
Proportion <9 years' education	−8.80				10.58	4.52		
	(5.28)				(3.26)	(1.72)		
Median income						0.09		
						(0.03)		
Living area a_{41}[a]	1.73							
	(0.63)							
Living area a_{42}[a]	1.58							
	(0.62)							

[a] Living area variables are: $a_{41} = 1$ when living area is 'urban, non-central city', $a_{41} = 0$ otherwise; $a_{42} = 1$ when living area is 'rural', $a_{42} = 0$ otherwise.

the additional ED characteristics in explaining the variation of the corresponding regression coefficients. The GLS γ-coefficients obtained at this stage are shown in Table 12.3. The displayed standard errors of the coefficients are the ordinary GLS estimates under the model.

Comparison of Table 12.3 with Table 12.1 shows that the design-based estimates are in most cases close to the model-based estimates. The only major differences are in the γ-coefficients of the models identified for the intercept and the race and lead coefficients which is clear considering the different ED characteristics included in the two analyses.

As to the estimates displayed in Table 12.3, we notice that the ED characteristic 'proportion of non-whites' has a negative marginal effect on both the regression coefficients of lead and race. Considering the lead coefficient for example, the negative effect of this characteristic can be interpreted as follows: with the other ED characteristics held fixed, the average marginal lead effect on hypertension decreases as the proportion of non-white people in an ED increases.

The other γ-coefficients can be interpreted in a similar way where each such coefficient represents the marginal effect of the corresponding ED characteristic with the other characteristics held fixed. In order to illustrate the joint effects of the various ED characteristics on the values of the regression coefficients, we calculated weighted averages of the empirical predictors of the regression coefficients in domains defined by these characteristics. (The empirical predictors were calculated by substituting the variance estimates for the unknown variances in formula (12.19) in the next section.) Focusing on the lead coefficient, we find that in the domain defined by (proportion of non-whites < median value, proportion with less than nine years of school > median value, living area = rural, poverty level = poverty stratum), the average effect of the empirical predictors is as high as 2.66 compared to the overall average effect of 1.56. The average effect drops to 1.81 when the proportion of non-whites is greater than the median value (with the other characteristics remaining unchanged) and it is as low as 0.04 in the domain defined by (proportion of non-whites > median value, proportion with less than nine years of school > median value, living area = urban, poverty level = poverty stratum).

In an attempt to explain the differences in the lead effects, we calculated the average log lead levels in the various domains. As it turned out, there are statistically significant differences in the mean lead levels between the domains. Furthermore, the estimated lead effects decrease almost systematically as the average log lead levels increase. For example, the average log lead levels in the three domains considered above are 2.61, 2.68 and 2.79 respectively with all pairwise differences being significant at the 0.01 level. Thus, our analysis has indicated the possibility of a non-linear marginal lead effect on hypertension with the higher effect operating for the lower lead levels.

12.7 PREDICTION OF DESIGN CLUSTER COEFFICIENTS

12.7.1 Definitions and Theoretical Formulae

In most circumstances the regression parameters γ and the components of variance will be the main items of interest. However, there may be circumstances in which we would wish to predict the values $\{\boldsymbol{\beta}_c\}$ for some or all clusters—for example if we wish to predict the value of Y for additional members of known clusters. As an extension to this, it is sometimes argued that a linear combination of $\{\boldsymbol{\beta}_c\}$ across all clusters in the population may be of direct interest, for example,

$$k(\boldsymbol{\beta}) = \sum_{c=1}^{M} k_c \boldsymbol{\beta}_c,$$

where M is the number of clusters in the finite population. Examples for such averages are:

1. The mean per cluster: $k_c = 1/M$;
2. The mean per unit: $k_c = N_c/N_o$, where N_c is the cluster size and $N_o = \sum_{c=1}^{M} N_c$;
3. The mean per cluster in a domain $U(d)$ of T clusters: $k_c = 1/T$ if $c \in U(d)$, $k_c = 0$ otherwise.

If the survey data were to be analysed by OLS regression, without regard to the clustered structure in the population then the coefficients obtained are, in a sense, an average of the $\boldsymbol{\beta}_c$. A similar analysis but using the selection probabilities to estimate p-weighted regression coefficients is also, in a sense, an averaging process. This was the form of analysis carried out by Harlan *et al.* (1985).

For given variances $\{\sigma_j^2\}$ and σ^2, the BLU predictors of the vectors $\boldsymbol{\beta}_c$ under the model are given by Pfeffermann (1984):

$$\hat{\boldsymbol{\beta}}_c = \begin{cases} \mathbf{A}_c\hat{\boldsymbol{\gamma}}_{\mathrm{GLS}} + \boldsymbol{\Delta}\mathbf{X}_c'\boldsymbol{\Sigma}_{cc}^{-1}(\mathbf{Y}_c - \mathbf{X}_c^*\hat{\boldsymbol{\gamma}}_{\mathrm{GLS}}) & c = 1,\ldots,m, \\ \mathbf{A}_c\hat{\boldsymbol{\gamma}}_{\mathrm{GLS}} & c = m+1,\ldots,M, \end{cases} \tag{12.19}$$

where $\boldsymbol{\Sigma}_{cc} = E_\xi(\mathbf{y}_c - \mathbf{X}_c^*\boldsymbol{\gamma})(\mathbf{y}_c - \mathbf{X}_c^*\boldsymbol{\gamma})' = \sigma^2 I + \mathbf{X}_c\boldsymbol{\Delta}\mathbf{X}_c'$. By unbiasedness we mean that $E_\xi(\hat{\boldsymbol{\beta}}_c - \boldsymbol{\beta}_c) = \mathbf{0}$ where the expectation is over the joint distribution of $\mathbf{y}$ and $\boldsymbol{\beta}' = (\boldsymbol{\beta}_1', \ldots, \boldsymbol{\beta}_m')$ given the sample of units. Notice from equation (12.19) that, for $c = m+1, \ldots, M$, the optimal predictors of the regression coefficients are the same as the optimal estimators of their expectations. For clusters in the sample, these estimators are corrected by taking into account the deviations of the observations $\mathbf{y}_c$ from their estimated mean. The BLU predictor of the $k(\boldsymbol{\beta})$ is obtained by averaging the BLU predictors $\hat{\boldsymbol{\beta}}_c$ from equation (12.19), i.e.

$$\hat{k}(\boldsymbol{\beta}) = \sum_{c=1}^{M} k_c\mathbf{A}_c\hat{\boldsymbol{\gamma}}_{\mathrm{GLS}} + \boldsymbol{\Delta}\sum_{c=1}^{m} k_c\mathbf{X}_c'\boldsymbol{\Sigma}_{cc}^{-1}(\mathbf{Y}_c - \mathbf{X}_c^*\hat{\boldsymbol{\gamma}}_{\mathrm{GLS}}). \tag{12.20}$$

It is clear intuitively, and can be shown more rigorously, that when averaging over a large number of clusters, most of which are not represented in the sample, the role of the correction factors corresponding to the selected clusters (equation 12.19) is diminished. This holds particularly in the practical situation where the sample sizes $\{n_c\}$ are small and m is large (see Table 12.4). Hence the predictor $\hat{k}(\boldsymbol{\beta})$ can be approximated in such situations by considering only the first term in the right-hand side of equation (12.20). Replacing $\hat{\boldsymbol{\gamma}}_{\mathrm{GLS}}$ by the p-weighted estimator $\hat{\boldsymbol{\gamma}}_{\mathrm{PW}}$ yields the alternative estimator

$$\tilde{k}(\boldsymbol{\beta}) = \left(\sum_{c=1}^{M} k_c\mathbf{A}_c\right)\hat{\boldsymbol{\gamma}}_{\mathrm{PW}} \tag{12.21}$$

The advantage of $\tilde{k}(\boldsymbol{\beta})$ is that its computation does not require the estimation of the unknown model variances. Although less efficient under the model than $\hat{k}(\boldsymbol{\beta})$, it is unbiased in the sense that $E_\xi[\tilde{k}(\boldsymbol{\beta}) - k(\boldsymbol{\beta})] = 0$ and is consistent under both an informative and a non-informative sampling design, provided that $\lim_{M\to\infty}\sum_{c=1}^{M} k_c^2 = 0$ which guarantees that $\tilde{k}(\boldsymbol{\beta})$ converges in probability to $\sum_{c=1}^{M} k_c\mathbf{A}_c\boldsymbol{\gamma}$ and hence that $[\tilde{k}(\boldsymbol{\beta}) - k(\boldsymbol{\beta})]$ converges in probability to zero.

The estimators considered so far assume that the weighted mean of cluster characteristics $\mathbf{A}_c$, $\mathbf{K}_A = \sum_{c=1}^{M} k_c\mathbf{A}_c$, is known. In practice, the information about the characteristics of clusters not in the sample may not be easily accessible.

In such cases, $\mathbf{K}_A$ may be estimated by

$$\hat{\mathbf{K}}_A = \sum_{c=1}^{m} k_c \mathbf{A}_c w_c^* \quad \text{where } w_c^* = M(1/\pi_c) \Big/ \sum_{c=1}^{M} (1/\pi_c).$$

The estimator $\hat{\mathbf{K}}_A$ is an example of the use of the sample selection probabilities in what is essentially a model-dependent analysis.

12.7.2 Empirical Results II: Prediction of Design Cluster Coefficients

In this section we consider the performance of the various predictors using again the NHANES data. We consider three prediction formulae:

1. The empirical predictors obtained by substituting the variance estimates for the unknown variances in equation (12.19).
2. The predictors $\hat{\boldsymbol{\beta}}_c^{\mathrm{G}} = \mathbf{A}_c \hat{\boldsymbol{\gamma}}_{\mathrm{GLS}}$ with $\mathbf{A}_c$ defined by the ED characteristics identified for each of the regression coefficients and reported in Table 12.3.
3. The predictors $\hat{\boldsymbol{\beta}}_c^{\mathrm{W}} = \tilde{\mathbf{A}}_c \hat{\boldsymbol{\gamma}}_{\mathrm{PW}}$ with $\tilde{\mathbf{A}}_c$ representing the ED characteristics identified in the initial *p*-weighted stage of the analysis and reported in Table 12.1.

The predictors $\hat{\boldsymbol{\beta}}_c^{\mathrm{G}}$ and $\hat{\boldsymbol{\beta}}_c^{\mathrm{W}}$ are model- and design-based estimators respectively of the model expectations of the coefficients, although the ED characteristics used for the two regression models differ.

Table 12.4 shows, for each of the prediction methods, weighted averages of the predictors calculated over all EDs in the sample. The weights are the ratios of the ED sizes to the ED sample inclusion probabilities so that the weighted averages are predictors of the population mean

$$k(\boldsymbol{\beta}) = \left(\sum_{c=1}^{M} N_c \boldsymbol{\beta}_c \right) \Big/ \sum_{c=1}^{M} N_c.$$

Also shown are the standard deviations between the predicted values in the sample EDs. For the haemoglobin variable, for example, the predicted coefficient is the same in every cluster since no cluster level variables were significant and the variance of the error term was estimated as zero (see Table 12.3). Thus the standard deviation between clusters for this coefficient is zero. For body mass, again no cluster level variables are needed and the standard deviation is again zero for $\boldsymbol{\beta}^{\mathrm{G}}$ and $\boldsymbol{\beta}^{\mathrm{W}}$. However, the error variance was estimated to be non-zero (see Table 12.3) and thus the empirical predictor shows a between-cluster variation for this coefficient and hence a non-zero standard deviation.

The last row of Table 12.4 displays the PW estimates of the (fixed) regression coefficients postulated by Harlan *et al.* (1985). These estimates can again be viewed as predictors of the population average $k(\boldsymbol{\beta})$.

The first notable result from Table 12.4 is that the weighted averages

Table 12.4 Weighted averages and standard deviations[a] (in brackets) of design- and model-based predictors of ED regression coefficients

	Regression coefficients ($\boldsymbol{\beta}_c$)							
Predictor	Intercept	Age	Age^2	Body mass	Race	Lead (log)	Haemog-lobin	Serum zinc (log)
Empirical predictor	45.39 (1.55)	0.55 (0.21)	−0.005 (0.002)	0.91 (0.04)	2.40 (2.09)	1.56 (1.23)	0.96 (—)	−4.06 (—)
Model, $\{\boldsymbol{\beta}^G\}$	45.39 (1.55)	0.55 (0.21)	−0.005 (0.002)	0.91 (—)	2.39 (1.87)	1.56 (1.23)	0.96 (—)	−4.06 (—)
Design, $\{\boldsymbol{\beta}^W\}$	38.49 (2.84)	0.54 (0.19)	−0.005 (0.002)	0.96 (—)	2.33 (1.32)	1.10 (1.57)	1.21 (—)	−3.26 (—)
Uniform regression	37.03 (—)	0.54 (—)	−0.005 (—)	0.95 (—)	2.35 (—)	1.39 (—)	1.25 (—)	−3.23 (—)

[a]The notation (—) indicates that the same predictors apply to every ED and hence that the SD vanishes.

displayed in the first two rows are essentially identical. This result illustrates the comment preceding equation (12.21). (The two averages have to coincide for every coefficient β_j for which the residual variance $\sigma_j^2 = 0$, since in this case the multiplier of the corresponding correction factor in formula 12.20 is nil.)

Perhaps the most interesting result in Table 12.4 is the similarity between the weighted averages of the design-based predictors and the corresponding coefficients of the uniform p-weighted regression model of Harlan *et al.* The similarity between these two sets of coefficients can be explained as follows. Under the model, the estimators obtained by postulating a uniform regression model are consistent for

$$\boldsymbol{\theta}_G = \left(\sum_{c=1}^{M} N_c \bar{\mathbf{G}}_c\right)^{-1} \left(\sum_{c=1}^{M} N_c \bar{\mathbf{G}}_c \mathbf{A}_c\right) \boldsymbol{\gamma} = \boldsymbol{\theta}^* \boldsymbol{\gamma}$$

(say): where $\bar{\mathbf{G}}_c = \sum_{t=1}^{N_c} (\mathbf{x}_{ct} \mathbf{x}'_{ct})/N_c$. The weighted averages of the predictors $\hat{\boldsymbol{\beta}}_c^W$ are consistent for

$$\boldsymbol{\phi}_N = \left(\sum_{c=1}^{M} N_c\right)^{-1} \left(\sum_{c=1}^{M} N_c \mathbf{A}_c\right) \boldsymbol{\gamma} = \boldsymbol{\phi}^* \boldsymbol{\gamma}$$

(say). Now $\boldsymbol{\theta}_G \simeq \boldsymbol{\phi}_N$ for all $\boldsymbol{\gamma}$ iff $\boldsymbol{\theta}^* \simeq \boldsymbol{\phi}^*$. For the latter condition to hold, the distribution of the matrices $\{\bar{\mathbf{G}}_c\}$ should be unrelated to that of the matrices $\{\mathbf{A}_c\}$ and the sizes $\{N_c\}$. This will happen in particular when the regressor variables assume similar values in the various design groups. The largest relative difference between corresponding elements of the matrices $\boldsymbol{\theta}^*$ and $\boldsymbol{\phi}^*$, observed in the present study, was 12%.

The overall conclusion from Tables 12.3 and 12.4 is that the common use of uniform regression models can indeed be sufficient for studying the 'average effects' of the regressors included in the model. However, as the analysis of the results in Table 12.3 and the standard deviations between the ED predictors in Table 12.4 show, the uniform model fails to represent the distinct regression relationships holding in small groups or subdomains of the population.

The substantive conclusion of this study is that the level of lead has indeed a significant marginal effect on hypertension, but the size of the effect is not constant and depends on socioeconomic and demographic characteristics. Clearly, these characteristics do not have a direct causal effect on the size of the lead effect. Rather, they represent latent environmental conditions and other factors affecting the value of the lead effect. Further research is needed to identify these factors and model their relationship to the lead effect.

12.8 MODEL MISSPECIFICATION AND *p*-WEIGHTED ESTIMATORS

The analysis throughout this chapter has been essentially model-based although the selection probabilities have been used to obtain *p*-weighted estimators for regression parameters at some stages. The arguments for using the model-based GLS estimators are that when the model holds these estimators are efficient, and this is confirmed by the smaller standard errors exhibited in Table 12.3 compared to the corresponding standard errors exhibited in Table 12.1.

However, there are several arguments in favour of *p*-weighted estimators, which are drawn together here:

1. The *p*-weighted estimators are still consistent and in many situations the loss of efficiency is relatively small. Also it is usually the case in surveys that the sample size is relatively large.
2. Another important consideration in the use of the sampling design as a source of randomness is that it offers an inference mode which is more robust and in some ways simpler to perform than an inference based on rigorous model assumptions.
3. The essential assumption for the model-based inference is that the model is applicable to both the population and the sample data achieved. This may not be so if the sampling scheme is informative. This point has been developed in Part B of the book, but suppose, for example, that not all the design variables used to select the sample are included in the set of cluster characteristics or that the sampling schemes employed within the selected clusters are such that they tend to over-sample units with large values of the dependent variable. In such cases the model defined in equation (12.4) does not apply to the sample data and estimators such as $\hat{\gamma}_{GLS}$ and $\hat{\gamma}_{OLS}$ are biased.

Notice, on the other hand, that by including the design variables used for the selection of the clusters as part of the cluster characteristics, the informativeness of the design can be eliminated at least with respect to the model holding for the regression coefficients. In such cases the use of the sampling weights becomes redundant.

The use of the sampling weights to protect against selection bias has proved to perform satisfactorily in empirical studies carried out by Holt *et al.* (1980), Nathan and Holt (1980), Hausman and Wise (1981) and Pfeffermann and Holmes (1985). Assuming certain linear relationships between the variables included in the regression model and the variables used for the design, more efficient estimators which are again free of selection bias can be constructed. Examples for such estimators can be found in the preceding references and are described in Part B of the book.

12.9 ACKNOWLEDGEMENTS

Much of the work on this study was carried out while the first author was visiting the Research Triangle Institute, NC, USA, and it was supported by research contract NO1-HD-5-2932 from the National Institute of Health and the National Institute of Child Health and Human Development, US Department of Health. Thanks are due to Barry Grabard, Ralph Folsom and Babu Shah for helpful comments.

APPENDIX: PROBABILITY WEIGHTED GLS ESTIMATION OF γ

The GLS estimator of γ can be written as

$$\hat{\gamma}_{\text{GLS}} = (\mathbf{X}^{*\prime}\mathbf{\Sigma}^{-1}\mathbf{X}^{*})^{-1}\mathbf{X}^{*\prime}\mathbf{\Sigma}^{-1}\mathbf{Y} = \left\{\sum_{c=1}^{m}(\mathbf{X}_c^{*\prime}\mathbf{\Sigma}_{cc}^{-1}\mathbf{X}_c^{*})\right\}^{-1}\sum_{c=1}^{m}\mathbf{X}_c^{*\prime}\mathbf{\Sigma}_{cc}\mathbf{Y}_c \quad (12.\text{A}1)$$

where $\mathbf{\Sigma}_{cc} = E(\mathbf{u}_c\mathbf{u}_c') = \sigma^2\mathbf{I}_{n_c} + \mathbf{X}_c\mathbf{\Delta}\mathbf{X}_c'$ so that

$$\mathbf{\Sigma}_{cc}^{-1} = \frac{1}{\sigma^2}\mathbf{I}_{n_c} - \frac{1}{\sigma^2}\mathbf{X}_c(\mathbf{X}_c'\mathbf{X}_c + \sigma^2\mathbf{\Delta}^{-1})^{-1}\mathbf{X}_c' \quad (12.\text{A}2)$$

using a well-known identity in matrix algebra. Inserting equation (12.A2) in the right-hand side of equation (12.A1) gives

$$\hat{\gamma}_{\text{GLS}} = \sum_{c=1}^{m}[\mathbf{X}_c^{*\prime}\mathbf{X}_c^{*} - \mathbf{X}_c^{*\prime}\mathbf{X}_c\mathbf{Q}_c^{-1}\mathbf{X}_c'\mathbf{X}_c^{*}]^{-1}\sum_{c=1}^{m}[\mathbf{X}_c^{*\prime}\mathbf{Y}_c - \mathbf{X}_c^{*\prime}\mathbf{X}_c\mathbf{Q}_c^{-1}\mathbf{X}_c'\mathbf{Y}_c] \quad (12.\text{A}3)$$

where $\mathbf{Q}_c = \mathbf{X}_c'\mathbf{X}_c + \sigma^2\mathbf{\Delta}^{-1}$. Let $\mathbf{W}_c = \text{diag}(w_{c1}, \ldots, w_{cn_c})$ be the matrix of sampling weights corresponding to sampling within the cth selected cluster and define $\mathbf{Q}_{c,w} = [\mathbf{X}_c'\mathbf{W}_c\mathbf{X}_c + \sigma^2\mathbf{\Delta}^{-1}]$. Using this notation, the desired estimator of γ is obtained as

$$\hat{\gamma}_{\text{GLS}}^{\text{W}} = \sum_{c=1}^{m} \frac{1}{\pi_c} [\mathbf{X}_c^{*\prime}\mathbf{W}_c\mathbf{X}_c^* - \mathbf{X}_c^{*\prime}\mathbf{W}_c\mathbf{X}_c\mathbf{Q}_{c,w}^{-1}\mathbf{X}_c'\mathbf{W}_c\mathbf{X}_c^*]^{-1}\mathbf{R}$$

where

$$\mathbf{R} = \sum_{c=1}^{m} \frac{1}{\pi_c} [\mathbf{X}_c^{*\prime}\mathbf{W}_c\mathbf{Y}_c - \mathbf{X}_c^{*\prime}\mathbf{W}_c\mathbf{X}_c\mathbf{Q}_{c,w}^{-1}\mathbf{X}_c'\mathbf{W}_c\mathbf{Y}_c] \tag{12.A4}$$

The estimator $\hat{\gamma}_{\text{GLS}}^{\text{W}}$ is approximately p-unbiased and consistent for the GLS census estimator of γ.

CHAPTER 13

Logistic Models for Contingency Tables

D. Holt and P. D. Ewings

13.1 INTRODUCTION

This chapter describes the use of logistic models for contingency table analysis when the contingency table is based upon survey data. We shall assume that a multidimensional contingency table is formed where one of the variables, the dependent variable, is of particular interest and the other dimensions of the contingency table are made up of explanatory variables that are used to help us understand the variation in the dependent variable.

For simplicity we shall generally assume that the dependent variable is dichotomous, but the models and methods proposed are readily generalizable to the case of a dependent variable with more than two categories and the extension will be indicated where appropriate.

Standard procedures for contingency table analysis assume that the data are independent and identically distributed (IID) and that sampling distributions such as the Poisson, multinomial or product multinomial are appropriate. In practice, of course, if we consider the case of a two-stage sample, observations from the same cluster usually exhibit positive intra-cluster correlation and the relatively simple distributional assumptions above are inappropriate. We know that the sample proportion, $\bar{y}$, for example when derived from a two-stage sample will not have variance $\mu(1-\mu)/n_o$, where μ is the population proportion and n_o is the sample size, but that this will be inflated because of the positive correlation between observations in the same cluster. The result is a larger variance, and estimators of variance which do not contain this inflation will be misleading. This may result in confidence intervals that are too narrow and test statistics that appear to be more significant than they really are.

In contingency table analysis, as we have seen in Chapter 4, the same sort of effects occur and standard procedures such as the Pearson test, X^2_{P}, or the likelihood ratio test, X^2_{LR}, based upon the IID assumptions, will not actually be distributed as χ^2 with the appropriate degrees of freedom and will usually appear to be more significant than they are. In higher dimension contingency tables there is a wide variety of logistic models that can be formulated and hence hypotheses to be tested, and there is a need for a general framework for model selection, estimation and interpretation which is suitable for use with survey data and which is comparable to the framework in the IID case.

In Chapter 4, methods are described for the analysis of contingency tables using survey data. Modifications to the ordinary χ^2 test statistics are described to take account of the clustering effects. These methods are relatively easy to compute, require only the design effects for the proportions in various marginal tables and seem to work well most of the time. If the objective of the analysis is simply to fit logistic or loglinear models to the contingency table, and to arrive at a model which fits the table adequately taking into account the cluster effects, then we believe that the methods in Chapter 4 are generally the best available although problems can occur with small numbers of clusters. Indeed, even if the methods described in this chapter are to be followed we would recommend that the approach in Chapter 4 be adopted as a preliminary analysis to identify likely models of interest.

In this chapter we put forward a class of logistic models which specifically allow for cluster effects and we describe estimation and testing procedures for contingency table analysis. If Chapter 4 is a satisfactory approach to the problem it is fair to ask what additional benefit can be derived from this chapter, particularly since the mathematics and the computing procedures are both rather complex. There are two benefits, of very different kinds, which can flow from this modelling approach.

1. An understanding of the effect of data structure and design on the standard procedures, and
2. A deeper understanding of the underlying social processes which may lead to the cluster effects which we observe.

The first benefit is a methodological one. We observe in some cases that χ^2 procedures are hardly affected by some kinds of cluster effects and yet in other cases the effects may be severe. There are a number of factors which contribute to this situation. The combined effect of these is complex, but by putting forward a plausible model of the data structure we can use computer simulations or analytic methods to understand which factors or combinations of factors lead to situations where the χ^2 statistics are affected.

The second, and arguably the more important benefit, is a substantive one. The approach in Chapter 4 is to take account of cluster effects by modifying the χ^2 statistic accordingly. It is as if the cluster effects were a nuisance that prevented us from using standard IID procedures. Actually the cluster effects are part of the population structure, and this chapter describes an approach to modelling these effects and allows us to conjecture about the underlying social processes which may be the source of the cluster effects which occur. The approach is somewhat different from that described in Chapters 11 and 12.

13.2 THE STANDARD LOGISTIC FRAMEWORK

Suppose that the dependent variable, Y, is dichotomous and that the contingency table contains $2J$ cells so that there are J combinations of the independent variables. For the jth combination of the independent variables there are two cells corresponding to the two categories of Y. The standard logistic model is given by

$$\boldsymbol{\lambda} = \mathbf{X}\boldsymbol{\beta}, \tag{13.1}$$

where $\boldsymbol{\lambda}$ is a $J \times 1$ vector of logits of the conditional probabilities

$$P\{Y=1|j\} = \theta_{1j}, \qquad j = 1,\ldots,J, \tag{13.2}$$

$$P\{Y=0|j\} = \theta_{0j} = 1 - \theta_{1j},$$

and

$$\lambda_j = \log \frac{\theta_{1j}}{\theta_{0j}}.$$

Here $\mathbf{X}$ is a $J \times q$ matrix of dummy variables which reflects the particular set of main effects and interactions included in the logistic model, and $\boldsymbol{\beta}$ is a $q \times 1$ vector of logistic regression parameters; $\boldsymbol{\lambda}$ cannot be observed and is replaced by a vector of empirical logits $\mathbf{l} = \{l_j = \log(\hat{\theta}_{1j}/\hat{\theta}_{0j})\}$ derived from sample data. Hence

$$\mathbf{l} = \mathbf{X}\boldsymbol{\beta} + \boldsymbol{\varepsilon} \tag{13.3}$$

where $E(\boldsymbol{\varepsilon}) = \mathbf{0}$, $\mathrm{Var}(\boldsymbol{\varepsilon}) = \mathbf{U}$ and where $\mathbf{U}$ is derived from the binomial distribution using equation (13.2), so that $\mathbf{U}$ is $(J \times J)$:

$$\mathbf{U} = \mathrm{diag}\left[\frac{1}{n_j}\{\theta_{0j}^{-1} + \theta_{1j}^{-1}\}\right],$$

where n_j is the number of observations for the jth combination of explanatory variables.

More generally, if the dependent variable is polytomous with $I + 1$ categories $\{0, 1, 2, \ldots, I\}$ and the vector of empirical logits is written in the natural order

$$\mathbf{l} = \{l_{11}, \ldots, l_{I1} \cdots l_{1j}, \ldots, l_{Ij} \cdots l_{1J}, \ldots, l_{IJ}\},$$

then $\mathbf{U}$ is block diagonal, $(IJ \times IJ)$:

$$\mathbf{U} = \oplus \frac{1}{n_j} \mathbf{U}_j, \tag{13.4}$$

where

$$\mathbf{U}_j = \begin{bmatrix} \theta_{0j}^{-1} + \theta_{1j}^{-1} \cdots \theta_{0j}^{-1} \cdots \theta_{0j}^{-1} \\ \vdots & & \vdots \\ \theta_{0j}^{-1} & \cdots\cdots & \theta_{0j}^{-1} + \theta_{Ij}^{-1} \end{bmatrix}. \tag{13.5}$$

The model (13.3) is in the general linear framework, but **U** has no unknown scalar multiplier and is simply a function of the probabilities $\{\theta_{ij}\}$, or equivalently the logistic parameters $\boldsymbol{\beta}$. Since there is no scalar multiplier, the standard F-tests for hypotheses testing need no denominator and are distributed as χ^2. For estimation purposes, an iterative form of generalized least squares is needed since the variance matrix is unknown. The usual procedure is to choose an initial set of values, estimate the logistic regression parameters, subsequently refine the variance matrix and iteratively estimate $\boldsymbol{\beta}$ in this way. This is known as *iterative generalized least squares* (GLS). A full discussion of the model (13.3) is given for the dichotomous case in Cox (1970) and for the polytomous case in Nerlove and Press (1973).

13.3 MODIFICATION TO ALLOW FOR CLUSTER EFFECTS

The model (13.3) assumes IID sampling and needs to be modified to take account of clustering in survey data. A common approach is to add a random-effect term η_c to the model for empirical logits derived from the cth cluster $c = 1,\ldots,m$, such that $E(\eta_c) = 0$; $\text{var}(\eta_c) = w^2$, and $\text{cov}(\eta_c, \eta_{c'}) = 0$, $c \neq c'$.

Thus if the vector of empirical logits is calculated for each cluster separately so that for the dichotomous case **l** is now of length Jm (and for the polytomous case of length IJm) then equation (13.3) becomes

$$\mathbf{l} = \mathbf{X}\boldsymbol{\beta} + \boldsymbol{\eta} + \boldsymbol{\varepsilon}, \tag{13.6}$$

where $E(\boldsymbol{\eta}) = \mathbf{0}$, $E(\boldsymbol{\varepsilon}) = \mathbf{0}$, $\text{Var}(\boldsymbol{\eta}) = \mathbf{W}$ and $\text{Var}(\boldsymbol{\varepsilon}) = \mathbf{U}$.

Now **X** ($Jm \times q$) is made up of m identical submatrices ($J \times q$) as described after equation (13.2). For the dichotomous case the vector $\boldsymbol{\eta}$ of random-effect components is made up of subvectors of length J for each cluster and each having a constant value η_c, $c = 1,\ldots,m$. Hence $\boldsymbol{\eta}' = (\eta_1\mathbf{1}'_J,\ldots,\eta_c\mathbf{1}'_J,\ldots,\eta_m\mathbf{1}'_J)$ where $\mathbf{1}_J$ is a vector of unit values of length J. For the polytomous case there would be mJ subvectors $(\eta_{1c},\ldots,\eta_{Ic})'$, each of length I and each being repeated J times within each cluster c, $c = 1\cdots m$, i.e.

$$\boldsymbol{\eta}' = [\mathbf{1}'_J \otimes (\eta_{11}\cdots\eta_{I1})\cdots\mathbf{1}'_J \otimes (\eta_{1c}\cdots\eta_{Ic})\cdots\mathbf{1}'_J \otimes (\eta_{1m}\cdots\eta_{Im})].$$

We refer to this model as the *constant cluster effect model.* This simple extension may be appropriate in some circumstances, but it does have obvious limitations which make it inadequate in general. The model requires that the variability of logits between clusters is the same for all combinations of the independent variables and this may be too strong an assumption. Secondly, the model requires that logits from the same cluster are affected in the same way for all combinations of the independent variables by a common cluster effect η_c (or η_{ic} in the polytomous case). This too is a very strong assumption which may be inappropriate in many situations, as illustrated by the following simple example.

Consider the attitude towards voting reform (in favour, against) as the dichotomous dependent variable and party affiliation (parties A, B) as the single independent variable. It is easy to imagine that in particular clusters (polling districts) the additive cluster effect on logits for party A could be negatively correlated with the effect on logits for party B. If the supporters of party A see a particular local advantage in voting reform then the logit in favour of reform will be locally higher for this party than the national average, but the supporters of party B will recognize the same local considerations as a disbenefit of voting reform and the logit in favour for party B supporters will be depressed compared to their national average.

A more flexible approach is obtained by removing the restriction that the cluster effect subvector for each cluster should be the same value η_c for all combinations of the explanatory variables within each cluster. More generally, for a dichotomous response, we allow the cluster effects to vary across the combinations of the explanatory variables within the cluster and to be correlated with each other. Thus $\boldsymbol{\eta}' = (\boldsymbol{\eta}_1', \ldots, \boldsymbol{\eta}_c', \ldots, \boldsymbol{\eta}_m')$, where $\boldsymbol{\eta}_c' = (\eta_{1c}, \ldots, \eta_{Jc})$, and where

$$\begin{aligned} \operatorname{cov}(\eta_{jc}, \eta_{j'c'}) &= w_{jj'}, \qquad c = c', \\ &= 0, \qquad c \neq c', \end{aligned}$$

and

$$\operatorname{cov}(\eta_{jc}, \varepsilon_{j'c'}) = 0, \quad \text{for all } j, j', c \text{ and } c'.$$

Thus for the dichotomous response variable

$$\mathbf{l} = \mathbf{X}\boldsymbol{\beta} + \boldsymbol{\eta} + \boldsymbol{\varepsilon}, \tag{13.7}$$

where

$$\operatorname{Var}(\boldsymbol{\varepsilon}) = \mathbf{U} = \bigoplus_{j,c} \mathbf{U}_{jc}/n_{jc},$$

$$\operatorname{Var}(\boldsymbol{\eta}) = \mathbf{W} = \bigoplus_{c} \mathbf{W}_c,$$

and n_{jc} is the number of observations for the jth combination of the explanatory variables in the cth cluster.

Here $\mathbf{U}_{jc}$ is a scalar and hence $\mathbf{U}$ is diagonal of order Jm. Also $\mathbf{W}_c = (w_{jj'})$ is the covariance matrix of cluster effects for each cluster and $\mathbf{W}$ is block diagonal with m blocks each of order J and the blocks are identical repetitions. For the polytomous response variable with $I + 1$ response categories the basic formulation is the same, but $\mathbf{U}_{jc}$ is now a matrix of order $I \times I$ and $\mathbf{U}$ is block diagonal, with a typical block:

$$\mathbf{U}_{jc} = \begin{bmatrix} \theta_{0jc}^{-1} + \theta_{1jc}^{-1} \cdots \theta_{0jc}^{-1} \\ \vdots \\ \theta_{0jc}^{-1} \cdots\cdots \theta_{0jc}^{-1} + \theta_{Ijc}^{-1} \end{bmatrix}. \tag{13.8}$$

The definition of $\mathbf{W}_c$ is extended in the natural way to allow cluster effects η_{ijc} to be correlated within each cluster across the levels of the explanatory variables. Thus $\mathbf{W}_c$ is a matrix of order IJ and $\mathbf{W}$ still retains a block diagonal form.

The method of estimation for this more general model is still iterative GLS. However, the covariance matrix has a strong structure with two block diagonal components and with the blocks of $\mathbf{W}$ being repetitions. In practice it is simpler to approximate by using the same values in $\mathbf{U}_{jc}$ for every cluster as the weights in the iterative GLS procedure. Thus we use

$$\mathbf{U}_j = \begin{bmatrix} \theta_{0j.}^{-1} + \theta_{1j.}^{-1} \cdots & \theta_{0j.}^{-1} \\ \cdot & \\ \cdot & \\ \theta_{0j.}^{-1} & \theta_{0j.}^{-1} + \theta_{Ij.}^{-1} \end{bmatrix}$$

in place of

$$\mathbf{U}_{jc} = \begin{bmatrix} \theta_{0jc}^{-1} + \theta_{1jc}^{-1} \cdots & \theta_{0jc}^{-1} \\ \cdot & \\ \cdot & \\ \theta_{0jc}^{-1} & \cdots \theta_{0jc}^{-1} + \theta_{Ijc}^{-1} \end{bmatrix}.$$

At first glance the two variance components η_{jc} and ε_{jc} may seem indistinguishable and hence whilst $\mathbf{V} = \mathbf{U} + \mathbf{W}$ may be estimable the two separate covariance matrices may not be. However, using $\mathbf{U}_j$ for $\mathbf{U}_{jc}$ we see that $\mathbf{U}$ is determined from the logistic regression parameters $\boldsymbol{\beta}$ rather than as a function of the logistic regression residuals and hence the two variance components $\mathbf{U}$ and $\mathbf{W}$ may be disentangled.

13.4 ESTIMATION AND HYPOTHESIS TESTING

The model is contained within the framework of the general linear model and, as we have already indicated, estimates may be obtained using the iterative GLS procedure.

$$\begin{gathered} \mathbf{l} = \mathbf{X}\boldsymbol{\beta} + \boldsymbol{\eta} + \boldsymbol{\varepsilon}, \\ \mathrm{Var}(\boldsymbol{\eta}) = \mathbf{W}; \qquad \mathrm{Var}(\boldsymbol{\varepsilon}) = \mathbf{U}; \qquad \mathrm{Var}(\mathbf{l}) = \mathbf{V} = \mathbf{W} + \mathbf{U}. \end{gathered} \tag{13.9}$$

Tests of hypotheses may be derived from the standard approach to generalized linear models. If we wish to test the hypothesis $H_0:\mathbf{C\beta}=\mathbf{r}$ then

$$X^2=(\mathbf{C}\hat{\boldsymbol{\beta}}-\mathbf{r})'[\mathbf{C}(\mathbf{X}'\hat{\mathbf{V}}^{-1}\mathbf{X})^{-1}\mathbf{C}']^{-1}(\mathbf{C}\hat{\boldsymbol{\beta}}-\mathbf{r})\sim\chi^2 \tag{13.10}$$

with the appropriate number of degrees of freedom. Thus if $\mathbf{C}$ is $q\times p$, $\boldsymbol{\beta}$ is $p\times 1$ and $\mathbf{r}$ is $q\times 1$ then, X^2, the test statistic is distributed as χ^2_q. For contingency tables most hypotheses are of the form $H_0:\boldsymbol{\beta}_0=\mathbf{0}$, where $\boldsymbol{\beta}_0$ is a subvector of $\boldsymbol{\beta}$.

Thus

$$X^2=\hat{\boldsymbol{\beta}}_0'[(\mathbf{X}'\hat{\mathbf{V}}^{-1}\mathbf{X})_0^{-1}]^{-1}\hat{\boldsymbol{\beta}}_0$$

where $\hat{\boldsymbol{\beta}}_0$ is the appropriate subvector of $\hat{\boldsymbol{\beta}}$ and $(\mathbf{X}'\hat{\mathbf{V}}^{-1}\mathbf{X})_0^{-1}$ is the corresponding submatrix of $(\mathbf{X}'\hat{\mathbf{V}}^{-1}\mathbf{X})^{-1}$.

An important practical consideration is that the model, as formulated, requires the empirical logits to be calculated separately for each combination of the explanatory variables and for each cluster. In many situations there will be particular combinations of the explanatory variables for which, in some clusters, there are zero observations in one or more of the categories of the response variable and so the empirical logit will be undefined. To overcome this problem we choose to calculate the empirical logits collapsed across the clusters

$$l_j=\log\left[\sum_c n_{jc}\hat{\theta}_{1jc}\Big/\sum_c n_{jc}\hat{\theta}_{0jc}\right] \tag{13.11}$$

Thus, for the dichotomous response variable, $\mathbf{l}=\{l_j\}$ is a vector of length J and for the polytomous case of length IJ. By using Taylor series methods, $\mathrm{Var}(\mathbf{l})$ may be derived in terms of $\mathbf{U}$ and $\mathbf{W}$ and estimation methods for the logistic regression parameters and $\hat{\mathbf{U}}$ and $\hat{\mathbf{W}}$ may be derived (Ewings, 1985).

For the dichotomous dependent variable

$$\mathrm{var}(l_j)=\frac{1}{d_jn_j\theta_{1j}(1-\theta_{1j})}+\frac{w_{jj}}{d_j}\sum_c\frac{n_{jc}(n_{jc}-1)}{n_j^2}, \tag{13.12}$$

where $d_j=[1+\frac{1}{2}(1-2\theta_{1j})^2w_{jj}]$ and

$$\mathrm{cov}(l_j,l_{j'})=\frac{w_{jj'}}{d_jd_{j'}}\sum_c\frac{n_{jc}n_{j'c}}{n_jn_{j'}},\qquad j\neq j'. \tag{13.13}$$

In many practical situations $d_j\simeq 1$.

Estimation of the components of variance $w_{jj'}$ is a difficult issue, but after extensive numerical comparisons of alternatives Ewings (1985) recommends the following; define

$$d_{jj'}=\tilde{\theta}_{1j}(1-\tilde{\theta}_{1j})\tilde{\theta}_{1j'}(1-\tilde{\theta}_{1j'})\sum_c\left(\frac{n_{jc}}{n_j}\right)\left(\frac{n_{j'c}}{n_{j'}}\right)$$

$$+\delta_{jj'}\tilde{\theta}_{1j}\frac{(1-\tilde{\theta}_{1j})[\frac{1}{2}-3\tilde{\theta}_{1j}(1-\tilde{\theta}_{1j})]}{n_j} \tag{13.14}$$

where $\delta_{jj'}$ is the Kronecker delta and $\tilde{\theta}_{1j}$ is the estimator of θ_{1j} from the fitted model.

Then let

$$A_{1jj'} = \frac{1}{d_{jj'}} \sum_{c} \frac{n_{jc}}{n_j} \frac{n_{j'c}}{n_{j'}} (\hat{\theta}_{1jc} - \hat{\theta}_{1j})(\hat{\theta}_{1j'c} - \hat{\theta}_{1j'}), \tag{13.15}$$

$$A_{2jj'} = \frac{\delta_{jj'}}{d_{jj'}} \frac{\tilde{\theta}_{1j}(1 - \tilde{\theta}_{1j})}{n_j}, \tag{13.16}$$

and

$$\hat{w}_{jj'} = A_{1jj'} - A_{2jj'}. \tag{13.17}$$

To estimate the variance–covariance matrix of $\mathbf{l}$ for the iterative GLS procedure θ_{1j} is replaced by $\tilde{\theta}_{1j}$ and $w_{jj'}$ by $\hat{w}_{jj'}$ in equations (13.12) and (13.13). The extension to a polytomous Y-variable is given in Ewings (1985).

A second important practical consideration is that a modification is required to the above procedures. The essential block submatrix of $\mathbf{W}$ is either $J \times J$ if the response variable is dichotomous or $IJ \times IJ$ in the polytomous case. In multidimensional contingency tables with a number of categories for each explanatory variable, J can easily be a moderately large number compared to the number of clusters in the survey data. Essentially $\mathbf{W}$ is estimated by subtracting two matrices $\mathbf{A}_1$ and $\mathbf{A}_2$, say, and may not be non-negative definite

$$\hat{\mathbf{W}} = \mathbf{A}_1 - \mathbf{A}_2 = \mathbf{A}_2^{1/2}\mathbf{Q}(\mathbf{\Lambda} - \mathbf{I})\mathbf{Q}'\mathbf{A}_2^{1/2}$$

where $\mathbf{Q}$ and $\mathbf{\Lambda}$ are the eigenvectors and eigenvalues of $\mathbf{A}_2^{-1/2}\mathbf{A}_1\mathbf{A}_2^{-1/2}$.

Let $\mathbf{D} = \text{diag}(\max\{\lambda_1, 1\}, \max\{\lambda_2, 1\}, \ldots, \max\{\lambda_m, 1\})$.

Then

$$\tilde{\mathbf{W}} = \mathbf{A}_2^{1/2}\mathbf{Q}(\mathbf{D} - \mathbf{I})\mathbf{Q}'\mathbf{A}_2^{1/2}$$

is non-negative definite.

We refer to $\tilde{\mathbf{W}}$ as the 'corrected cluster effects covariance matrix' and this adjustment has been used throughout the numerical results which follow. Extensive numerical work reported in Ewings (1985) shows that an adjustment for non-negative definiteness is essential for stability of results.

The iterative GLS procedure remains as before, with the exception that the $w_{jj'}$ in equations (13.12) and (13.13) are replaced at each step by $\tilde{w}_{jj'}$, the terms of the corrected $\tilde{\mathbf{W}}$. This provides an estimate $\tilde{\mathbf{V}}$ of $\mathbf{V} = \text{Var}(\mathbf{l})$ that uses a non-negative definite estimate of the cluster effects covariance component $\mathbf{W}$. The test statistic is now

$$\tilde{X}^2 = \hat{\boldsymbol{\beta}}_0'[(\mathbf{X}'\tilde{\mathbf{V}}^{-1}\mathbf{X})_0^{-1}]^{-1}\hat{\boldsymbol{\beta}}_0 \tag{13.18}$$

where $\boldsymbol{\beta}_0$ is still of length q but, since we use logits collapsed across clusters, $\mathbf{X}$

is now $J \times q$ and $\tilde{\mathbf{V}}$ is $J \times J$ for the dichotomous case. For the polytomous case $\mathbf{X}$ is $IJ \times q$ and $\tilde{\mathbf{V}}$ is $IJ \times IJ$.

13.5 THE IMPACT OF POPULATION STRUCTURE ON THE χ^2 TEST

If the main purpose of the analysis is to fit loglinear or logistic models to survey data and the cluster effects are regarded as a complicating nuisance then, as we stated earlier, the methods described in Chapter 4 are recommended. The ordinary X_{P}^2 and X_{LR}^2 test statistics may be adjusted by a factor derived from the data and the resulting adjusted test statistic is approximately distributed as a χ^2 variable with the appropriate degrees of freedom.

There are, however, two reasons for building up a more detailed model of the population structure. The main reason is that we may regard the cluster effects, not as an added complication but as an integral part of the population structure that may tell us about the underlying social process. In a standard loglinear or logistic analysis the whole of the interest is focused on the deterministic part of the model $\mathbf{L} = \mathbf{X}\boldsymbol{\beta}$. The error structure is of no direct interest except as a source of residuals for diagnostic plots and checking and adequacy of the model fit. In the more detailed model of the population structure, interest should focus on both the deterministic and error structure parts of the model. In particular the latter may help us to understand the social processess which underpin the phenomenon under investigation. We return to this point in Section 13.6.

The other purpose in building up a more detailed model of the population structure is to better understand the various factors which influence the properties of the ordinary χ^2 procedures. The standard fitting and testing procedures using the Pearson X_{P}^2 test or the asymptotically equivalent likelihood ratio test, X_{LR}^2, are readily accessible with software available in most of the major statistical computer packages. The modification described in Chapter 4, and certainly the more detailed models described in this chapter, are much less accessible. In some circumstances the standard procedures are adequate. If we better understand the impact of the various factors on the standard procedures we would be better able to judge the need for more sophisticated approaches.

Two methods can be used to explore these questions; one is to examine the true distribution of the unadjusted test statistic, the other is to use computer simulations.

Assuming our model to be correct, the asymptotic distribution of the ordinary statistic under the null hypothesis follows from results on quadratic forms as described in Chapter 4. For a hypothesis of the form $H_0\!:\boldsymbol{\beta}_0 = \mathbf{0}$, let $\hat{\mathbf{b}}_0$ be the estimator of $\boldsymbol{\beta}_0$ under IID assumptions, with corresponding covariance matrix $(\mathbf{X}'\mathbf{U}^{-1}\mathbf{X})_0^{-1}$. Then the asymptotic distribution of the unadjusted test statistic

$\hat{\mathbf{b}}_0'[(\mathbf{X}'\hat{\mathbf{U}}^{-1}\mathbf{X})_0^{-1}]^{-1}\hat{\mathbf{b}}_0$ is a weighted sum of χ_1^2 random variables, with the weights being the eigenvalues of the matrix

$$[(\mathbf{X}'\mathbf{U}^{-1}\mathbf{X})_0^{-1}]^{-1}[\mathrm{Var}(\hat{\mathbf{b}}_0)].$$

The first matrix in this product is the covariance matrix of $\hat{\mathbf{b}}_0$ under IID assumptions, whilst the second is the 'true' covariance matrix which, under our model, is a function of $\mathbf{X}$, $\mathbf{U}$ and $\mathbf{W}$. For given $\mathbf{X}$, $\mathbf{U}$ and $\mathbf{W}$ we can therefore define the distribution of the unadjusted statistic under H_0, and estimate coverage probabilities using the methods described in Chapter 4. Since the distribution depends on $\mathbf{W}$ and the n_{jk} (through $\mathbf{U}$), we can choose various values for these factors of interest and examine the corresponding effect on the unadjusted procedure.

We begin with the simplest case of a 2×2 table where there are two levels of the independent variable and $\mathbf{W}$ is block diagonal and comprised of m identical 2×2 blocks. Table 13.1 contains results for the test of independence in the 2×2 table. The true level of the nominal 5% X_{LR}^2 test is given for various values of the 2×2 component block $\mathbf{W}_c$ and various sample allocations. All of the cluster-effect matrices assume that there is zero covariance between the cluster effects for the two levels, j, of the independent variable. The first five cases have cluster effects with increasing variances which are the same for both levels of j. The last two cases have cluster effects with different variances. The first five sample allocations show decreasing cross-classing of the independent variables. In the first case all clusters contain equal numbers of observations at both levels of the independent variable. The cases (b) through to (e) show clusters with greater homogeneity for the categories of the independent variable. In the extreme case (e) clusters either have observations in one category of the independent variable or the other, but not both. Case (f) shows a constant number of observations of 25 in category 1 of $\mathbf{X}$ and 5 in category 2 across all clusters.

In Table 13.1 we notice that the level of significance of the nominal 5% test gets further away from 5% as the cluster effects become stronger. For the cases where the cluster-effect variances are unequal the impact on the X_{LR}^2 procedure is very similar to the case where both cluster effects have equal variance at the mean of the separate cluster-effect variances. Thus, for example, the row with cluster-effect variances 0.1 and 0.5 is very similar to the row with cluster-effect variances 0.3 and 0.3. We note also that the more extreme sample allocations, with less cross-classing have a greater impact on the test. We note with interest that case (f) with uniformly unequal allocation has less impact on the X_{LR}^2 procedure than the cases where the allocation varies from cluster to cluster. In fact the impact is less on the procedure for this case than the first case, with equal allocation across the levels of the independent variable in every cluster. Table 13.1 contains results for X_{LR}^2 but X_{P}^2 exhibits essentially the same behaviour. Thus in the whole of this discussion and what follows X_{P}^2 could be substituted for X_{LR}^2.

Table 13.1 True level of significance (%) for the 5% X^2_{LR} test for various values of $\mathbf{W}_c$ and sample allocations $\{n_{jc}\}$ for 2 × 2 tables

$\mathbf{W}_c$	Sample allocations					
	(a)	(b)	(c)	(d)	(e)	(f)
$\begin{pmatrix} 0.1 & 0 \\ & 0.1 \end{pmatrix}$	9	10	11	12	13	7
$\begin{pmatrix} 0.2 & 0 \\ & 0.2 \end{pmatrix}$	13	14	17	19	21	9
$\begin{pmatrix} 0.3 & 0 \\ & 0.3 \end{pmatrix}$	17	18	22	25	27	11
$\begin{pmatrix} 0.4 & 0 \\ & 0.4 \end{pmatrix}$	20	22	26	29	32	13
$\begin{pmatrix} 0.5 & 0 \\ & 0.5 \end{pmatrix}$	23	25	30	33	36	16
$\begin{pmatrix} 0.1 & 0 \\ & 0.3 \end{pmatrix}$	13	14	17	19	21	9
$\begin{pmatrix} 0.1 & 0 \\ & 0.5 \end{pmatrix}$	17	18	22	25	27	11

Note: Sample allocations:
(a) 100 clusters of $n_{1c} = 15$ $n_{2c} = 15$
(b) 50 clusters of $n_{1c} = 20$ $n_{2c} = 10$, 50 clusters of $n_{1c} = 10$, $n_{2c} = 20$
(c) 50 clusters of $n_{1c} = 25$ $n_{2c} = 5$, 50 clusters of $n_{1c} = 5$, $n_{2c} = 25$
(d) 50 clusters of $n_{1c} = 28$ $n_{2c} = 2$, 50 clusters of $n_{1c} = 2$, $n_{2c} = 28$
(e) 50 clusters of $n_{1c} = 30$ $n_{2c} = 0$, 50 clusters of $n_{1c} = 0$, $n_{2c} = 30$
(f) 100 clusters of $n_{1c} = 25$ $n_{2c} = 5$

In Table 13.2 we keep the variance for each cluster effect fixed at 0.3 but allow the covariance to vary between $+0.3$ and -0.3. The various sample allocations are the same as in Table 13.1. For sample allocation (e) all of the values in the column are the same (27%) since the covariance term will have no impact when the sample allocation is such that clusters contain only the first level of **X** or the second level but not both. In this case the intra-cluster correlation of the explanatory variable **X** is 1.0 ($\tau_x = 1$). For the most extreme covariance structure $w_{12} = -0.3$ all of the sample allocations (a)–(e) have the same impact on the X^2_{LR} test. When $w_{12} = 0.3$ we have what we term the constant cluster effect. The correlation between the two cluster effects is 1.0, and the variances are equal. When one cluster effect is high, so too is the other and vice versa. In this case, and when there is no change in the sample allocation across levels of **X** for each cluster ($\tau_x = 0$), there is no impact on the significance level

Table 13.2 True level of significance (%) for the 5% X^2_{LR} test for various values of $\mathbf{W}_c$ and sample allocations $\{n_{jc}\}$ for 2 × 2 tables

	Sample allocations					
$\mathbf{W}_c$	(a)	(b)	(c)	(d)	(e)	(f)
$\begin{pmatrix} 0.3 & 0.3 \\ & 0.3 \end{pmatrix}$	4	7	16	22	27	4
$\begin{pmatrix} 0.3 & 0.2 \\ & 0.3 \end{pmatrix}$	8	11	18	23	27	7
$\begin{pmatrix} 0.3 & 0.1 \\ & 0.3 \end{pmatrix}$	12	15	20	24	27	9
$\begin{pmatrix} 0.3 & 0 \\ & 0.3 \end{pmatrix}$	17	18	22	25	27	11
$\begin{pmatrix} 0.3 & -0.1 \\ & 0.3 \end{pmatrix}$	21	21	24	25	27	14
$\begin{pmatrix} 0.3 & -0.3 \\ & 0.3 \end{pmatrix}$	27	27	27	27	27	19

Note: Sample allocations same as Table 13.1.

of the X^2_{LR} test. As the covariance gets smaller, so that the correlation between the two cluster effects becomes smaller and eventually negative, so the impact on the test becomes stronger. This is a confirmation of the property mentioned in Section 3.3.3 that when domain membership cuts across clusters ($\tau_x = 0$), the design effect of the difference in domain means is small (the Kish conjecture). In our case, domain membership is defined by the categories of the explanatory variable **X**, and a negligible design effect for **X** implies that the X^2_{LR} test will be unaffected, no matter how strong the cluster effects are for the response variable Y, for each category of **X** separately. However, Table 13.2 also shows that the Kish conjecture does not hold when the model with constant cluster effects does not hold. When the covariance between the cluster effects is zero, for example ($w_{12} = 0$), the impact on the significance level of the χ^2 test is almost as severe for sample allocation (a) with equal sample allocations to the two categories of the **X** variable as it is for other allocations.

For three-way tables, there are a number of possible models to fit and corresponding hypotheses to test. Consider for example a 2 × 2 × 2 table for which there are two dichotomous explanatory variables. The design matrix **X** comprises dummy variables to indicate the two categories of each of the explanatory variables and the saturated logistic model is given by

$$E(l) = \beta_0 + \beta_1 x_1 + \beta_2 x_2 + \beta_{12} x_{12}$$

where $x_1 = 1$ for the second category of the first explanatory variable,
$= 0$ for the first category of the first explanatory variable;
$x_2 = 1$ for the second category of the second explanatory variable,
$= 0$ for the first category of the second explanatory variable;
$x_{12} = x_1 x_2 = 1$ for the second category of both explanatory variables,
$= 0$ otherwise.

In this case **X** is a matrix containing four columns of which the first is a column of 1's and the remaining three are the values of x_1, x_2 and $x_{12} = x_1 x_2$.

The logistic regression parameter β_{12} represents the interaction between the two explanatory variables (and is equivalent to a three-factor interaction in the corresponding loglinear model).

We may wish to test hypotheses against the saturated model such as $\beta_{12} = 0$ or $\beta_1 = \beta_{12} = 0$. This is used to test the overall goodness of fit of the model comprising the remaining terms. Alternatively, one may wish to test conditional hypotheses such as $\beta_1 = 0$ conditional on the assumption $\beta_{12} = 0$. In this case the likelihood ratio test may be based on two models such as

$$E(l) = \beta_0 + \beta_1 x_1 + \beta_2 x_2$$

and

$$E(l) = \beta_0 + \beta_2 x_2.$$

The appropriate hypotheses are listed in Table 13.3. Note that for three-way tables with more categories in each dimension the terms β_1, β_2, β_{12} would be vectors of parameters. Tests 1a–1d are for testing the fit of a given model relative to the saturated model. Tests 2a–2c test the same models as 1a–1c, but relative to the model that has no three-factor interaction in the loglinear model or two-factor effect β_{12} in the logit model.

The results in Table 13.4 show a similar pattern to previous tables.

Table 13.3 Various hypotheses for logit and log-linear models in three-way tables

Hypotheses	Parameters in logit model
1a	β_0 (i.e. $\beta_1 = \beta_2 = \beta_{12} = 0$)
1b	β_0, β_2 (i.e. $\beta_1 = \beta_{12} = 0$)
1c	β_0, β_1 (i.e. $\beta_2 = \beta_{12} = 0$)
1d	β_0, β_1, β_2 (i.e. $\beta_{12} = 0$)
2a	1a conditional on $\beta_{12} = 0$
2b	1b conditional on $\beta_{12} = 0$
2c	1c conditional on $\beta_{12} = 0$

Table 13.4 True level of significance (%) for the 5% X^2_{LR} test for various values of $\mathbf{W}_c$ and sample allocations for $2 \times 2 \times 2$ tables

$\mathbf{W}_c$	Hypothesis	d.f.	Sample allocation (a)	(b)	(c)
	1a	3	4	8	21
	1b	2	4	10	18
	1c	2	4	4	18
$\mathbf{W}_{1c}$	1d	1	4	4	13
	2a	2	4	10	18
	2b	1	4	13	13
	2c	1	4	4	13
	1a	3	15	20	29
	1b	2	13	17	24
	1c	2	13	17	24
$\mathbf{W}_{2c}$	1d	1	11	13	17
	2a	2	13	17	24
	2b	1	11	13	17
	2c	1	11	13	17
	1a	3	28	31	36
	1b	2	23	23	29
	1c	2	23	29	29
$\mathbf{W}_{3c}$	1d	1	17	21	21
	2a	2	23	23	29
	2b	1	17	12	21
	2c	1	17	21	21

Cluster-effect covariance matrices:

$$\mathbf{W}_{1c} = \begin{bmatrix} 0.3 & 0.3 & 0.3 & 0.3 \\ & 0.3 & 0.3 & 0.3 \\ & & 0.3 & 0.3 \\ & & & 0.3 \end{bmatrix} \qquad \mathbf{W}_{2c} = 0.3\mathbf{I}$$

$$\mathbf{W}_{3c} = \begin{bmatrix} 0.3 & -0.3 & -0.3 & -0.3 \\ & 0.3 & -0.3 & -0.3 \\ & & 0.3 & -0.3 \\ & & & 0.3 \end{bmatrix}$$

Sample allocations:

(a) 100 clusters of $n_{11c} = 6$, $n_{12c} = 7$, $n_{21c} = 8$ and $n_{22c} = 9$

(b) 50 clusters of $n_{11c} = 12$, $n_{12c} = 12$, $n_{21c} = 3$ and $n_{22c} = 3$
50 clusters of $n_{11c} = 3$, $n_{12c} = 3$, $n_{21c} = 12$ and $n_{22c} = 12$

(c) 25 clusters of $n_{11c} = 21$, $n_{12c} = 3$, $n_{21c} = 3$ and $n_{22c} = 3$
25 clusters of $n_{11c} = 3$, $n_{12c} = 21$, $n_{21c} = 3$ and $n_{22c} = 3$
25 clusters of $n_{11c} = 3$, $n_{12c} = 3$, $n_{21c} = 21$ and $n_{22c} = 3$
25 clusters of $n_{11c} = 3$, $n_{12c} = 3$, $n_{21c} = 3$ and $n_{22c} = 21$

Cluster-effect matrix $\mathbf{W}_{1c}$ shows a pattern of constant cluster effects. The sample allocation (a) is the same for all clusters so that there is no intra-cluster correlation for the explanatory variable X ($\tau_x = 0$). Thus the ordinary X^2_{LR} test is satisfactory. Similarly with allocation (b) there are no cluster effects for X for some tests since the appropriate margin has a constant allocation:

50 clusters	12	12 | 24		50 clusters	3	3 | 6	
	3	3 | 6			12	12 | 24	
	—	— | —			—	— | —	
	15	15 | 30			15	15 | 30	

The marginal distribution for the column sums is (15, 15) for all 100 clusters and the corresponding tests 1c, 1d and 2c are unaffected. Other tests show that the cluster effects have a large impact on the significance level of the X^2_{LR} test. We note that on most occasions the effect on the conditional tests 2a–2c is smaller than the corresponding unconditional tests 1a–1c. However, the difference is small in all cases and both tests are badly affected.

For larger tables, where the average number of observations per cell per cluster is smaller, the impact on the X^2_{LR} test is smaller than that shown in Table 13.4. However, these results are not presented here, but given in Ewings (1985).

The other method that can be used to gain insight into the various factors influencing the χ^2 procedure is that of computer simulations. One advantage of this method is that the behaviour of other test statistics can be studied, in addition to the unadjusted procedure.

We briefly describe the computer simulation procedure. Suppose that we require data under a particular logistic model $\mathbf{L} = \mathbf{X}\boldsymbol{\beta}$, with fixed $\boldsymbol{\beta}$, a known cluster-effect covariance matrix $\mathbf{W}$ and with fixed sample sizes, n_{jc}, in each combination of the independent variables, j, for each cluster, c. (*NB.* These could be different in different clusters.) For the cluster effects we generate m independent vectors, $\boldsymbol{\eta}_{jc}$, from a multivariate normal distribution with covariance matrix $\mathbf{W}$. For each cluster these are added to the overall logits $\mathbf{L}$ to obtain cluster-specific values of the logits. From these, for each cluster, and for each combination of the independent variables j, we may obtain θ_{ijc} the probability of the response variable falling into each of the $I+1$ categories. For the dichotomous response variable case we have θ_{1jc} and θ_{0jc} ($\theta_{0jc} + \theta_{1jc} = 1$). Then for the fixed number of observations n_{jc}, binomial random variables may be generated with $P\{Y=1\} = \theta_{1jc}$. Hence the sample for all combinations of the independent variables in each cluster may be generated.

We concentrate on the performance of various test statistics calculated under the null hypothesis. Thus, repeated sets of data are generated using an original model structure $\mathbf{L} = \mathbf{X}\boldsymbol{\beta}$ that corresponds to the nominal hypothesis being tested. A total of 500 simulations is used in each case. Basing the empirical results on

***Table 13.5** Estimated levels of significance (%) for nominal 5% tests using two test statistics from the loglinear framework (X^2_{LR} and the modified statistic $X^2_{\text{P}}/\bar{\delta}$) and the modified logit model test statistic ($\tilde{X}^2$). Levels based on 500 simulations in each case, using various values of $\mathbf{W}_c$ and sample allocations for 2 × 2 × 2 tables*

$\mathbf{W}_c$	Sample allocation	Test	d.f.	Estimated levels X^2_{LR}	Estimated levels $X^2_{\text{P}}/\bar{\delta}$	Estimated levels $\tilde{X}^2$
$\mathbf{W}_{1c}$	(a)	1a	3	10	6	6
		1b	2	9	7	7
		1c	2	7	5	5
	(b)	1a	3	9	4	4
		1b	2	9	5	5
		1c	2	8	6	5
	(c)	1a	3	14	6	5
		1b	2	13	6	4
		1c	2	13	6	5
$\mathbf{W}_{2c}$	(a)	1a	3	15	6	7
		1b	2	12	6	7
		1c	2	13	7	7
	(b)	1a	3	18	6	6
		1b	2	15	4	5
		1c	2	13	6	6
	(c)	1a	3	23	8	8
		1b	2	21	7	7
		1c	2	20	6	5

Cluster-effect covariance matrices:

$$\mathbf{W}_{1c} = 0.2\mathbf{I} \qquad \mathbf{W}_{2c} = \begin{bmatrix} 0.4 & -0.1 & -0.1 & -0.1 \\ & 0.4 & -0.1 & -0.1 \\ & & 0.4 & -0.1 \\ & & & 0.4 \end{bmatrix}$$

Sample allocations:

(a) 100 clusters of $n_{11c} = 6$, $n_{12c} = 7$, $n_{21c} = 8$ and $n_{22c} = 9$

(b) 50 clusters of $n_{11c} = 12$, $n_{12c} = 12$, $n_{21c} = 3$ and $n_{22c} = 3$
 50 clusters of $n_{11c} = 3$, $n_{12c} = 3$, $n_{21c} = 12$ and $n_{22c} = 12$

(c) 25 clusters of $n_{11c} = 21$, $n_{12c} = 3$, $n_{21c} = 3$ and $n_{22c} = 3$
 25 clusters of $n_{11c} = 3$, $n_{12c} = 21$, $n_{21c} = 3$ and $n_{22c} = 3$
 25 clusters of $n_{11c} = 3$, $n_{12c} = 3$, $n_{21c} = 21$ and $n_{22c} = 3$
 25 clusters of $n_{11c} = 3$, $n_{12c} = 3$, $n_{21c} = 3$ and $n_{22c} = 21$

500 simulations means that, if the true significance of the test under study is 5%, then the simulations will yield a value between 3% and 7% with probability 0.95.

The results presented here all relate to $2 \times 2 \times 2$ contingency tables generated with no main effects or interaction effects (i.e. $\beta_1 = \beta_2 = \beta_{12} = 0$). The procedures examined are the modified logit model developed in this chapter (i.e. the test statistic $\tilde{X}^2$ defined in equation (13.18)), the modification to the usual χ^2 procedure proposed in Chapter 4 (i.e. a test statistic of the form $X_{\mathrm{P}}^2/\bar{\delta}$) and the unadjusted X_{LR}^2 test statistic itself. Table 13.5 gives actual levels of nominal 5% tests (as estimated from simulations) for the three test statistics, using two values for the cluster-effect covariance matrix $\mathbf{W}$ and three different sample allocations. It can be seen that the unadjusted procedure can have distorted significance levels while the two modified procedures behave well. Extensive results reported in Ewings (1985) show this pattern to hold under a wide variety of conditions; distortion of the significance level is confirmed for the unadjusted X_{LR}^2 while $X_{\mathrm{P}}^2/\bar{\delta}$ and $\tilde{X}^2$ generally have a level close to the nominal 5%. Note that whilst $X_{\mathrm{P}}^2/\bar{\delta}$ is given in these tables the corresponding modification to X_{LR}^2 has similar properties.

Under certain unusual circumstances, however, the statistic developed in Chapter 4 is found to behave aberrantly. These circumstances arise when the number of clusters involved is very small. An example of this is given in Table 13.6. Three values of $\mathbf{W}$ have been used in the simulations, with either six or twelve clusters in total and two different sample allocations in each case. The hypothesis being tested here is $\beta_1 = \beta_2 = \beta_{12} = 0$. First, note the usual phenomenon of distorted significance levels for the unadjusted X_{LR}^2 when cluster effects are present. In the case of the first cluster-effects matrix, $\mathbf{W}_1 = \mathbf{0}$ (i.e. there are no cluster effects in this case) the unadjusted procedure is satisfactory. The levels for $\tilde{X}^2$ are a little erratic for these data, but the performance of $X_{\mathrm{P}}^2/\bar{\delta}$ is more disturbing. When the sample allocation varies between clusters, the level for $X_{\mathrm{P}}^2/\bar{\delta}$ is often distorted to a greater extent than that for the unadjusted X_{LR}^2. (When the sample allocation is constant across clusters the performance improves, but this is a reflection of the model used to generate the data and the hypothesis being tested.) The reason that $X_{\mathrm{P}}^2/\bar{\delta}$ behaves worse than $\tilde{X}^2$ is probably related to the use of a correction for non-negative definiteness in estimating the cluster-effects matrix $\mathbf{W}$ before calculation of $\tilde{X}^2$. If no such correction is made, the significance levels for $\tilde{X}^2$ can also be seriously distorted (although results are not presented here). It appears that, when there are few clusters, the estimates of variance based on between-cluster sums of squares become unstable. The total variance in the model (including that due to the cluster effects as well as the binomial error) is often underestimated under these circumstances, leading to large test statistics and actual significance levels considerably in excess of the nominal 5%. Including the correction for non-negative definiteness makes the variance estimates more stable, producing

***Table 13.6** Estimated levels of significance (%) for nominal 5% tests using two test statistics from the loglinear framework (X^2_{LR} and the modified statistic $X^2_P/\bar{\delta}$) and the modified logit model test statistic ($\tilde{X}^2$). Levels based on 500 simulations in each case, using various values of $\mathbf{W}_c$ and sample allocations for $2 \times 2 \times 2$ tables*

			Estimated levels		
$\mathbf{W}_c$	m	Sample allocation	X^2_{LR}	$X^2_P/\bar{\delta}$	$\tilde{X}^2$
0	6	(a)	5	9	3
0	6	(b)	4	20	1
0	12	(c)	6	7	3
0	12	(d)	5	11	1
0.2**I**	6	(a)	11	8	8
0.2**I**	6	(b)	17	26	6
0.2**I**	12	(c)	9	6	7
0.2**I**	12	(d)	18	13	6
0.4**I**	6	(a)	20	9	11
0.4**I**	6	(b)	32	32	9
0.4**I**	12	(c)	18	7	11
0.4**I**	12	(d)	31	14	9

Sample allocations:

(a) 6 clusters 8, 7/7, 8

(b) 2 clusters 21, 3/3, 3
1 cluster 3, 21/3, 3
1 cluster 3, 3/21, 3
2 clusters 3, 3/3, 21

(c) 12 clusters 8, 7/7, 8

(d) 4 clusters 21, 3/3, 3
2 clusters 3, 21/3, 3,
2 clusters 3, 3/21, 3
4 clusters 3, 3/3, 21

levels closer to the nominal figure. It should, however, be emphasized that under most typical conditions (i.e. when there are many clusters in the sample), both $\tilde{X}^2$ and $X^2_P/\bar{\delta}$ behave in a similar fashion, achieving good results in the simulation studies.

The overall conclusion is that both the distribution of the explanatory variables across clusters and the covariance matrix of the cluster effects are important. In situations where a constant cluster effect might be appropriate (e.g. where a variable has been omitted from the analysis that varies between clusters and affects all logits in the same way in each cluster) then explanatory variables that have low intra-cluster correlation will lead to corresponding χ^2 tests which are relatively unaffected. However, in general this will not be the case and analyses of real data, to be presented in the next section, show that the impact on the ordinary X^2_{LR} test can be severe. A modification to the X^2_{LR} test, such as described in Chapter 4, is strongly recommended, except in special circumstances.

13.6 THE INTERPRETATION OF THE CLUSTER-EFFECT COVARIANCE MATRIX

In this section we turn to the possible interpretation of the cluster-effect covariance matrix **W**, to help understand the social processes which may underlie cluster effects. We use as an illustration the analysis of ever married women living in urban areas and aged over 30 years at the time of the survey. The data source is the 1974 Fiji Fertility Survey which is described in detail in the Fiji Fertility Survey Report (World Fertility Survey 1976). The dependent variable is dichotomous: whether or not the woman was first married at age under 20 years. The explanatory variables used are

1. Ethnicity: (a) Fijian; (b) Indian;
2. Education: (a) none or lower primary; (b) upper primary or secondary; and
3. Work experience before marriage: (a) no; (b) yes.

There was a large decline in fertility in the 1960s in Fiji which started earlier and was much more pronounced amongst the Indian community. The general fertility rate fell from 240 births per 1000 in 1957 to about 170 per 1000 in 1965. One of the important demographic factors in this decline was the reduction in teenage marriages, particularly amongst the Indian population.

The contingency table is thus a 2^4 table and was, in fact, derived from a larger 2^5 table involving both urban and rural places of residence. For the larger table, involving urban and rural women, a simple additive logistic model using the main effects of each explanatory variable was an adequate fit to the data. The $\tilde{X}^2$ statistic, adjusted for the cluster effects was 12.98 with 11 degrees of freedom. It is of interest to note that the usual X^2_{LR} statistic, taking no account of the cluster effects, for the same model is 24.12 with 11 degrees of freedom which erroneously suggests the possible need for a more complicated model. This is the misleading effect of using the unadjusted X^2_{LR} statistic.

Table 13.7 contains the estimated covariance matrix for the cluster effects for the urban part of the sample; the fitted proportion of women married in their teens in each category and also the correlation matrix for the cluster effects. We will concentrate on the interpretation of the cluster effects, but from the fitted proportions we see the higher incidence of teenage marriage among Indians, those with no or little education and those with no work experience before marriage.

The covariance matrix shows the extent to which logits vary between clusters for the various categories of women. For Fijian women, for example, with no work experience before marriage, those with little or no education have a variance of 1.78 and those with more education have a smaller variance of 0.29. The positive covariance of 0.45 between these two categories shows that to some extent the logits of the two groups of women vary in the same way across clusters; they are positively correlated.

Table 13.7 Cluster-effect matrices for logistic models of age at marriage less than 20 years for women aged over 30 years in urban dwellings

(a) Variance–covariance matrix of cluster effects								
Ethnicity:	Fijian				Indian			
Education:	Low		High		Low		High	
Work exp.:	No	Yes	No	Yes	No	Yes	No	Yes
	1.78	−0.15	0.45	0.12	0.11	0.97	0.46	−0.01
		0.41	0.14	−0.01	0.26	−0.23	0.24	0.03
			0.29	0.18	0.24	0.25	0.09	0.13
				0.36	0.18	0.17	−0.22	0.29
					0.29	0.02	0.08	0.16
						0.63	0.05	0.07
							0.54	−0.19
								0.24
Fitted proportion married	0.74	0.61	0.58	0.44	0.83	0.73	0.70	0.58
(b) Correlation matrix								
	1.00	−0.18	0.63	0.15	0.15	0.92	0.47	−0.02
		1.00	0.41	−0.03	0.75	−0.45	0.51	0.10
			1.00	0.56	0.83	0.58	0.23	0.49
				1.00	0.56	0.36	−0.50	0.99
					1.00	0.05	0.20	0.61
						1.00	0.09	0.18
							1.00	−0.53
								1.00

Note: The categories of education are:
low—none or lower primary,
high—upper primary or secondary.

If we first consider the variances we note that in general these are larger for women with no education than for women with otherwise similar characteristics but with more education. The only exception is Indian women with no work experience. It might be argued that women with no education (and perhaps particularly those with no work experience before marriage either) are much more influenced by purely local factors, whereas those with education are less influenced in this way. The latter group may be more influenced by their education and the wider sources of information available to them. To the extent that the local factors are constant for all women in a cluster one would expect to see larger between-cluster variation for women with no education and hence a larger variance. The wider sources of information for educated women would tend to ameliorate cluster differences. One might expect the same reasoning to apply to work experience since this too broadens the experience of the young women before marriage. However, there is no such systematic pattern in the matrix. For Fijians with no education and Indians with secondary education the variances are smaller for those with work experience, but the reverse is the case for the other categories.

However, this may not be so surprising. The term 'work experience' can mean very different things and this is likely to vary considerably between clusters and between categories of women. For some women 'work experience' will mean the start of a career and for others it may mean no more than a short period of employment between education and marriage as a means of supplementing the household income. Also for many women, work experience will still be a purely local influence since the added exposure will be to other people, largely women, drawn from the same locality.

We note that the entries in the covariance matrix have nothing to do with whether the logit itself is generally high or low for any category of women. Indians with no education and no work experience have a high average logit (a probability of marriage before age 20 of 0.83) but the variance of the cluster effect is relatively small. This presumably shows that the cultural influence of early marriage as the norm is very strong for this group and does not vary greatly between clusters.

The second aspect of the analysis of the cluster effects is to study further the covariation of the cluster effects. In principle this could involve the covariance matrix or the correlation matrix and we choose the latter. A multivariate analysis of the covariance matrix would be dominated by the large variances, and in some situations this may give useful and interpretable results. As an alternative our rationale is that there may be underlying factors which cause between-cluster variation, but the impact of the factor may cause larger between-cluster variation for one category than another. Thus suppose, for example, a factor existed for all Fijian women but caused only a small variance for educated Fijians and a larger variance for uneducated Fijians. We use the correlation matrix to seek to identify the existence of this factor and the individual logit variances as a separate measure of the size of the impact of the factor on the categories of women. As with many multivariate analyses of this kind the terms 'cause' and 'identify the factor' overstate the power of the procedure. In fact we seek to understand cluster effects within a framework which may suggest underlying social processes or may highlight effects of hypotheses which should be the subject of further research.

The constant cluster-effect model described in Section 13.3 would result in correlations of 1.0 for all pairs of categories. More generally one might expect the cluster-effect correlation matrix to exhibit some structure. The simplest way to explore this is through a principal components analysis of the correlation matrix. For the correlation matrix in Table 13.7 the first three principal components and their eigenvalues are given in Table 13.8. The joint variation

Table 13.8 Principal components of cluster effects for logistic models of age at first marriage: urban dwellers aged over 30

	Component		
	I	II	III
Fijian			
Low education			
No work experience	0.28	−0.47	−0.33
Work experience	0.16	−0.12	0.64
High education			
No work experience	0.50	−0.19	0.06
Work experience	0.44	0.35	−0.12
Indian			
Low education			
No work experience	0.44	−0.00	0.38
Work experience	0.30	−0.27	−0.50
High education			
No work experience	0.00	−0.60	0.25
Work experience	0.41	0.42	−0.01
Eigenvalues	3.45	2.24	2.13
% variance	43.1	28.0	26.7
Cumulative	43.1	71.1	97.8

of the cluster effects may be accounted for by three principal components of which the first accounts for 43.1% of the total variation. Unfortunately this approach is not always helpful since the principal components analysis takes no account whatsoever of the factorial structure of the explanatory variables and may be very difficult to interpret. It may be better to sacrifice some of the variance explained to identify factors which relate to the explanatory variables and are interpretable. In this example the first principal component is suggestive of a general effect although the weights given to different categories of women are quite variable. We will try a constant cluster effect as one component of the variance structure. The second principal component is essentially a contrast between (a) women with both secondary education and work experience before marriage regardless of ethnicity, and (b) the remainder. One might conjecture that those with secondary education and work experience before marriage enter the labour force with some expectations of career development which will impact upon the age of marriage. The third principal component essentially involves only women with no education and is a contrast between those with and those without work experience before marriage. However, the signs on the coefficients are reversed for Indians compared to Fijians so that this third factor is an interaction between ethnicity and work experience for women with no education only. Once again the actual coefficients obtained in the principal components analyis vary for different categories of women.

An alternative approach is to set up a matrix of orthogonal contrasts, which will not be so successful as principal components in accounting for variation, but will reflect the implied structure of the principal components, while giving equal weight to similar categories of women and being more interpretable. In this way, we can decompose the correlation matrix into a new set of variables that reflect the factorial structure of the explanatory variables.

The vector of cluster effects $\boldsymbol{\eta}_c$ has variance–covariance matrix $\mathbf{W}_c$, where the $\mathbf{W}_c$ are the identical block diagonal elements of $\mathbf{W}$ as described in Section 13.3. The elements of $\boldsymbol{\eta}_c$ are $\{\eta_{1c}, \ldots, \eta_{Jc}\}$ and we seek linear combinations of these elements which capture the main sources of the cluster-effect variance and which are interpretable. Thus we choose a suitable set of orthogonal transformations $\boldsymbol{\eta}^* = \boldsymbol{\Gamma}'\boldsymbol{\eta}$ such that $\boldsymbol{\Gamma}'\boldsymbol{\Gamma} = \mathbf{I}$ and $\text{Var}(\boldsymbol{\eta}^*) = \boldsymbol{\Gamma}'\mathbf{W}\boldsymbol{\Gamma}$. The matrix $\boldsymbol{\Gamma}$ is chosen to be interpretable in terms of the factorial structure and to reflect the patterns seen in the principal components analysis. In the example,

$$\boldsymbol{\Gamma} = \begin{bmatrix} 1 & 1 & 1/3 & 1/2 & 1 & 1/3 & 1/2 & 1 \\ 1 & 1 & 1/3 & 1/2 & -1 & 1/3 & 1/2 & -1 \\ 1 & 1 & 1/3 & -1 & 0 & 1/3 & -1 & 0 \\ 1 & 1 & -1 & 0 & 0 & -1 & 0 & 0 \\ 1 & -1 & 1/3 & 1/2 & 1 & -1/3 & -1/2 & -1 \\ 1 & -1 & 1/3 & 1/2 & -1 & -1/3 & -1/2 & 1 \\ 1 & -1 & 1/3 & -1 & 0 & -1/3 & 1 & 0 \\ 1 & -1 & -1 & 0 & 0 & 1 & 0 & 0 \end{bmatrix} \boldsymbol{\Delta},$$

$$\boldsymbol{\Delta}^2 = \text{diag}(1/8, 1/8, 3/8, 1/3, 1/4, 3/8, 1/3, 1/4),$$

where $\boldsymbol{\Delta}$ is a diagonal matrix chosen to normalize the columns of $\boldsymbol{\Gamma}$. The first column represents a grand mean (i.e. the constant cluster effect); the second is an ethnicity main effect; the third is the contrast between women with secondary education and work experience and the remainder and so on. In terms of the principal components analysis the columns of interest are the first, third and eighth.

Applying this transformation to the correlation matrix of cluster effects reveals the variances of the transformed cluster effects, $\boldsymbol{\Gamma}'\mathbf{W}\boldsymbol{\Gamma}$ are as follows:

3.02 0.04 1.69 0.12 0.08 0.04 0.96 2.05

Thus the three main transformed variables reflect the structure of the principal components analysis although now all comparable categories are given equal weight. The first variable accounts for 38% of the total (compared to 43.1% for the first principal component) and the total variation explained by the three main components is 84.5% (compared to 97.8%). We note that the third main component is a little stronger than the second (i.e. eigenvalue $2.05 > 1.69$).

The overall interpretation is that there appear to be three main components influencing the cluster effects. The first is a general cluster effect for all women in the cluster regardless of their category of ethnicity, education or work experience. This suggests a common local influence which transcends any social divisions. The second separates women with both secondary education and work experience before marriage from the remainder. It is not simply that women in the former category are less likely to marry as teenagers, it is that when the logit for this group is lower in any cluster than the general average across all clusters, then the logits for remaining categories of women in the same cluster are higher than average. In social terms this suggests local influences which reach only one section of the community. This social division is strong, although not as strong as the general cluster effect which reaches all women in the cluster. We note that heterogeneity of the population could cause this effect. If, for example, in some clusters women with secondary education have only limited long-term work opportunities whereas in other clusters the term 'work experience before marriage' means the opportunity to develop a career , then this heterogeneity could show itself as a cluster effect of the kind observed. Thus the 'local influence' may be due to the imprecise nature of the interpretation of the term 'work experience' rather than any direct consequence of work experience on the women. The third important term, being an interaction, is harder to interpret although it only applies to women with little or no education. The effect is a contrast between those with and without work experience although it is reversed for the other ethnic group.

To a large extent the analysis of the cluster effects raises questions rather than supplying answers. The main use will be to suggest further avenues to explore in understanding the cluster variations in age at first marriage and thence to identify the most appropriate strategies for population programmes.

CHAPTER 14

Conclusions

14.1 OVERVIEW

The questions the survey analyst should attempt to answer and the ones which we have tried to pose throughout this book are: which parameters do we really want to estimate and why? The answers to these questions will depend on the problem being analysed and on the structure of the population. The methods will depend additionally on the relationship between the sample and the population. In some cases there may be a simple aggregate parameter of major interest, but more commonly there will be no single predetermined parameters of interest and the analysis will evolve in the iterative manner of Figure 1.1. Frequently this evolution will be from a simple aggregate model to far more complex disaggregated models which attempt to represent the underlying causal mechanism which generated the population data. This evolutionary analysis is reflected in the structure of the book which itself reflects the way our thinking has evolved.

In Chapter 1 we proposed three objectives for this book:

1. To investigate the effects of complex designs and population structure on standard procedures.
2. To explore methods of making simple adjustments to standard procedures to restore their properties.
3. To develop new statistical methods more appropriate to complex survey data.

Our broad conclusion on (1) is clear. Standard procedures based upon SRS and IID assumptions are generally inappropriate for complex survey data. There are several reasons for this. As discussed in Part A, conventional standard errors and test procedures can be misleading, especially with clustered data. As discussed in Part B, conventional point estimators of parameters can be severely biased under disproportionate stratification or other forms of non-epsem selection. And finally, as discussed in Part C, the parameters of IID-based statistical models may simply fail to reflect the actual complex structure of the population and hence be unsuitable for analytic interpretation.

The traditional survey-sampling approach to objective (2) has been to adjust point estimators by p-weighting and to adjust standard errors and test procedures by the use of complex variance estimation procedures. Whilst we have considered this approach for various statistical methods, especially in Part A, we have also argued that there is much more to the analysis of complex surveys than just p-weighting and variance estimation for non-linear statistics. The fact that IID assumptions are unsatisfactory is a symptom of the deeper fact that the survey variables are related in some way to the structure of the population as represented by strata, clusters and so on. If survey analysts claim to investigate 'causal' relations and to define model parameters which have explanatory interpretations then they should investigate the relation of the survey variables to this complex structure by appropriate modelling. This leads to the adjustment approach of Part B and the disaggregated approach of Part C.

As argued in Chapter 1, however, there are limits to the process of disaggregation, and in practice a compromise between a purely aggregated and a purely disaggregated approach is necessary. Thus the issues raised in Parts A and B are still relevant within the general context of statistical modelling. In particular, for non-epsem designs there remains the issue of whether or not to weight unit values by the inverse inclusion probabilities. This is a difficult issue for which there is no simple resolution. If the differences between the inclusion probabilities are a function only of known design variables, such as stratum identifiers or size measures, and if these design variables are included as independent variables in the model and the model is correctly specified, then unweighted model-based inference is appropriate. Even if the design variables are not included as independent variables in the model but a correct model of the selection mechanism is specified, as in Part B, then unweighted model-based inference is still appropriate. The issue of weighting arises only when the correctness of either the model of the population structure or the model of the selection mechanism is open to question.

In the traditional survey-sampling approach, it is argued that weighted estimates should be used since they are robust to model misspecification, being design-consistent for population 'parameters' which define the best approximation, in a given sense, of the specified model to the true model. The major problem with this approach is that the parameters of the best approximating model may still not be relevant for analytic interpretation. Certainly a vital element in the model-building process is to check on the adequacy of the model specified. We note that a comparison of both weighted and unweighted parameter estimates offers a useful diagnostic check of model adequacy. If both estimates are close (say the difference is non-significant) then the choice of weighted or unweighted estimate is not of great practical importance, although of the two we would tend to recommend the unweighted estimates on the grounds that the diagnostic check had not offered any evidence against the specified model. If the two sets of estimates differ significantly then this is evidence that the model is misspecified and the analyst should consider elaborating the model, further, say by including quadratic and interaction terms involving the design variables. The weighted and unweighted estimates may then be compared again and the process repeated. If finally the two sets of estimates are still significantly different then the analyst concludes that the two sets of estimates are estimating different 'parameters', which define effectively two different approximations to the underlying true model. In this situation one possibility is that the analyst reports both sets of estimates, accompanied if possible by some speculation on the possible forms of model misspecification which might account for the differences.

14.2 IMPEDIMENTS TO PROGRESS

We would not claim that the work described in this book is complete, or that the methods proposed and techniques described will not be improved upon and eventually replaced. However, enough is known to suggest that the issues raised are of importance to anyone who analyses survey data and that methods based upon IID assumptions, for example, should be replaced. Alternative methods are scattered widely throughout the statistical literature and by bringing some of them together in this book we hope that we have contributed to a needed change in statistical methodology.

A major impediment to this change is the lack of availability of suitable computer software. Some specialist software does exist (see, for example, Wolter, 1985, Appendix E). However, a major step forward would be taken if one of the major software packages such as SPSS X, SAS or BMDP made available some of the procedures described. In the long run we feel that separate specially written software is not the answer; rather we look to the continued development of existing packages.

Another major impediment to progress is the use of 'disclosure control techniques' on data tapes issued to secondary analysts by survey agencies (Duncan and Lambert, 1986). On grounds of protecting confidentiality, the unit labels which identify important population groupings such as clusters and strata may be omitted from tapes. If this is the case then a secondary analysis cannot take into account population structure and the methods we have advocated cannot be used. This is a major restriction and if surveys are to be analysed properly then the extent to which data labels can be made available to analysts for statistical purposes should be reviewed.

14.3 AREAS FOR FURTHER WORK

As noted in Section 14.2, there is much room for further work of which the most obvious area is the empirical evaluation of alternative methods. To take just one example, it would be useful to extend the empirical study of Holt *et al.* (1980) to the methods for loglinear models discussed in Chapter 4. Empirical studies need to be conducted carefully, however, if their results are to be generalizable to actual survey practice. There is room for both Monte Carlo studies and case studies.

A second general area which we have only touched upon is the use of model-checking procedures, such as residual analysis, measures of leverage and influence and the analysis of transformations. Special techniques are required for hierarchical data sets, for unequal inclusion probabilities and to take advantage of the large numbers of observations often present in survey data (Smith, 1987).

The range of statistical methods we have covered has been limited to contingency tables analysis, regression analysis and multivariate analysis. Many other methods remain to be explored. In particular we have almost wholly concentrated on methods for the analysis of cross-sectional surveys, whereas there are many useful analytical techniques applicable to repeated survey data involving a longitudinal component.

There are a number of other important aspects of complex surveys which lie outside the ambit of this book but which are nevertheless related.

1. There is the general problem of incomplete data arising from factors such as non-response and discussed at length in Madow *et al.* (1983). This work is particularly related to the discussion of methods in Part B.
2. There is the issue of response errors and the impact that these may have on the various methods proposed. Modelling response errors can be linked to analytical techniques such as factor analysis. Koch *et al.* (1975) present an integrated discussion of response errors and complex surveys.
3. Finally there is the issue of survey design for analytic surveys, which has been discussed, for example, by Sedransk (1965). For example, if the effect of clustering is to inflate the variance of estimators then this will result in less efficient procedures and loss of power for tests. For techniques such as regression the same issues apply as for other applications of regression analysis, but the specific choice of independent variable values to reduce the variance of regression parameter estimators or to achieve orthogonality will emphasize the distinction between unweighted modelling techniques and the use of p-weighted estimators. Indeed, if the values of the independent variables are chosen by design as opposed to random selection then the distinction is irreconcilable. Finally, we note that for the modelling methods of Part C the choice of a good design is not clear. If cluster-level variables are to be used as well as individual-level variables in multilevel regression models then the choice of the number of clusters selected and the number of observations per cluster will have different effects on different coefficients.

When we began to work on methods for the analysis of complex surveys, there was comparatively little recent literature in the journals. However, during the last ten years there has been a large growth in published articles and an increasing number of researchers throughout the world have turned their attention to these issues. The distribution of publication dates in our list of references bears testament to this development. We are conscious also of parallel developments in areas such as econometrics, psychometrics, epidemiology and survival analysis.

Our view, which we have tried to express throughout the book, is that the issues raised are not mere technical issues which have only a marginal esoteric interest for would-be analysts of survey data. On the contrary our experience suggests that the use of analytic procedures which take account of the population structure and the sample selection mechanism can change the objectives of the analysis and can have a substantial impact on the subsequent interpretation of the results. We hope that this book will help to encourage survey analysts to establish these procedures in their daily practice.

References

Agresti A. (1984) *Analysis of Ordinal Categorical Data*. New York: Wiley.

Aitkin M. and Longford N. (1986) Statistical modelling issues in school effectiveness studies (with discussion). *Journal of the Royal Statistical Society* **A149** 1–43.

Altham P. M. E. (1970a) The measurement of association of rows and columns for an $r \times s$ contingency table. *Journal of the Royal Statistical Society* **B32** 63–73.

Altham P. M. E. (1970b) The measurement of association in a contingency table: three extensions of the cross-ratios and metric methods. *Journal of the Royal Statistical Society* **B32** 395–407.

Altham P. M. E. (1976) Discrete variable analysis for individuals grouped into families. *Biometrika* **63** 263–9.

Anderson J. A. (1972) Separate sample logistic discrimination. *Biometrika* **19** 19–35.

Anderson S., Auquier A., Hauck W. W., Oakes D., Vandaele W. and Weisberg H. I. (1980) *Statistical Methods for Comparative Studies*. New York: Wiley.

Anderson T. W. (1957) Maximum likelihood estimates for a multivariate normal distribution when some observations are missing. *Journal of the American Statistical Association* **52** 200–3.

Anderson T. W. (1973) Asymptotically efficient estimation of covariance matrices with linear structure. *The Annals of Statistics* **1** 135–41.

Anderson T. W. (1984) *An Introduction to Multivariate Statistical Analysis* 2nd edn. New York: Wiley.

Barrell G. V., Holt D. and MacKean J. M. (1987) A comparison of the reasons for marathon participation of British and Australian non-elite runners. *Comparative Physical Education and Sport* **5** 131–8.

Barrett J. P. and Goldsmith L. (1976) When is n sufficiently large? *American Statistician* **20** 67–70.

Bayless D. L. (1968) Variance estimation in sampling from finite populations. Ph.D. thesis, Texas A & M University.

Bean J. A. (1975) Distribution and properties of variance estimators for complex multistage probability sample. *Vital and Health Statistics*, Series 2, No. 65. Washington DC: National Center for Health Statistics.

Bebbington A. C. and Smith T. M. F. (1977) The effect of survey design on multivariate analysis. In: *The Analysis of Survey Data* vol. 2 *Model Fitting* eds C. A. O'Muircheartaigh and C. Payne. New York: Wiley, 175–92.

Bedrick E. J. (1983) Adjusted chi-squared tests for cross-classified tables of survey data. *Biometrika* **70** 591–6.

Berk R. A. (1983) An introduction to sample selection bias in sociological data. *American Sociological Review* **48** 387–98.

Binder D. A. (1983) On the variances of asymptotically normal estimators from complex surveys. *International Statistical Review* **51** 279–92.

Birnbaum Z. W., Paulson E. and Andrews F. C. (1950) On the effect of selection performed on some coordinates of a multi-dimensional population. *Psychometrika* **15** 191–204.

Bishop Y. M. M., Fienberg S. E. and Holland P. W. (1975) *Discrete Multivariate Analysis: Theory and Practice.* Cambridge, Mass.: MIT Press.

Blalock H. M. (1985) Cross level analyses. In: *The Collection and Analysis of Community Data*, ed. J. B. Casterline. International Statistical Institute, World Fertility Survey.

Box G. E. P. (1954) Some theorems on quadratic forms applied in the study of analysis of variance problems: I. Effect of inequality of variance in the one-way classification. *Annals of Mathematical Statistics* **25** 290–302.

Box G. E. P. and Jenkins G. M. (1976) *Time Series Analysis, Forecasting and Control* revised edn. San Francisco: Holden-Day.

Breslow N. E. and Day N. E. (1980) *Statistical Methods in Cancer Research* vol. 1 *The Analysis of Case-Control Studies.* Lyon: International Agency for Research on Cancer.

Brier S. E. (1980) Analysis of contingency tables under cluster sampling. *Biometrika* **67** 591–6.

Brown K. G. and Burgess M. A. (1983) On maximum likelihood and restricted maximum likelihood approaches to estimation of variance components. *Journal of Statistical Computation and Simulation* **4** 1–18.

Browne M. W. (1984) Asymptotically distribution-free methods for the analysis of covariance structures. *British Journal of Mathematical and Statistical Psychology* **37** 62–83.

Bureau of the Census (1978) *The Current Population Survey: Design and Methodology.* Washington: US Bureau of the Census.

Campbell C. (1977) Properties of ordinary and weighted least squares estimators of regression coefficients for two stage samples. *Proceedings of the American Statistical Association, Social Statistics Section* 800–5.

Campbell C. and Meyer M (1978) Some properties of *t* confidence intervals for survey data. *Proceedings of the American Statistical Association, Section on Survey Research Methods* 437–42.

Cassel C. M., Sarndal C. E. and Wretman J. H. (1977) *Foundations of Inference in Survey Sampling.* New York: Wiley.

Chambless L. E. and Boyle K. E. (1985) Maximum likelihood methods for complex sample data: Logistic regression and discrete proportional hazards models. *Communications in Statistics-Theory and Methods* **14** 1377–92.

Choudhry G. H. and Lee H. (1987) Variance estimation for the Canadian Labour Force Survey. *Survey Methodology* **13** 147–61.

Christensen R. (1984) A note on ordinary least squares methods for two-stage sampling. *Journal of the American Statistical Association* **79** 720–1.

Christensen R. (1987) The analysis of two-stage sampling data by ordinary least squares. *Journal of the American Statistical Association* **82** 492–8.

Cochran W. G. (1977) *Sampling Techniques* 3rd edn. New York: Wiley.

Cox D. R. (1969) Some sampling problems in technology, In: *New Developments in Survey Sampling* eds N. L. Johnson and H. Smith. New York: Wiley.

Cox D. R. (1970) *The Analysis of Binary Data.* London: Methuen.

Cox D. R. and Hinkley D. V. (1974) *Theoretical Statistics.* London: Chapman and Hall.

Cox D. R. and Oakes D. O. (1984) *Analysis of Survival Data.* London: Chapman and Hall.

Dalen, J. (1986) Sampling from finite populations: actual coverage probabilities for confidence intervals on the population mean. *Journal of Official Statistics* **2** 13–24.

DeMets D., and Halperin M. (1977) Estimation of simple regression coefficients in samples arising from sub-sampling procedures. *Biometrics* **33** 47–56.

Deming W. E. (1950) *Some Theory of Sampling*. New York: Dover.

Deming W. E. (1956) On simplification of sampling design through replication with equal probabilities and without stages. *Journal of the American Statistical Association* **51** 24–53.

Dempster A. P., Laird N. and Rubin D. B. (1977) Maximum likelihood from incomplete data via the EM algorithm. *Journal of the Royal Statistical Society* **B39** 1–38.

Dippo C. S., Fay R. E. and Morganstein D. H. (1984) Computing variances from complex samples with replicate weights. *Proceedings of the American Statistical Association, Section on Survey Research Methods* 489–94.

Dixon W. J. (ed.) (1981), *BMDP Statistical Software 1981*. Berkeley: University of California Press.

Draper N. R. and Smith H. (1981) *Applied Regression Analysis*, 2nd edn. New York: Wiley.

DuMouchel W. H. and Duncan G. J. (1983) Using sample survey weights in multiple regression analyses of stratified samples. *Journal of the American Statistical Association* **78** 535–43.

Duncan G. T. and Lambert D. (1986) Disclosure-limited data dissemination. *Journal of the American Statistical Association* **81** 10–28.

Durbin, J. (1953) Some results in sampling theory when the units are selected with unequal probabilities. *Journal of the Royal Statistical Society* **B15**, 262–9.

Durbin J. (1967) Design of multi-stage surveys for the estimation of sampling errors. *Applied Statistics* **16** 152–64.

Edwards A. W. F. (1963) The measure of association in a 2 × 2 table. *Journal of the Royal Statistical Society* **A126** 109–14.

Ewings P. D. (1985) Logit models for the analysis of contingency tables derived from survey data. Unpublished Ph.D. thesis, University of Southampton.

Fay R. E. (1979) On adjusting the Pearson chi-square statistic for clustered sampling. *Proceedings of the American Statistical Association, Social Statistics Section*, 402–6.

Fay R. E. (1984) Application of linear and log-linear models to data from complex samples. *Survey Methodology* **10** 82–96.

Fay R. E. (1985) A jackknifed chi-squared test for complex samples. *Journal of the American Statistical Association* **80** 148–57.

Fay R. E. (1987) Additional evaluation of chi-squared methods for complex samples. *Proceedings of the American Statistical Association, Survey Research Section*, 680–5.

Fears T. R. and Brown C. C. (1986) Logistic regression methods for retrospective case-control studies. *Biometrics* **42** 955–60.

Fellegi I. P. (1980) Approximate tests of independence and goodness of fit based on stratified multistage samples. *Journal of the American Statistical Association* **75** 261–8.

Ferber R. (1980) *Readings in the Analysis of Survey Data*. Chicago: American Marketing Association.

Fichtenbaum R. and Shahidi H. (1988) Truncation bias and the measurement of income inequality. *Journal of Business and Economic Statistics* **6** 335–337.

Frankel M. R. (1971) *Inference from Survey Samples: an Empirical Investigation*. Michigan: Institute for Social Research.

Freeman D. H., Freeman J. L. and Brock D. B. (1977) Modularization for the analysis of complex sample survey data. *Bulletin of the International Statistical Institute* **47** Book 3 3–20.

Fuller W. A. (1973) Regression analysis for sample surveys. Paper presented at 1st meeting of International Association of Survey Statisticians, Vienna.

Fuller W. A. (1975) Regression analysis for sample surveys. *Sankhya* **C37** 117–32.
Fuller W. A. (1984) Least squares and related analyses for complex survey designs. *Survey Methodology* **10** 97–118.
Fuller W. A. (1986) *PC CARP*. Ames, Iowa: Statistical Laboratory, Iowa State University.
Fuller W. A. (1987a) Estimators of the factor model for survey data. In: *Applied Probability, Stochastic Processes and Sampling Theory*, eds I. B. MacNeill and G. J. Umphrey. Dordrecht: D. Reidel 265–84.
Fuller W. A. (1987b) *Measurement Error Models*. New York: Wiley.
Fuller W. A. and Battese G. E. (1973) Transformations for estimation of linear models with nested-error structure. *Journal of the American Statistical Association* **68** 626–32.
Gilks W. R. (1986) The relationship between birth history and current fertility in developing countries. *Population Studies* **40** 437–55.
Glock C. Y. (ed.) (1967) *Survey Research in the Social Sciences*. New York: Sage.
Goldstein H. (1986a) Multilevel mixed linear model analysis using iterative generalized least squares. *Biometrika* **73**, 43–56.
Goldstein H. (1986b) Efficient statistical modelling of longitudinal data. *Annals of Human Biology* **13** 129–142.
Goldstein H. (1987a) Multilevel covariance component models. *Biometrika* **74** 430–1.
Goldstein H (1987b) *Multilevel Models in Educational and Social Research*. London: Griffin.
Goodman L. A. (1979) Simple models for the analysis of cross-classifications having ordered categories. *Journal of the American Statistical Association* **74** 537–52.
Goodman L. A. and Kruskal W. H. (1954) Measures of association for cross classifications. *Journal of the American Statistical Association* **49** 732–64.
Goodman L. A. and Kruskal W. H. (1959) Measures of association for cross classifications II: further discussion and references. *Journal of the American Statistical Association* **54** 123–63.
Goodman L. A. and Kruskal W. H. (1963) Measures of association for cross classifications III: Approximate sampling theory. *Journal of the American Statistical Association* **58** 310–64.
Goodman L. A. and Kruskal, W. H. (1972) Measures of association for cross classifications IV: Simplification of asymptotic variances. *Journal of the American Statistical Association* **67** 415–21.
Gross W. F. (1984) A note on 'chi-squared tests with survey data'. *Journal of the Royal Statistical Society* **B46** 270–2.
Hajek J. (1960) Limiting distributions in simple random sampling from a finite population. *Pub. Math. Inst. Hung. Acad. Sci.* **5** 361–74.
Hajek J. (1964) Asymptotic theory of rejective sampling with varying probabilities from a finite population. *Annals of Mathematical Statistics* **35** 1491–1523.
Hall B. H., Schnake R. and Cummins C. (1987) *Time Series Processor, Version 4.1, User's Manual*. Palo Alto: TSP International.
Hansen M. H. and Hurwitz W. N. (1943) On the theory of sampling from finite populations. *Annals of Mathematical Statistics* **14** 333–62.
Hansen M. H., Hurwitz W. N. and Madow W. G. (1953) *Sample Survey Methods and Theory*, vol. II. New York: Wiley.
Harlan W. R., Landis J. R., Schmouder R. L., Goldstein N. G. and Harlan L. C. (1985) Blood lead and blood pressure, *Journal of the American Medical Association* **253** 530–4.
Harville D. A. (1977) Maximum likelihood approaches to variance component estimation and to related problems. *Journal of the American Statistical Association* **72** 320–40.
Hausman J. A. and Wise D. A. (1977) Social experimentation, truncated distributions and efficient estimation. *Econometrica* **45** 319–39.

Hausman J. A. and Wise D. A. (1981) Stratification on endogenous variables and estimation: The Gary Income Maintenance Experiment. In: *Structural Analysis of Discrete Data with Econometric Applications* eds C. F. Manski and D. McFadden. Cambridge, Mass.: MIT Press.

Heckman J. (1979) Sample selection bias as a specification error. *Econometrica* **47** 153–61.

Hidiroglou M. A., Fuller W. A. and Hickman R. D. (1980) *SUPERCARP* 6th edn. Ames, Iowa: Statistical Laboratory, Iowa State University.

Hidiroglou M. A. and Rao J. N. K. (1987a) Chi-squared tests with categorical data from complex surveys: Part I—simple goodness-of-fit, homogeneity and independence in a two-way table with applications to the Canada Health Survey (1978–1979). *Journal of Official Statistics* **3** 117–32.

Hidiroglou M. A. and Rao J. N. K. (1987b) Chi-squared tests with categorical data from complex surveys: Part II—independence in a three-way table with applications to the Canada Health Survey (1978–1979). *Journal of Official Statistics* **3** 133–40.

Hill M. S. (1981) Some illustrative design effects: proper sampling errors versus simple random sample assumptions. In: *Five Thousand American Families—Patterns of Economic Progress* vol. IX eds M. S. Hill, D. H. Hill, J. N. Morgan. Michigan: Institute of Social Research 457–65.

Hirschi T. and Selvin H. C. (1967) *Delinquency Research: An Appraisal of Analytic Methods*. New York: Free Press.

Holmes D. J. (1987) The effect of selection on the robustness of multivariate methods. Unpublished Ph.D. thesis, University of Southampton.

Holst L. (1973) Some limit theorems with applications to sampling theory. *Annals of Statistics* **1** 644–58.

Holt D. and Scott A. J. (1981) Regression analysis using survey data. *The Statistician* **30** 169–78.

Holt D., Scott A. J. and Ewings P. D. (1980) Chi-squared tests with survey data. *Journal of the Royal Statistical Society* **A143** 303–20.

Holt D. and Smith T. M. F. (1976) The design of surveys for planning purposes. *Australian Journal of Statistics* **28** 37–44.

Holt D. and Smith T. M. F. (1979) Post stratification. *Journal of the Royal Statistical Society* **A142** 33–46.

Holt D., Smith T. M. F. and Winter P. D. (1980) Regression analysis of data from complex surveys. *Journal of the Royal Statistical Society* **A143** 474–87.

Holt M. M. (1977) SURREGR: Standard errors of regression coefficients from sample survey data. Research Triangle Institute, Research Triangle Park, NC.

Horvitz D. G. and Thompson D. J. (1952) A generalization of sampling without replacement from a finite universe. *Journal of the American Statistical Association* **47** 663–85.

Hyman H. H. (1955) *Survey Design and Analysis*. New York: Free Press.

IFIR and CBS (1979) *The Profitability of the Wine-Grape Branch in 1976/77 and 1977/78*. Institute of Farm Income Research and Central Bureau of Statistics, Israel, Series H, No. 2 (in Hebrew).

Isaki C. T. and Fuller W. A. (1982) Survey design under the regression super-population model. *Journal of the American Statistical Association* **77** 89–96.

Jewell N. P. (1985) Least squares regression with data arising from stratified samples of the dependent variable. *Biometrika* **72** 11–21.

Johnson N. L. and Pearson E. S. (1969) Tables of percentage points of non-central chi. *Biometrika* **56** 255–72.

Jowell R. and Airey C. (1985) *British Social Attitudes—the 1984 Report*. Aldershot: Gower.

Jowell R., Witherspoon S. and Brook L. (1986) *British Social Attitudes—the 1986 Report*. Aldershot: Gower.

Kalton G. (1977) Practical methods for estimating survey sampling errors. *Bulletin of the International Statistical Institute* **47**(3) 495–514.

Kemsley W. F. F., Redpath R. U. and Holmes M. (1980) *Family Expenditure Survey Handbook*. London: HMSO.

Kendall P. L. and Lazarsfeld P. F. (1950) Problems of survey analysis. In: *Continuities in Social Research: Studies in the Scope and Method of 'The American Soldier'* eds R. K. Merton and P. F. Lazarsfeld. Chicago: Free Press.

Keyfitz N. (1957) Estimates of sampling variances where two units are selected from each stratum. *Journal of the American Statistical Association* **52** 503–10.

King M. L. and Evans M. A. (1986) Testing for block effects in regression models based on survey data. *Journal of the American Statistical Association* **81** 677–9.

Kish L. (1965) *Survey Sampling*. New York: Wiley.

Kish L. (1987) *Statistical Design for Research*. New York: Wiley.

Kish L. and Frankel M. R. (1974) Inference from complex samples (with discussion). *Journal of the Royal Statistical Society* **B36** 1–37.

Kish L., Groves R. M. and Krotki K. (1976) Sampling errors for fertility surveys, *WFS Occasional Paper No. 17*. The Hague: International Statistical Institute.

Kish L. and Hess I. (1959) On variances of ratios and their differences in multistage samples. *Journal of the American Statistical Association* **54** 416–446.

Koch G. G., Freeman D. H. and Freeman J. L. (1975) Strategies in the multivariate analysis of data from complex surveys. *International Statistical Review* **43** 59–78.

Koch G. G. and Lemeshow S. (1972) An application of multivariate analysis to complex sample survey data. *Journal of the American Statistical Association* **67** 780–2.

Kotchen J., Kotchen T. A., Guthrie G. P. Jr, Cottrill C. M. and McKean H. E. (1980) Correlates of adolescent blood pressure at five-year follow-up. *Hypertension* **2** (supp. 1) 124–9.

Kovar J. G., Rao J. N. K. and Wu C. F. J. (1988) Bootstrap and other methods to measure errors in survey estimates. To appear in *Canadian Journal of Statistics*.

Kovar M. G. and Johnson C. (1986) Design effects from the Mexican American portion of the Hispanic Health and Nutrition Examination Survey: a strategy for analysts. *Proceedings of the American Statistical Association, Section on Survey Research Methods*, 396–9.

Krewski D. (1978) Jackknifing U-statistics in finite populations. *Communications in Statistics—Theory and Methods* **A7** 1–12.

Krewski D. and Rao J. N. K. (1981) Inference from stratified samples: properties of the linearization, jackknife and balanced repeated replication methods. *Annals of Statistics* **9** 1010–19.

Krzanowski W. J. (1979) Between-group comparison of principal components. *Journal of the American Statistical Association* **74** 703–7.

Krzanowski, W. J. (1984) Principal component analysis in the presence of group structure. *Applied Statistics* **33** 164–8.

Larntz K. (1978) Small-sample comparisons of exact levels for chi-squared goodness-of-fit statistics. *Journal of the American Statistical Association* **76** 221–30.

Lawley D. N. (1943) A note on Karl Pearson's selection formula. *Proceedings of the Royal Society of Edinburgh* **A62** 28–30.

Layard M. W. J. (1972) Large sample tests for the equality of two covariance matrices. *Annals of Mathematical Statistics* **43** 123–41.

Layard M. W. J. (1974) A Monte Carlo comparison of tests for equality of covariance matrices. *Biometrika* **61** 461–5.

Lemeshow S. and Stoddard A. M. (1984) A comparison of alternative variance estimation strategies for estimating the slope of a linear regression in sample surveys. *Communications in Statistics—Simulation and Computation* **13** 153–68.

Lepkowski J. M. and Landis J. R. (1980) Design effects for linear contrasts of proportions and logits. *Proceedings of the American Statistical Association, Section of Survey Research Methods* 224–9.

Light R. J. and Margolin B. H. (1971) An analysis of variance for categorical data. *Journal of the American Statistical Association* **66** 534–44.

Little R. J. A. (1982a) Models for nonresponse in sample surveys. *Journal of the American Statistical Association* **77** 237–50.

Little R. J. A. (1982b) Sampling errors of fertility rates from the WFS. *WFS Technical Bulletin No. 10.* The Hague: International Statistical Institute.

Lord F. and Novick H. (1968) *Statistical Theories of Mental Test Scores.* Reading, Mass.: Addison-Wesley.

McCarthy P. J. (1969) Pseudo replication: half samples. *Internationl Statistical Review* **37** 239–64.

McCarthy P. J. (1976) The use of balanced half-sample replication in cross-validation studies. *Journal of the American Statistical Association* **71** 596–604.

McCullagh P. and Nelder J. A. (1983) *Generalized Linear Models.* London: Chapman and Hall.

McDowell A., Engel A., Massey J. T. and Maurer K. (1981) *Plan and Operation on the Second National Health and Nutrition Examination Survey 1976–80.* DHHS Publication No. (PHS) 81–1317, US Department of Health and Human Survices, National Center for Health Statistics, Hyattsville, MD.

McHugh R. B. and Mielke P. W. (1968) Negative variance estimates and statistical dependence in nested sampling. *Journal of the American Statistical Association* **63** 1000–3.

McKennell A. C. (1969) Methodological problems in a survey of aircraft noise annoyance. *The Statistician* **19** 1–29.

Maddala G. S. (1983) *Limited-dependent and Qualitative Variables in Econometrics.* Cambridge: Cambridge University Press.

Madow W. G. (1948) On the limiting distribution of estimates based on samples from finite universes. *Annals of Mathematical Statistics* **19** 535–45.

Madow W. G., Olkin I. and Rubin D. B. (1983) *Incomplete Data in Sample Surveys, Vol. 2, Theory and Bibliographies.* New York: Academic Press.

Mahalanobis P. C. (1946) Recent experiments in statistical sampling in the Indian Statistical Institute. *Journal of the Royal Statistical Society* **109** 325–70.

Manski C. F. and McFadden D. (eds) (1981) *Structural Analysis of Discrete Data with Econometric Applications.* Cambridge Mass.: MIT Press.

Mardia K. V., Kent J. T. and Bibby J. M. (1979) *Multivariate Analysis.* London: Academic Press.

Mellor R. W. (1973) Subsample replication variance estimation. Unpublished Ph.D. thesis, Harvard University.

Midzuno H. (1951) On the sampling systems with probability proportional to sum of sizes. *Annals of the Institute of Statistical Mathematics* **23** 99–108.

Miller R. G. (1981) *Simultaneous Statistical Inference* 2nd edn. New York: Springer-Verlag.

Molina C. E. A. (1982) Measures of association for cross-classifications under complex sampling designs. Unpublished Ph.D. thesis, University of Southampton.

Molina C. E. A. and Smith T. M. F. (1986) The effect of sample design on the comparison of association. *Biometrika* **73** 23–33.

Molina C. E. A. and Smith T. M. F. (1988) The effect of sampling on operative measures of association with a ratio structure. *International Statistical Review* **56** 235–42.

Mosteller F. (1968) Association and estimation in contingency tables. *Journal of the American Statistical Association* **63** 1–28.

Mosteller F. and Tukey J. W. (1977) *Data Analysis and Regression*. Reading, Mass.: Addison-Wesley.

Murthy M. N. and Roy A. S. (1975) Development of the sample design of the Indian National Sample Survey during its first twenty-five rounds. *Sankhya* **C37** 1–42.

Nathan G. (1983) A simulation comparison of estimators for a regression coefficient under differential non-response. *Communications in Statistics: Theory and Methods* **12** 645–59.

Nathan G. and Holt D. (1980) The effect of survey design on regression analysis. *Journal of the Royal Statistical Society* **B42** 377–86.

Nerlove M. and Press S. J. (1973) *Univariate and multivariate log-linear and logistic models*. Santa Monica, Calif.: Rand Corporation.

Neyman J. (1934) On the two different aspects of the representative method: the method of stratified sampling and the method of purposive selection. *Journal of the Royal Statistical Society* **97** 558–606.

Ohlsson E. (1986) Asymptotic normality of the Rao–Hartley–Cochran estimator: an application of the martingale CLT. *Scandinavian Journal of Statistics* **13** 17–28.

Patil G. P. and Rao C. R. (1978) Weighted distributions and size biased sampling with applications, etc. *Biometrics* **34** 179–90.

Payne C. D. (1985) *The GLIM System Release 3.77 Manual*, Oxford: Numerical Algorithms Group.

Pearson K. (1903) On the influence of natural selection on the variability and correlation of organs. *Philosophical Transactions of the Royal Society* **A200** 1–66.

Pearson K. (1904) On the theory of contingency and its relation to association and normal correlation. *Draper's Co. Res. Mem. Biometric Ser. 1*. Reprinted (1948) in *Karl Pearson's Early Papers*, Cambridge University Press.

Pervaiz M. K. (1986) A comparison of tests of equality of covariance matrices with special reference to the case of cluster sampling. Unpublished Ph.D. Thesis, University of Southampton.

Pfeffermann D. (1982) Multivariate prediction in finite population sampling with missing data. Paper presented at the International Meeting on Analysis of Sample Survey Data and on Sequential Analysis, Jerusalem.

Pfeffermann D. (1984) On extension of the Gauss–Markov theorem to the case of stochastic regression coefficients. *Journal of the Royal Statistical Society* **B46** 139–48.

Pfeffermann D. and Holmes D. J. (1985) Robustness considerations in the choice of method of inference for the regression analysis of survey data. *Journal of the Royal Statistical Society* **A148** 268–78.

Pfeffermann D. and LaVange L. (1986) The analysis of survey data using stochastic regression coefficients with application to NHANES data. Research Triangle Institute, Research Triangle Park, NC.

Pfeffermann D. and Nathan, G. (1981) Regression analysis of data from complex samples. *Journal of the American Statistical Association* **76** 681–9.

Pfeffermann D and Nathan G. (1985) Problems in model identification based on data from complex sample surveys. *Bulletin of the International Statistical Institute* **51** 12.2.

Pfeffermann D. and Smith T. M. F. (1985) Regression models for grouped populations in cross-section surveys. *International Statistical Review* **53** 37–59.

Pirkle J. L., Schwartz J., Landis J. R. and Harlan W. R. (1985) The relationship between blood lead levels and blood pressure and its cardiovascular risk implications. *American Journal of Epidemiology* **121** 246–58.

Plackett R. L. (1965) A class of bivariate distributions. *Journal of the American Statistical Association* **60** 516–22.

Porter R. M. (1973) On the use of survey sample weights in the linear model. *Annals of Economic and Social Measurements* **2** 141–58.

Prentice R. L. and Pyke R. (1979) Logistic disease incidence models and case-control studies. *Biometrika* **66** 403–411.

Raj D. (1966) Some remarks on a simple procedure of sampling without replacement. *Journal of the American Statistical Association* **61** 391–96.

Rao J. N. K. (1975a) Unbiased variance estimation for multistage designs. *Sankya* **C 37**, 133–9.

Rao J. N. K. (1975b) Analytic studies of sample survey data. *Survey Methodology* **1** (Supplementary Issue).

Rao J. N. K. (1978) Sampling designs involving unequal probabilities of selection and robust estimation of a finite population total. In *Contributions to Survey Sampling and Applied Statistics* ed. H. A. David. New York: Academic Press, 69–87.

Rao J. N. K. and Lanke J. (1984) Simplified unbiased variance estimation for multistage designs. *Biometrika* **71** 387–96.

Rao J. N. K. and Scott A. J. (1979) Chi-squared tests for analysis of categorical data from complex surveys. *Proceedings of the American Statistical Association, Section on Survey Research Methods* 58–66.

Rao J. N. K. and Scott A. J. (1981) The analysis of categorical data from complex sample surveys: chi-squared tests for goodness-of-fit and independence in two-way tables. *Journal of the American Statistical Association* **76** 221–30.

Rao J. N. K. and Scott A. J. (1982) On chi-squared tests for multi-way tables with cell proportions estimated from survey data. Paper presented at Israeli Statistical Association International Meetings on Analysis of Sample Survey Data and on Sequential Analysis, Jerusalem, June 1982.

Rao J. N. K. and Scott A. J. (1984) On chi-squared tests for multi-way tables with cell proportions estimated from survey data. *Annals of Statistics* **12** 46–60.

Rao J. N. K. and Scott A. J. (1987) On simple adjustments to chi-square tests with sample survey data. *Annals of Statistics* **15** 385–97.

Rao J. N. K. and Thomas D. R. (1988) The analysis of cross-classified categorical data from complex sample surveys. *Sociological Methodology* **18** 213–69.

Rao J. N. K. and Wu C. F. J. (1985) Inference from stratified samples: second-order analysis of three methods for non-linear statistics. *Journal of the American Statistical Association* **80** 620–30.

Rao J. N. K. and Wu C. F. J. (1987) Methods for standard errors and confidence intervals from sample survey data: some recent work. *Bulletin of the International Statistical Institute* **52**(3) 5–21.

Rao J. N. K. and Wu C. F. J. (1988) Resampling inference with complex survey data. *Journal of the American Statistical Association* **83** 231–41.

Roberts G., Rao J. N. K. and Kumar S. (1987) Logistic regression analysis of sample survey data. *Biometrika* **74** 1–12.

Rosen B. (1972) Asymptotic theory for successive sampling with varying probabilities without replacement, I and II, *Annals of Mathematical Statistics* **43** 373–97, 748–76.

Rosen B. (1974) Asymptotic theory for Des Raj's estimator, I and II. *Scandinavian Journal of Statistics* **1** 71–83, 135–44.

Rosenberg M. (1968) *The Logic of Survey Analysis*. New York: Basic Books.

Royall R. M. (1976) The linear least-squares prediction approach to two-stage sampling. *Journal of the American Statistical Association* **71** 657–64.

Royall R. M. (1986) Model robust confidence intervals using maximum likelihood estimators. *International Statistical Review* **54** 221–6.

Royall R. M. and Cumberland W. G. (1981) An empirical study of the ratio estimator and estimators of its variance. *Journal of the American Statistical Association* **76** 66–77.

Royall R. M. and Herson J. (1973) Robust estimation in finite populations. *Journal of the American Statistical Association* **68** 880–93.

Rubin D. B. (1985) The use of propensity scores in applied Bayesian inference. In: *Bayesian Statistics 2*, eds. J. M. Bernardo, M. H. DeGroot, D. V. Lindley and A. F. M. Smith. Amsterdam: North-Holland.

Runciman W. G. (1972) *Relative Deprivation and Social Justice.* Harmondsworth: Penguin.

Rust K. (1985) Variance estimation for complex estimators in sample surveys. *Journal of Official Statistics* **1** 381–97.

Rust K. (1987) Practical problems in sampling error estimation. *Bulletin of the International Statistical Institute* **52**(3) 39–56.

SAS (1985) *SAS User's Guide: Statistics, Version 5 Ed.* Cary, NC: SAS Institute.

SAS/IML (1985) *SAS/IML User's Guide Version 5 Ed.* Carry, NC: SAS Institute.

Satterthwaite, F. E. (1946) An approximate distribution of estimates of variance components. *Biometrics* **2** 110–14.

Scott, A. J. (1977) Some comments on the problem of randomisation in survey sampling. *Sankhya* **C39** 1–9.

Scott A. J. and Holt D. (1982) The effect of two stage sampling on ordinary least squares methods. *Journal of the American Statistical Association* **77** 848–54.

Scott A. J. and Rao J. N. K. (1981) Chi-squared tests for contingency tables with proportions estimated from survey data. In: *Current Topics in Survey Sampling* eds D. Krewski, R. Platek and J. N. K. Rao. New York: Academic Press, 247–66.

Scott A. J., Rao J. N. K. and Thomas D. R. (1989) Weighted least squares and quasi-likelihood estimation for categorical data under singular models.

Scott A. J. and Wild C. J. (1986) Fitting logistic models under case control or choice-based sampling. *Journal of the Royal Statistical Society* **B48** 170–82.

Scott A. J. and Wu C. F. (1981) On the asymptotic distribution of ratio and regression estimators. *Journal of the American Statistical Association* **76** 98–102.

Sedransk J. (1965) Analytical surveys with cluster sampling. *Journal of the Royal Statistical Society* **B27** 264–78.

Shah B. V., Holt M. M. and Folsom R. F. (1977) Inference about regression models from sample survey data. *Bulletin of the International Statistical Institute* **41**(3) 43–57.

Shaper A. G. and Pocock S. J. (1985) Blood lead and blood pressure. *British Medical Journal* **291** 1147–9.

Sheil J. and O'Muircheartaigh I. (1977) The distribution of non-negative quadratic forms in normal variables. Algorithm AS106. *Applied Statistics* **26** 92–8.

Silvey S. D. (1970) *Statistical Inference.* London: Chapman and Hall.

Skinner C. J. (1982) Multivariate analysis of sample survey data. Unpublished Ph.D. thesis, University of Southampton.

Skinner C. J. (1983) Multivariate prediction from selected samples. *Biometrika* **70** 289–92.

Skinner C. J. (1984) On the geometric approach to multivariate selection. *Psychometrika* **49** 383–90.

Skinner C. J. (1986a) Regression estimation and post-stratification in factor analysis. *Psychometrika* **51**, 346–56.

Skinner C. J. (1986b) Design effects of two-stage sampling. *Journal of the Royal Statistical Society* **B48** 89–99.

Skinner C. J. (1986c) Probability proportional to size sampling. In *Encyclopedia of Statistical Sciences* Vol. 7 eds N. L. Johnson and S. Kotz. New York: Wiley, 237–41.

Skinner C. J. (1988) On conditioning for model-based inference in survey sampling. *Biometrika* **75** 275–86.

Skinner C. J., Holmes D. J. and Smith T. M. F. (1986) The effect of sample design on principal component analysis. *Journal of the American Statistical Association* **81** 789–98.

Smith T. M. F. (1981) Regression analysis for complex surveys. In: *Current Topics in Survey Sampling* eds D. Krewski, R. Platek and J. N. K. Rao. New York: Academic Press, 267–92.

Smith T. M. F. (1983) On the validity of inferences from non-random samples. *Journal of the Royal Statistical Society* **A146** 394–403.

Smith T. M. F. (1987) Influential observations in survey sampling. *Journal of Applied Statistics* **14** 143–52.

Smith T. M. F. (1988) To weight or not to weight, that is the question. In: *Bayesian Statistics 3* eds J. M. Bernardo, M. H. DeGroot, D. V. Lindley and A. F. M. Smith. Oxford: Oxford University Press, 437–51.

Snow J. (1855) *On the Mode of Communication of Cholera* 2nd edn. London: Churchill.

SPSS X (1986) *User's Guide* 2nd edn. Chicago: SPSS.

Steel D. (1985) Statistical analysis of populations with group structure. Unpublished Ph.D. thesis, University of Southampton.

Stenlund H. and Westlund A. (1975) A Monte-Carlo study of simple random sampling from a finite population. *Scandinavian Journal of Statistics* **2** 106–8.

Stenlund H. and Westlund A. (1976) On the asymptotic normality of the mean estimator. *Scandinavian Journal of Statistics* **3** 127–31.

Sugden R. A. and Smith T. M. F. (1984) Ignorable and informative designs in survey sampling inference, *Biometrika* **71** 495–506.

Sukhatme P. V., Sukhatme B. V. Sukhatme S. and Asok C. (1984) *Sampling Theory of Surveys with Applications* 3rd edn. Ames, Iowa: Iowa State University Press.

Thomas D. R. and Rao J. N. K. (1987) Small-sample comparisons of level and power for simple goodness-of-fit statistics under cluster sampling. *Journal of the American Statistical Association* **82** 630–6.

Van Eck N. (1979) *Osiris IV User's Manual* 5th edn. Ann Arbor, Michigan: Institute of Social Research.

Verma V., Scott C. and O'Muircheartaigh C. (1980) Sample designs and sampling errors for the World Fertility Survey. *Journal of the Royal Statistical Society* **A143** 431–73.

Wald A. (1943) Tests of statistical hypotheses concerning several parameters when the number of observations is large. *Transactions of the American Mathematical Society* **54** 426–82.

White H. (1980) A heteroscedasticity-consistent covariance matrix estimator and a direct test for heteroscedasticity. *Econometrica* **41** 733–50.

Wolter K. M. (1985) *Introduction to Variance Estimation.* New York: Springer-Verlag.

Woodruff R. S. (1952) Confidence intervals for medians and other position measures. *Journal of the American Statistical Association* **47** 635–46.

Woodruff R. S. (1971) A simple method for approximating the variance of a complicated estimate. *Journal of the American Statistical Association* **66** 411–14.

Wong G. W. and Mason W. M. (1985) The hierarchical logistic regression model for multivariate analysis. *Journal of the American Statistical Association* **80** 513–24.

World Fertility Survey (1976) *Fiji Fertility Survey, 1974, Principal Report.* International Statistical Institute, The Netherlands.

Wu C. F. J., Holt D. and Holmes D. J. (1988) The effect of two-stage sampling on the F statistic. *Journal of the American Statistical Association* **83** 150–9.

Yule G. U. (1912) On the methods of measuring association between two attributes. *Journal of the Royal Statistical Society* **75** 579–642.

Author Index

Subject Index

HALD • Statistical Tables and Formulas
HALD • Statistical Theory with Engineering Applications
HAND • Discrimination and Classification
HEIBERGER • Computation for the Analysis of Designed Experiments
HOAGLIN, MOSTELLER and TUKEY • Exploring Data Tables, Trends and Shapes
HOAGLIN, MOSTELLER, and TUKEY • Understanding Robust and Exploratory Data Analysis
HOCHBERG and TAMHANE • Multiple Comparison Procedures
HOEL • Elementary Statistics, *Fourth Edition*
HOEL and JESSEN • Basic Statistics for Business and Economics, *Third Edition*
HOGG and KLUGMAN • Loss Distributions
HOLLANDER and WOLFE • Nonparametric Statistical Methods
HOSMER and LEMESHOW • Applied Logistic Regression
IMAN and CONOVER • Modern Business Statistics
JESSEN • Statistical Survey Techniques
JOHNSON • Multivariate Statistical Simulation
JOHNSON and KOTZ • Distributions in Statistics
Discrete Distributions
Continuous Univariate Distributions—1
Continuous Univariate Distributions—2
Continuous Multivariate Distributions
JUDGE, HILL, GRIFFITHS, LÜTKEPOHL and LEE • Introduction to the Theory and Practice of Econometrics, *Second Edition*
JUDGE, GRIFFITHS, HILL, LÜTKEPOHL and LEE • The Theory and Practice of Econometrics, *Second Edition*
KALBFLEISCH and PRENTICE • The Statistical Analysis of Failure Time Data
KAUFMAN and ROUSSEEUW • Finding Groups in Data: An Introduction to Cluster Analysis
KISH • Statistical Design for Research
KISH • Survey Sampling
KUH, NEESE, and HOLLINGER • Structural Sensitivity in Econometric Models
KEENEY and RAIFFA • Decisions with Multiple Objectives
LAWLESS • Statistical Models and Methods for Lifetime Data
LEAMER • Specification Searches: Ad Hoc Inference with Nonexperimental Data
LEBART, MORINEAU, and WARWICK • Multivariate Descriptive Statistical Analysis: Correspondence Analysis and Related Techniques for Large Matrices
LINHART and ZUCCHINI • Model Selection
LITTLE and RUBIN • Statistical Analysis with Missing Data
McNEIL • Interactive Data Analysis
MAGNUS and NEUDECKER • Matrix Differential Calculus with Applications in Statistics and Econometrics
MAINDONALD • Statistical Computation
MALLOWS • Design, Data, and Analysis by Some Friends of Cuthbert Daniel
MANN, SCHAFER and SINGPURWALLA • Methods for Statistical Analysis of Reliability and Life Data
MARTZ and WALLER • Bayesian Reliability Analysis
MASON, GUNST, and HESS • Statistical Design and Analysis of Experiments with Applications to Engineering and Science
MIKÉ and STANLEY • Statistics in Medical Research: Methods and Issues with Applications in Cancer Research
MILLER • Beyond ANOVA, Basics of Applied Statistics
MILLER • Survival Analysis
MILLER, EFRON, BROWN, and MOSES • Biostatistics Casebook
MONTGOMERY and PECK • Introduction to Linear Regression Analysis
NELSON • Applied Life Data Analysis